Biomathematics

Volume 20

Managing Editor
S. A. Levin

Editorial Board
C. DeLisi   M. Feldman   J. Keller   R. M. May   J. D. Murray
A. Perelson   L. A. Segel

Joel E. Cohen   Frédéric Briand
Charles M. Newman

Community
Food Webs

Data and Theory

With a Contribution by Zbigniew J. Palka

With 46 Illustrations

Springer-Verlag
Berlin Heidelberg New York
London Paris Tokyo Hong Kong

Joel E. Cohen
Laboratory of Populations
Rockefeller University
New York, NY 10021/USA

Frédéric Briand
World Conservation Center
International Union for the
Conservation of Nature
CH-1196 Gland/Switzerland

Charles M. Newman
Department of Mathematics
University of Arizona
Tucson, AZ 85721/USA

Cover Figure: The three-dimensional frequency histogram in the lower left shows the observed number of food chains of each length from 1 to 9 links in 113 community food webs. Each slice parallel to the axis labeled "length" is the observed frequency distribution for one web. The height of a slice at each value of length is the observed number of chains of that length. The webs are ranked from the smallest number of species to the largest number of species. The dot-and-arrow diagrams in the upper right are four configurations of a web on six vertices that produce an induced 4-cycle in the trophic niche overlap graph (see Chapter III.6).

ISBN-13:978-3-642-83786-9   e-ISBN-13:978-3-642-83784-5
DOI: 10.1007/978-3-642-83784-5

Library of Congress Cataloging-in-Publication Data
Cohen, Joel E.
Community food webs : data and theory / Joel E. Cohen, Frédéric
Briand, Charles M. Newman.
(Biomathematics; v. 20)
Includes bibliographical references.
ISBN-13:978-3-642-83786-9 (alk. paper : U.S.)
1. Food chains (Ecology) – Mathematical models. I. Briand,
Frédéric. II. Newman, Charles M. III. Title. IV. Series.
QH541.15.M3C63   1990
574.5'3 – dc20

Dataconversion: Kurt G. Mattes, Heidelberg

2141/3020-543210   Printed on acid-free paper

Preface

Food webs hold a central place in ecology. They describe which organisms feed on which others in natural habitats. This book describes some recently discovered empirical regularities in real food webs. It proposes a novel theory that unifies many of these regularities. It offers researchers the most extensive available collection of edited data on community food webs.

The book is intended for graduate students, teachers and researchers primarily in ecology, especially community ecologists with a quantitative orientation. The theoretical portions of the book provide materials that could be useful to teachers of applied combinatorics, in particular random graphs. Researchers in the theory of random graphs will find some unsolved mathematical problems here.

The first portion of the book, a general introduction, reviews the empirical and theoretical discoveries about food webs presented here.

The second portion of the book shows that community food webs obey several striking phenomenological regularities. Some of these regularities unify; they apply to all webs, regardless of the kind of habitat in which they are observed. For example, the ratio of number of trophic links (feeding connections between a living consumer and a living resource) to the number of kinds of organisms is approximately independent of the total number of kinds of organisms in the web. Other regularities differentiate; they show that the habitat of a web significantly influences the structure of the food web. For example, food chains in habitats that are three-dimensional on the human scale, such as the open ocean or forest canopy, are longer than those in two-dimensional habitats, such as a rocky ocean shore.

The third portion of the book presents a theoretical analysis of some of the unifying empirical regularities. Several simple models, based on random directed graphs, are considered. All but one of the models are clearly rejected by the data. The sole survivor, called the cascade model, explains the major empirical regularities qualitatively and quantitatively. The cascade model predicts the proportions of top, intermediate and basal species in a web. The model gives the first exactly derived predictions of the frequency distribution of the length of food chains, and these predictions are in acceptable agreement with observations. The model explains the newest observations of the frequency of intervality in webs.

The second and third portions of the book are preceded by introductions that review the background of the following chapters.

The fourth portion of the book presents 113 community food webs. Collected from scattered sources and carefully edited, these webs are the empirical basis for the results in this volume. We believe they are the largest available set of data on community food webs. We hope they will provide a valuable foundation for future studies of community food webs. We welcome corrections of errors in these data and additions to the stock of webs. We hope that, by making the data easily and widely available, we can attract other scientists to the study of food webs and thereby accelerate the obsolescence of this book.

NewYork, N.Y., November 1989   Joel E. Cohen

Synopses of the Chapters

Chapter I. General Introduction

§ 0. Food Webs and Community Structure
Joel E. Cohen

A central problem of biology is to develop helpful concepts (e. g., genes) and tested quantitative models (e. g., Mendel's laws) to describe, explain and predict biological variation. This book describes recent discoveries, descriptive and explanatory, about variation in the food webs of ecological communities.

Chapter II. Empirical Regularities

§ 0. Untangling an Entangled Bank
Joel E. Cohen

Darwin wrote about food webs in a literary way. The systematic attempt to record all the feeding relations in a natural community apparently began in the twentieth century. Now many webs have been reported. The great variability of these webs invites description and explanation. Descriptions can unify (all webs share certain properties) or differentiate (certain webs differ systematically from other webs). The chapters in this section present some examples of both kinds of descriptions. These empirical generalizations raise questions about the quality of the underlying data and the appropriateness of the data for the analyses that are made of them. Suggestions for improving the quality and quantity of future food web data are offered.

A. General Regularities

§ 1. Ratio of Prey to Predators in Community Food Webs
Joel E. Cohen

In community food webs, the ratio of the number of kinds of prey (or living resources) to the number of kinds of predators (or consumers) displays no increasing or decreasing trend, over the observed range of numbers of kinds of organisms.

§ 2. Community Food Webs Have Scale-Invariant Structure
Frédéric Briand and Joel E. Cohen

In community food webs, the proportions of top, intermediate and basal trophic species are, on average, independent of the total number of trophic species. This scale-invariance explains the direct proportionality between the numbers of prey and predator trophic species.

§ 3. Trophic Links of Community Food Webs
Joel E. Cohen and Frédéric Briand

In community food webs, the mean number of trophic links is proportional to the total number of trophic species. The numbers of trophic links of each kind (e. g. from basal to intermediate species, or from intermediate to top species) are also roughly proportional to the total number of trophic species.

§ 4. Food Webs and the Dimensionality of Trophic Niche Space
Joel E. Cohen

If the trophic niche of a kind of organism is a connected region in niche space, then it is possible for trophic niche overlaps to be described in a one-dimensional niche space if and only if the trophic niche overlap graph is an interval graph. An analysis of 30 food webs, using the combinatorial theory of interval graphs, suggests that a niche space of dimension one suffices, with unexpectedly high frequency, to describe the trophic niche overlaps implied by real food webs in single habitats.

B. Differential Regularities

§ 5. Environmental Control of Food Web Structure
Frédéric Briand

In community food webs, the trophic connectance is lower in habitats with marked fluctuations of the physical environment than in webs with relatively constant physical habitats.

§ 6. Environmental Correlates of Food Chain Length
Frédéric Briand and Joel E. Cohen

In community food webs, the average lengths and the maximal lengths of food chains are independent of primary productivity, contrary to the hypothesis that longer food chains should arise when more energy is available at their base. Environmental variability alone also does not appear to constrain mean or maximal chain length. However, habitats that are three-dimensional or solid, like the forest canopy or the water column of the open ocean, have distinctly longer food chains than habitats that are two-dimensional or flat on the human scale, like a grassland or lake bottom.

Chapter III. A Stochastic Theory of Community Food Webs

§ 1. Theory: Circles of Complexity, Spherical Horses
Joel E. Cohen

Theories of food web structure have been strongly influenced by the tradition in physics of modeling dynamic processes by systems of differential equations. Such models may be linear or non-linear, and may have fixed or random parameters. All such models necessarily posit dynamic processes, and such processes cannot be tested against static food web data. The theoretical approach taken in the following chapters is tuned to the nature of most food web data, which are static and phenomenological. The aim is to find the simplest assumptions, with the least theoretical superstructure, that can unify the observed empirical regularities. Any phenomenological model, no matter how successful, remains only a partial, cross-sectional description of dynamic ecological processes. In spite of its apparent limitations, the cascade model presented in the following chapters offers the first quantitative description of some important features of food webs.

§ 2. Models and Aggregated Data
Joel E. Cohen and Charles M. Newman

Several simple models, based on random directed graphs, are proposed to explain the structure of food webs. Several are rejected for qualitative or quantitative failures to describe the data. A model called the cascade model is shown to predict the form and parameters of the observed scale-invariance in the numbers of kinds of species and kinds of links as a consequence of the observed scale-invariance in the ratio of links to species.

§ 3. Individual Webs
Joel E. Cohen, Charles M. Newman and Frédéric Briand

The cascade model is tested against data from individual webs. It shows a higher ratio of links to species for webs in constant habitats than for webs in fluctuating habitats.

§ 4. Predicted and Observed Lengths of Food Chains
Joel E. Cohen, Frédéric Briand, and Charles M. Newman

An exact quantitative theory for the expected numbers of chains of each length, the first such theory, is derived from the cascade model of community food webs, and is tested with considerable success against the observed numbers of chains in 113 webs.

§ 5. Theory of Food Chain Lengths in Large Webs
Charles M. Newman and Joel E. Cohen

The cascade model provides the first exact explanation of why the lengths of food chains are much less than the number of species in a community. According to the cascade model, the median value of the longest chain increases very slowly with the number of trophic species, remaining below 17 for up to one million trophic species. When the number of trophic species in a web becomes extremely large, the cascade model predicts that the mean length of chains approximately equals the mean number of predators plus prey of any species in the web; this prediction is apparently new, and is testable.

§ 6. Intervality and Triangulation in the Trophic Niche Overlap Graph
Joel E. Cohen and Zbigniew J. Palka

In 113 community food webs, the fraction of webs that are interval is strongly associated with the number of species in the webs, declining from one for small webs (16 or fewer species) toward zero for large webs (33 or more species). The cascade model predicts that, for small numbers of species, the probability that a web is interval is near one, while for large numbers of species, the probability that a web is interval declines extremely rapidly toward zero. The quantitative and qualitative agreement between the observed and predicted relative frequencies of interval webs is reasonable. The broad ecological interpretation is that the larger the number of species in a community, the less likely it is that a single dimension suffices to describe the community's trophic niche space.

Chapter IV. Data on 113 Community Food Webs

Assembled and edited by *Frédéric Briand and Joel E. Cohen*

Table of Contents

Table of Contents

Chapter I. General Introduction

§0. Food Webs and Community Structure

Joel E. Cohen

1. Introduction

A central problem of biology is to devise helpful concepts (such as genes) and tested quantitative models (such as Mendel's laws) to describe, explain and predict biological variation. The problem of characterizing variation arises in different guises in population genetics (genetic variation), demography (variation by age, sex, or location), epidemiology (variation by risk factors and disease status), and ecology (variation in species composition and interactions in communities). In each field, there is variation over time, in space, and among units of observation (individuals, populations, or comparable habitats).

This introduction reviews some recent efforts to describe, explain and predict variation in the food webs of ecological communities. There are many notions of an ecological community and many approaches to describing and understanding community ecology. Panoramic reviews of community ecology are available (such as Diamond and Case 1986; Kikkawa and Anderson 1986; National Research Council 1986; May 1986). For present purposes, a community is whatever lives in a habitat (lake, forest, sea floor) that some ecologist wants to study.

Once the physical boundaries of a habitat are defined, it is natural to study flows of matter and energy across and within the boundaries. A partial description of these flows is provided by food webs, which used to be called food-cycles (Elton 1927).

A food web describes which kind of organisms in a community eat which other kinds, if any. A community food web (hereafter simply "web") describes the feeding habits of a set of organisms chosen on the basis of taxonomy, location or other criteria without prior regard to the feeding habits among the organisms.

Webs were invented in the natural-historical approach to community ecology as a descriptive summary of which species were observed to eat which others.

If an ecological community is like a city, a web is like a street map of the city: it shows where road traffic can and does go. A street map usually omits many important details, such as the flow of pedestrian and bicycle traffic, how much traffic flows along the available streets, what kind of vehicular traffic it is, the reasons for the traffic, the laws governing traffic flow, rush hours, and the origin of the vehicles. By analogy, a web often omits small flows of food or predation on minor species, the quantities of food or energy consumed, the chemical composition of food flows, the behavioral and physical constraints on predation, temporal variations (periodic or stochastic) in eating, and the population dynamics of species involved. Thus a web gives at best very sketchy information about the functioning of a community. But just as a map provides a helpful framework for organizing more detailed information, a web helps picture how a community works.

Many approaches to studying webs are available. I will not attempt here a comprehensive review of food webs, since such reviews are available (see Pimm 1982 and in press; DeAngelis et al. 1983; Lawton in press). A difference of temperament, training, and language seems to divide those who prefer to study webs in physical and chemical terms (such as Lotka 1925; Lindemann 1942; Wiegert 1976; Budyko 1980; Margalef 1984; Remmert 1984) from those who prefer to study webs in terms of the natural history of species of living organisms (such as many authors in the collection by Hazen 1964). Here "natural history" comprises morphological, genetic, physiological, behavioral, and demographic characteristics of species. Recent natural-historical approaches have focused on combinatorial aspects of web structure (Cohen 1978; Sugihara 1982, 1983, 1984), on the theory of interactions between web structure and the stability of dynamic models (May 1973; Pimm and Lawton 1977; Pimm 1982, 1984; Sugihara 1982), and on empirical generalizations (Paine 1980; Briand 1983; Beaver 1983, 1985).

Fortunately, nature is serenely indifferent to the prejudices ecologists bring her. It will eventually be necessary to integrate the physico-chemical and natural-historical approaches to community ecology. I hope that the food web models reviewed here will help bring about that integration.

This introduction reviews some recent discoveries about webs, suggests opportunities for further empirical and theoretical study, and sketches some uses for actual and potential knowledge about webs. I attempt to give here an informal description of the discoveries that are presented more technically in the following chapters.

So far as I know, webs were first described in scientific detail at the beginning of this century. Simplifications that they were, the webs appeared forbiddingly complex relative to the concepts available for understanding them. The webs differed strikingly from one habitat to another. Now enough webs have been patiently observed and recorded to demonstrate that ensembles or collections of webs display simple general properties that are not evident from any single web. Building on a collection of webs that I initiated (Cohen 1978), F. Briand assembled and edited 113 community webs from 89 distinct published studies.

Thus many field ecologists contributed to the discoveries reviewed here. Most of the world's biomes are represented among these webs. There are 55 continental (23 terrestrial and 32 aquatic), 45 coastal, and 13 oceanic webs, ranging from arctic to antarctic regions. The sources and major characteristics of these webs are listed in Chap. II.6. The webs are fully documented in Chapter IV of this book.

In what follows, I will illustrate what a web is and how a web is described. I will present some recent quantitative empirical generalizations about webs. Then I will present a simple model, called the cascade model, that unifies the quantitative generalizations. Though this model does not purport to represent everything field ecologists know is happening in webs, no other model at present connects and explains quantitatively what is observed. The cascade model also makes novel predictions that can be tested. Then I will describe problems from other parts of ecology that can be analyzed using the cascade model and the facts on which it is based. Finally, I will sketch some potential uses of facts and theories about webs.

2. Terms

Let me introduce some terms and illustrate them with an example. A trophic species is a collection of organisms that have the same diets and the same predators. This definition combines Sugihara's definitions (1982, p. 19) that resources are trophically equivalent if they have identical consumers and that consumers are trophically equivalent if they have identical resources. A trophic species will sometimes, but not always, be a biological species in the usual sense of biological species: a collection of organisms with shared genetics. A trophic species may be a biological species of plant or animal, or several species, or a stage in the life cycle of one biological species. Hereafter the word "species" without further specification means "trophic species".

Independently of Sugihara (1982, p. 19), Briand and I (Chap. II.2) introduced the concept of trophic species to find out if there was merit in a criticism that Pimm (1982, p. 168) made of my earlier finding (Chap. II.1) that webs generally had about 4 (biological) species of predators for every 3 (biological) species of prey. Pimm suggested that ecologists distinguish among species with fur or feathers, which are likely to be consumers, more often than among species with more difficult taxonomy, such as many plants, microorganisms and insects, which are likely to be consumed. The excess of predators, he suggested, could be an artifact of the interests and knowledge of ecologists.

To test that possibility, Briand and I devised an automated lumping procedure that puts together those biological species or other biological units of a web that eat the same kinds of prey and have the same kinds of predators. We call each equivalence class that results from such lumping a trophic species. Our intent was to apply a uniform rule to distinguishing among the units of a web in order to see if this uniform rule altered the ratio of predators to prey. Indeed it did! A slight excess of predators remains, but the ratio of predators to prey counting lumped or trophic species is much nearer 1:1 than the ratio based on the original data (Chap. II.2). Pimm's criticism had merit. We believe that using trophic species, as we shall do henceforth in this introduction, corrects a bias of ecologists and gives a more realistic picture of the trophic structure of communities.

A web is a collection of trophic species, together with their feeding relations. Each arrow in a web goes from food to eater, or from prey to predator. I call each arrow a "link", short for "trophic link".

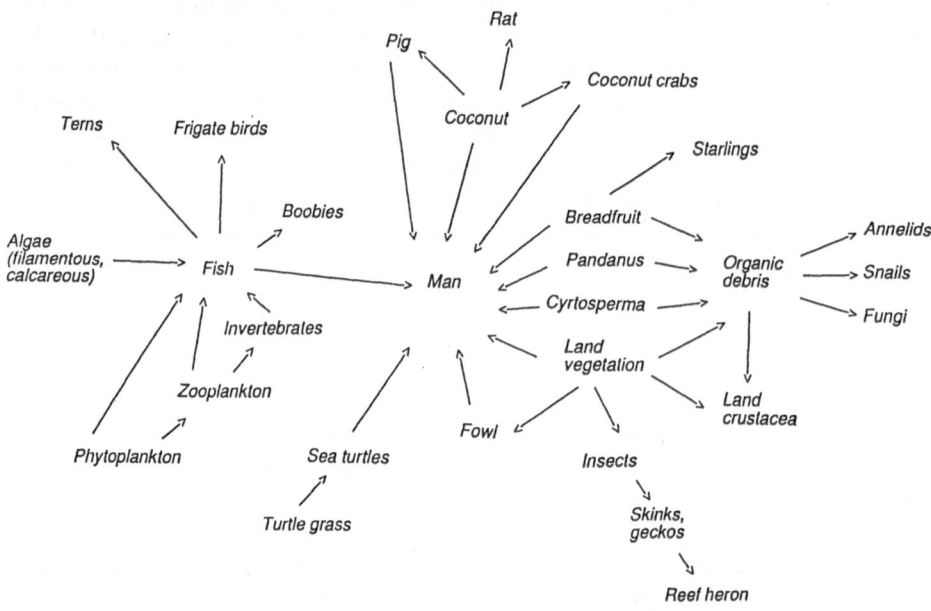

Fig. I.0.1. Food web in the Kapingamarangi Atoll. Redrawn from p. 157 of Niering 1963. As reported by Niering, the biological units in this figure range taxonomically from individual biological species (man, pig) to very large aggregates of species (phytoplankton, land vegetation), and do not necessarily correspond to trophic species

Fig. I.0.1 (redrawn from Niering 1963) pictures the unlumped web on an island in the Pacific Ocean. Some species are top, meaning that no other species in the web eats them, such as reef heron, starlings. Some species are intermediate, meaning that at least one species eats them, and they eat at least one species, such as insects, skinks, fish. Some species are basal, meaning that they eat no other species, such as algae, phytoplankton. The web omits decomposers. A crude way to quantify the structure of webs is to count the numbers of species that are top, intermediate and basal.

These three kinds of species specify four kinds of links: basal-intermediate links, such as phytoplankton to zooplankton; basal-top links, such as coconut to man; intermediate-intermediate links, such as zooplankton to fish; and inter-

mediate-top links, such as fish to frigate birds. Additional information about
structure is given by the numbers of links of each of these four kinds.

A chain is a path of links from a basal species to a top species, such as from
phytoplankton to fish to terns. The length of a chain is the number of links in
it. In Fig. I.0.1, the longest chain has only four links, and there is only one chain
of length four.

·A cycle is a directed sequence of one or more links starting from, and ending
at, the same species. A cycle of length 1 describes cannibalism, in which a species
eats itself. Cannibalism is common in nature. But ecologists report cannibalism
so unreliably that we have suppressed it from all the data even where it is
reported. A cycle of length 2 means that A eats B and B eats A. In this example,
as in most webs, there are no cycles of length 2 or more.

In summary, the terms just defined are trophic species, including top, intermediate and basal; links, including basal-intermediate, basal-top, intermediate-
intermediate and intermediate-top; and chains, length (the number of links) and
cycles.

In what follows, the terms "observed web" or "real web" mean a web edited
to eliminate obvious errors, inconsistencies and oversights, in which the original
ecologist's biological units are replaced by trophic species, and in which cannibalism and isolated species (species without feeding relations to any others) are
excluded. It is useful to ask in what sense such webs are "real".

Clearly the processed data are more constrained by reality than, for example,
webs constructed a priori as model ecosystems. As a relative term, "real" means,
not that the data are perfect, but that they are not invented.

I think it may eventually be possible to claim much more for edited webs based
on trophic species. By analogy, chemists have learned that it is more useful and
economical to describe chemical "reality" in terms of chemical elements, which
were once considered hypothetical, than in terms of gross phenomenology like
color, taste and density. Geneticists have learned that it is more useful and
economical to describe the factors affecting inheritance in terms of genes, which
were once considered hypothetical, than in terms of the gross phenomenology
of certain macroscopic characters. I suggest that a web in which the units are
trophic species may prove to be a more useful and economical description of the
trophic organization of ecological communities than a description in terms of
taxonomic phenomenology. Whether trophic species are closer to reality than the
full glory of a naturalist's notebook will have to be determined by the eventual
usefulness of the empirical and theoretical generalizations that develop using
trophic species.

3. Laws

Here are five laws or empirical generalizations about webs.

First, excluding cannibalism, cycles are rare. This generalization, without
detailed supporting data, has been known for a long time (Gallopín 1972). Of
113 webs, three webs each contain a single cycle of length 2, and there are no
other cycles (Chaps. III.2, III.4).

The rarity of cycles is not an artifact of using trophic species instead of the original units of observation, such as biological species, size classes, or aggregates of species. The reason is that the lumping procedure does not alter the connectivity of the web: the trophic species containing unit A is trophically linked to the trophic species containing unit B if and only if A was originally trophically linked to B. It follows that any cycle present in the original web must be represented by a cycle of the same length in the lumped web. Therefore, excluding cannibalism, if 110 of 113 lumped webs have no cycles, then 110 of the original webs had no cycles. The remaining three of the original webs had no cycles longer than length 2. There is no evidence that cycles occur in more webs if biological species are used instead of trophic species.

Second, chains are short (Hutchinson 1959). If one finds the maximum chain length within each web, then the median of this maximum in the 113 webs is four links and the upper quartile of the maximum chain length is five links (Chap. III.4). The longest chains in all 113 webs had ten links, and only one web had chains that long.

The last three laws deal with scale invariance (Chaps. II.1–3). Scale invariance means that webs of different size have constant shape, in some sense.

Our third law is scale invariance in the proportions of all species that are top species, intermediate species and basal species (see Fig. A.2.2a–c). There is evidently no increasing or decreasing trend in these proportions as the number of species increases (Chap. II.2). Here scale invariance describes the observation that as the number of species in 62 webs varies from 0 to 33, the proportions of top, intermediate and basal species apparently remain invariant. This scale invariance explains my earlier observation (Chap. II.1) that the ratio of number of predators to number of prey has no systematic increasing or decreasing trend when webs with different numbers of species are compared. The number of predators is the sum of the numbers of top plus intermediate species, while the number of prey is the sum of the numbers of intermediate plus basal species. Mithen and Lawton (1986) and Tilman (1986) have developed other explanations for the same finding.

Our fourth law is scale invariance in the proportions of the different kinds of links. In Fig. A.3.2a, for example, the abscissa is the number of species and the ordinate is the proportion of basal-intermediate links among all links. There is no clear evidence of an increasing or decreasing trend. The proportions of different kinds of links, like the proportions of species, are approximately scale-invariant.

The fifth law is that the ratio of links to species is scale-invariant. In Chap. III.4, Fig. 4 plots the observed number of links in each web against the observed number of species, for 113 webs. The data are approximated well by a straight line with slope about 2. That means that a web of 25 species has on average about 50 links. We first came across this generalization with 62 webs (Chap. II.3). Then Briand collected an additional 51 webs, and we found that the new data superimpose beautifully on the old data (Chap. III.4). Several other investigators independently arrived at equivalent conclusions (Chap. II.0). So far, this scale-invariant ratio of links to species is a consistent feature of nature.

In summary, I have reviewed evidence for five "laws"of webs. Qualitatively, these laws state that cycles are rare, chains are short, and there is scale-invariance in the proportions of different kinds of species, in the proportions of different kinds of links, and in the ratio of links to species. Each of these laws may be stated quantitatively.

By constructing hypothetical examples, it is not too hard to see that each of.these laws may fail to hold while the remaining laws continue to hold. This means that the laws are logically independent. That all five laws characterize observed webs suggests that the laws are not empirically independent, and that it might be possible to find fewer than five assumptions which could explain and unify the five laws.

I make no claim that these are the only important empirical "laws" of webs. For example, I have omitted my finding (Chap. II.4) that the trophic niches of predators in webs may be usually represented by intervals of a line (see also Chap. III.6), and Sugihara's findings (1982, 1983, 1984) on the rarity of homological holes and the high frequency of rigid circuits. I selected the five "laws" reviewed above because they are phenomenologically important and because a simple model can connect them qualitatively and quantitatively.

4. Models

I turn now to a model that shows how the five empirical regularities described in the preceding section are related.

Let S denote the number of trophic species and L the number of links. List all the species along both the rows and columns of a "predation matrix," a square table of numbers with S rows and S columns. Name the matrix A. Put a 1 in the intersection of row i and column j (element a_{ij} of the matrix A) if the species labeled j eats the species labeled i, and a 0 if species j does not eat species i. Since cannibalism is excluded from the data, all the diagonal elements (where $i = j$) are set equal to 0. In terms of this predation matrix, the total number of links is the sum of the elements of A. The sum picks up a 1 if there is a link from prey i to predator j and a 0 if there is no link.

The predation matrix also tells whether a species is top. If a species is top, then nobody eats it. That means that the row of that species should be all 0's. So a 0-row corresponds to a top species. Similarly, a 0-column corresponds to a basal species because the species eats nothing. A species that has neither a 0-row nor a 0-column is intermediate.

I now describe the cascade model, but not the calculations required to squeeze results out of it. Some limnologists (such as Carpenter et al. 1985) use the term "cascade" with a different meaning, to describe the dynamics of limiting nutrients in webs. When the term "cascade" appears, it seems advisable to look for a definition. In this book, "cascade" refers only to the model in the next paragraph.

First, the cascade model supposes that nature numbers the S species in the community from 1 to S (without showing us the numbering), and that the numbering specifies a pecking order for feeding, as follows. Any species j in this hierarchy or cascade can feed on any species i with a lower number $i < j$ (which doesn't mean that j does feed on i, only that j can feed on i). However, species j cannot feed on any species with a number k at least as large, $k \geq j$. Second, the cascade model assumes that each species actually eats any species below it according to this numbering with probability d/S, independently of whatever else is going on in the web. Thus the probability that species j does not eat species $i < j$ is $1 - d/S$.

The assumptions of an ordering of species, of a probability of feeding proportional to $1/S$ that is the same for all possible feeding relations, and of independence among feeding relations, are all there is to the cascade model. In the predation matrix A, a_{ij} is 0 always if $i \geq j$. The predation matrix in the cascade model is strictly upper triangular, i.e., every element on or below the main diagonal is 0. An element above the diagonal ($i < j$) is 1 with probability d/S and is 0 with probability $1 - d/S$, and all elements are independent.

As is conventional, I use E to denote the average or expected number. I now show how to compute $E(L)$, the expected number of links, according to the cascade model. The expected number of links is the expectation of the sum of the predation matrix elements. There are S^2 elements in the predation matrix A and the probability is d/S that an element a_{ij} ($i < j$) above the main diagonal equals 1. All other elements of A are 0 by construction. Since there are $S(S-1)/2$ elements above the main diagonal, the expected sum of the elements of A is $S(S-1)/2 \times d/S = d(S-1)/2 = E(L)$. Thus $E(L)$ is a linear function of S with slope $d/2$.

Since at present I have no theory to predict the slope, I have to estimate the slope from the data in Fig. 4 of Chap. III.4. The slope of the line there is approximately 2, so I take $d = 4$ approximately. That's the only curve-fitting in this model. Everything else is derived. Thus $E(L) = 2(S-1) = 2S - 2$. Among webs with 26 species, the average number of links is predicted to be 50. Since the number of species ranges from 3 to 48 in our data, the constant term -2 in this equation is negligible compared to the term $2S$ proportional to S. Qualitatively, the cascade model reflects the observation that the expected number of links is nearly proportional to the number of species. Quantitatively, the link-species scaling law fits because I made it fit by taking $d = 4$.

Roughly speaking $d/2$ (more exactly, $d(S-1)/(2S)$) is the average number of predators per species and roughly $d/2$ is the average number of prey per species. Here the average is taken over all webs with a given number of species and, more importantly, over all species within a web. Obviously, a species at the top of the cascade has no predators, while a species at the bottom of the cascade has no prey. However, averaged over all positions in the cascade, an average species has about 2 predators and about 2 prey.

As the number of species becomes large, the cascade model predicts 26 percent top species, 48 percent intermediate species and 26 percent basal species. Thus the model predicts a 1:1 ratio of predators to prey. We observe 29 percent top species, 53 percent intermediate and 19 percent basal (see Fig. A.2.2), giving roughly a 1.1:1 ratio of predators to prey. The model predicts the following percentages of basal-intermediate, basal-top, intermediate-intermediate and intermediate-top links: 27, 13, 33, and 27. We observe, correspondingly, 27, 8, 30, and 35 (see Fig. A.3.2).

It is nice that the cascade model reproduces all the laws of scale-invariance qualitatively, but far more striking that the cascade model gives a remarkable quantitative agreement between observed and predicted proportions. We put one number d into the cascade model and get out five independent numbers (because the three species proportions have to add up to 1 and the four link proportions have to add up to 1). I emphasize that these predictions use only the observed ratio of links to species.

For a finite number of species, we calculated from the cascade model the expected fraction of top species and the predicted variance. In Chap. III.2, Fig. 1 shows that the cascade model predicts not only the means but also the variability in the proportion of top species. I don't know whether the cascade model can predict the variability in proportions of links because I don't know how to calculate analytically what variability the cascade model predicts and have yet to do appropriate numerical simulations.

The cascade model was built to, and does, explain qualitatively and quantitatively the mean proportions of different kinds of species and links. Can the cascade model describe the number of chains of each length, counting all the possible routes from any basal species to any top species?

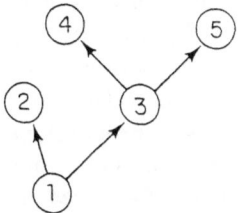

Fig. I.0.2. Hypothetical food web to illustrate how the frequency distribution of chain lengths is counted. There is one chain length 1 (from species 1 to species 2) and there are two chains of length 2 (from species 1 to species 4 and from species 1 to species 5)

Let me illustrate with an artificial example (Fig. I.0.2) how to get a frequency histogram of chain length from a web. The link from 1 to 2 is a chain of length 1. The path 1, 3, 4 is a chain of length 2, and the path 1, 3, 5 is another chain of length 2. A numerical summary of the chain length distribution of the web in Fig. I.0.2 is that it has one chain of length 1, two chains of length 2 and no longer chains.

In Chap. III.4, Fig. 1 shows the expected number of chains of each length, according to the cascade model, using parameters of a typical web, namely 17 species and d near 4. The same figure also shows the results of one hundred computer simulations of the model using the same parameters. The sample mean numbers of chains of each length agree well with the theoretically expected number calculated from the model. That agreement is evidence that both the calculations and the simulations are right.

How well does the cascade model predict the observed distribution of chain length of a real web? To find out, we generated random webs according to the cascade model with the parameters of the observed web. We measured how often the chain length distribution of a random web was further from the chain length distribution predicted by the cascade model than the real observed chain length distribution was from the predicted distribution. We used two measures of goodness of fit: the sum of squares of differences and a measure like Pearson's chi-squared. If the discrepancy between the observed and the expected frequency distributions was not larger than most of the discrepancies between webs randomly generated according to the cascade model and the mean frequency distribution expected from the model, we said the fit was good. If the discrepancy between observed and predicted chain length distributions was bigger than most simulated discrepancies, we said the fit was bad.

Have no illusions about what a good fit means. In Chap. III.4, Table 2, food web 18 illustrates a good fit while food web 37 illustrates a poor fit. Food web 18 is the Kapingamarangi Atoll food web (see Fig. I.0.1 above) of Niering (1963). For food web 18, Table 2 of Chap. III.4 shows four chains of length 4 while Fig. I.0.1 has one chain of 4 links. The reason for this discrepancy is that Cohen (1978) added to the predation matrix for this web links that Niering (1963) described in his text but omitted from his figure.

Of 62 webs in Briand's original collection, the chain length distributions of 11 or 12 (depending on the measure of goodness of fit used) were badly described by the cascade model. The model's success with the chain length distributions of 50 or 51 of these webs made us afraid that we had overfitted the model to the data. Perhaps by constructing the cascade model to explain the mean proportions of top, intermediate and basal species and the proportions of different kinds of links, we had used so much information from the data that there was no possibility for the fits to the chain length distribution to be bad, even though they were not used to build the model. This worried us. So Briand found and edited 51 additional webs which we had never analyzed before. The ratio of links to species was roughly the same for these new webs as for the old webs, as I mentioned already. With these fresh data, we found only five webs with poor fits to the cascade model's predicted frequency distribution of chain length. The proportion of poor fits, 5 of 51 webs, was smaller among the new webs than it had been among the original webs (Chap. III.4).

The cascade model uses no information about chain length to predict the frequency distributions of chain length. The predictions derive solely from the number of species and the number of links. No other parameters are free.

Apparently, the niche overlap graph of most webs is an interval graph, i.e., the overlaps of trophic niches revealed by most webs are consistent with the trophic niches being 1-dimensional (Chaps. II.4, III.6). Unexpectedly, the cascade model predicts the conditions under which intervality is common or rare. This prediction uses no fitted parameters.

The cascade model needs to be tested further, tested until it fails, as it surely will. How well can the cascade model predict the moments of chain length (as Stuart Pimm has asked), or patterns of omnivory? Can the cascade model relate to the combinatorial web models of Sugihara (1982, 1983, 1984)? Much testing remains to be done.

The cascade model makes new predictions. In large webs ($S > 17$), the cascade model implies a novel rule of thumb: The mean length of a chain should equal the mean number of prey species plus the mean number of predators of an average species (Chap. III.5). Both should equal a number near 4. This purported rule is open to empirical test.

The cascade model explains qualitatively why the longest chains in webs are typically short. Newman and I (Chap. III.5) derived the relative expected frequency of various chain lengths as the number of species goes to infinity, according to the cascade model, and found that, with a realistic value of d, practically no chains have length 8, 9, or 10. The cascade model predicts that, in very large webs, the length of the longest chain grows like $(\log S)/(\log \log S)$. That is very slow growth. In a web with 10^8 species, which is probably an upper bound for the world, the cascade model predicts that the longest chain will almost never have more than 20 links.

5. Connections

The cascade model connects with quantitative questions and theories elsewhere in ecology. I will sketch the connection of the cascade model with three topics: the species-area curve, the relative importance of predation and competition in communities, and allometric equations for the effects of body size.

First, one of the best known quantitative empirical generalizations of ecology is the species-area curve (MacArthur and Wilson 1967; Schoener 1976, 1986; Diamond and May 1981). In its simplest form, the species-area curve asserts that the number of biological species on an island is proportional to the area of the island raised to some power near 1/4. (When examined in detail [Schoener 1986], species-area curves are vastly more complicated.) The cascade model predicts, among other things, how the mean or maximal length of chains depends on the number of trophic species in a community. If the number of trophic species can be assumed or demonstrated (by a future empirical study of actual webs) to be proportional to the number of biological species, then a combination of the species-area curve and the cascade model predicts how chain length should vary on islands of different areas.

Without going into the details of the formulas, it is evident that if the number of species on an island increases very slowly with area, and if the maximal or mean chain length in a web increases very slowly with the number of species in a community, then the maximal or mean chain length should increase extremely slowly, or be practically constant, with increasing island area. The combination of the species-area curve and the cascade model explains, qualitatively at least, why there is not a known relation between the area a community occupies and the mean or maximal chain length of its web.

An alternative explanation, suggested by Robert T. Paine, is that there is no known relation between the area of a community and the mean or maximal chain length of its web because nobody has looked for such a relation. If the cascade model provokes an ecologist to examine the relation empirically, the model will have served a useful purpose.

Second, the cascade model relates to the roles of competition and predation in ecological communities. Hairston, Smith and Slobodkin (1960), as described succinctly by Schoener (1982, p. 590), "argued that competition should prevail among top predators, whereas predation should prevail among organisms of intermediate trophic status, mainly herbivores. Because the herbivores are held down by competing top carnivores, competition should prevail again among the herbivore's [sic] food species, green plants." Menge and Sutherland (1976, p. 353) proposed, by contrast, that as trophic position goes from high to low within a community, the relative importance of predation should increase monotonically while the relative importance of competition should decline monotonically. Connell (1983) and Schoener (1983) reviewed at length field experiments on interspecific competition which bear on these generalizations, and Schoener (1985) analyzed the points of agreement and disagreement in the two reviews.

Predation and competition can be interpreted in terms of quantities computable from the cascade model. It is then possible to examine whether these quantities behave according to the generalizations of Hairston et al. (1960) or Menge and Sutherland (1976). For example, a natural measure of the amount of predation on trophic species i in the cascade model is the expected (or average) number of predators on trophic species i, which is easily seen to be $d(S-i)/S$. There are $S-i$ species above species i in the trophic pecking order, and the probability that any one of them will feed on species i is d/S, so the expected number of predators on species i is the product $d(S-i)/S$. Since $i = 1$ is the lowest trophic position in the cascade model and $i = S$ is the highest, the cascade model implies that this measure of predation should increase linearly as trophic position goes from high to low within a community, exactly as proposed by Menge and Sutherland (1976). As the generalizations of Hairston et al. and Menge and Sutherland pertain to the relative importance of competition and predation, the behavior of a measure of predation needs to be related to the behavior of a measure of competition, such as one used by Briand (Chap. II.5).

Third, physical interpretations of the ordering of trophic species assumed in the cascade model may make it possible to connect the study of webs with the study of allometry and physiological ecology. The combination might be called "ecological allometry". For example, extending to entire webs a qualitative suggestion of Elton (1927, pp. 68–70) for individual chains, suppose that each trophic species consists of individuals more or less homogeneous with respect to size or mass, and that the larger the species' label $i = 1, 2, \ldots, S$ in the cascade model (i.e., the higher the trophic position), the larger the mass of each individual in that species. (Food chains of parasites generally follow the opposite rule: parasites are much smaller than their hosts [Elton 1927, Chap. 6].)

The assumption that body mass increases with a species' label i in the cascade model can be tested empirically, since it implies that no (nonparasitic) trophic species can eat a species larger than itself. When trophic species in real webs are ordered by body mass, is the predation matrix generally upper triangular, as assumed by the cascade model? [Since this introduction was written, the identical question occurred independently to Warren and Lawton (1987). In the food web of an acid pond community, when trophic links were determined by laboratory tests (not in the field), the predation matrix was largely, but not entirely, upper triangular.]

If size-ranked predation matrices are generally upper triangular, the cascade model can connect facts about food webs with quantitative empirical generalizations that physiological ecologists have discovered about body size (Peters 1983; Calder 1984; Peterson et al. 1984; Peters and Raelson 1984; Vézina 1985; May and Rubinstein 1985). From preliminary calculations, it appears that several empirical ecological generalizations, which have previously lacked a physical explanation, may be derived from a combination of the cascade model with assumptions or facts about body size.

6. Applications

This work may eventually contribute to human well-being in four ways.

First, environmental toxins cumulate along food chains. "Eating 0.5 kg of Lake Erie fish can cause as much PCB [polychlorinated biphenyl] intake as drinking 1.5×10^6 L of Lake Erie water"(National Research Council 1986). An understanding of the distribution of the length of food chains is necessary, though not sufficient, for understanding how toxins are concentrated by living organisms.

Second, people have not been very successful at anticipating all the consequences of introducing or eliminating species. Such perturbations of natural ecosystems are being practiced with increasing frequency in programs of biological control. An understanding of the invariant properties of webs is essential for anticipating the consequences of species' removals and introductions. For example, a perturbation that eliminated most of the top trophic species, or most of the basal trophic species, could be expected to be followed by major changes in the structure of the web if the community adjusts to reestablish invariant proportions of top, intermediate and basal species. The cascade model or its successors may eventually make it possible to derive more quantitative predictions.

Third, an understanding of webs will help in the design of nature reserves and of those future ecosystems that will be required for long-term manned spaceflight and extra-terrestrial colonies. A nature reserve with all top species would be expected to have trouble, according to the cascade model. For humans to survive and to be fed in space, we need to know more about the care and feeding of webs.

Fourth, and finally, since some webs include man, an understanding of webs may give us a better understanding of man's place in nature, here on earth. We have not detected any consistent differences between webs that contain man and webs that do not. Of course, we have not looked yet at webs of agricultural ecosystems strongly influenced by man. When we look at new classes of webs, we may expect to see new patterns.

Chapter II. Empirical Regularities

§0. Untangling an Entangled Bank

Joel E. Cohen

> ... plants and animals, most remote in the scale of nature,
> are bound together by a web of complex relations.
>
> Darwin (1859, p. 73)

The chapters in the empirical portion of this book are part of a funny story. At least, the story is funny if viewed from sufficient distance. In cartoon form, the story has three panels. In the first panel, country folks (the field ecologists) happily record the glories of nature in their notebooks and publish summaries in the form of food webs. In the second panel, naive city folks (the theoretical ecologists) assemble and analyze the food webs. They trumpet to the world general patterns that emerge from the collected food webs. (That's what this empirical part of the book is about.) In the third panel, the field ecologists, some puzzled, some aroused, rear back and dig in their heels: "Wait a minute! We didn't expect anybody to use our food webs as *data.*" Meanwhile, in the background, a few mice busily build a scaffolding of theory to hold together the general patterns found in panel two. (That's what the theoretical portion of the book, Chap. III, is about.)

1. Natural History

Let us return to the first panel. Food webs figure, in literary garb, in one of the most famous paragraphs in biology, the last paragraph of Darwin's *On the Origin of Species* (1859). That paragraph begins:

It is interesting to contemplate an entangled bank, clothed with many plants of many kinds, with birds singing on the bushes, with various insects flitting about, and with worms crawling through the damp earth, and to reflect that these elaborately constructed forms, so different from each other, and dependent on each other in so complex a manner, have all been produced by laws acting around us.

Darwin summarizes his theory of evolution and resumes:

Thus, from the war of nature, from famine and death, the most exalted object which we
[Darwin speaks anthropocentrically here] are capable of conceiving, namely, the production of
the higher animals, directly follows.

The study of food webs is the study of that war of nature, and of the laws acting
around us which govern it.

Darwin's literary account of food webs presaged and soon motivated empirical
studies. In papers published from 1876 onward, Forbes (1977 reprint) described
in detail the diets of birds, fishes and insects. These papers may contain the first
sink food webs. In his 1878 paper on the food of Illinois fishes, Forbes explained
the purposes of his investigations, and emphasized that "We ought also to gain,
by this means, some addition to our knowledge of the causes of variation, of the
origin and increase, the decline and extinction of species ... What groups crowd
upon each other in the struggle for subsistence? Do closely allied species, living
side by side, ever compete for food?" Forbes did not cite Darwin explicitly here,
but Darwin's ideas appeared clearly.

Camerano (1880), in a paper generously sent me by Stuart Pimm, initiated
more abstract descriptions of the entangled bank. His hypothetical tree-like
diagrams show feeding relations among different classes of organisms. Unlike
Forbes's sink webs, Camerano's sketch was intended as a description of an entire community. The diagrams are remarkably similar in form to Darwin's illus-
tration of divergence of character (1859, lithograph inserted between pages 116
and 117), though with different labels. In function, Camerano's diagrams closely
resemble schematic webs seen in elementary textbooks of ecology today. Camerano distinguished vegetation, herbivores, and carnivores. Among carnivores, he
distinguished predators, parasites and endoparasites. He even initiated a mathematical formalism to describe or explain equilibrium in complex communities.

The earliest food web graphs in English that I know of are Shelford's (1913)
hypothetical descriptions of communities. It remained for British empiricism to
produce a real description of a whole community. Summerhayes and Elton (1923)
reported a detailed food web of Bear Island. Another detailed food web centered
on the herring and plankton community (Hardy 1924) and a web of the animals
that live on pine (Richards 1926) quickly followed. An industry was born. The
industry continues to this day, with considerable improvements in technology
and product.

As for the technology of inferring the existence of trophic links, Forbes collected consumers and examined macroscopically the contents of their stomachs.
Many observers determined trophic links simply by macroscopic observation of
one living organism eating another. The techniques of natural history have not
lost their value, but have recently been joined by more sophisticated techniques.
The stomachs of certain marine organisms contain a gray-green paste from which
it is not easy, or even possible, visually to identify the diet. Recently, antibodies
have been applied to that paste to identify proteins that are specific to individual prey species. In addition, the isotopic composition of the tissue of organisms
living around hot-water vents in the deep ocean floor has been carefully analyzed
to determine whether the tissue was based on sun-based detritus chains or on

chains that derived energy and nutrients from the earth's core (Van Dover et al. 1988). (See also Hart 1989.)

As for product, only recently have accounts of food webs improved on the reporting, more than a century ago, of Forbes (1977 reprint). Forbes gave detailed matrices, listing consumers by species in column headings and prey often by species in row labels, with actual numbers of specimens of each consumer and each prey item. Some of his predation matrices were classified by month of observation to bring out seasonal changes in feeding habits. Unfortunately, Forbes's high standard was not maintained. In the hands of many later reporters, a food web became little more than an error-prone diagram with boxes and arrows. By happy contrast, a recent tropical rocky intertidal web is presented as a large matrix accompanied by the number of individuals of each kind observed feeding, the mean wet weight of each kind of consumer, various exclosure experiments to determine the regulatory role, if any, played by consumers, percent cover of certain sessile species, rates of biomass accumulation and change in abundance of various species in the web (Menge et al. 1986). In another recent collection of webs, the detailed variations in time and space of feeding relations in tree holes of various sizes are accompanied by the abundance of each species (Kitching 1987).

Chapter IV of this book presents 113 community food webs in the form of predation matrices, with a simple identification of each species. In many cases, the original sources provide additional information. Based on the numerous other smaller collections of webs assembled by other investigators, I guess that probably over a thousand community food webs, and perhaps equal numbers of source webs and sink webs (Cohen 1978), have been observed, though many remain unpublished.

Just as molecular biology has benefited from computerized banks of protein sequences and nucleic acid sequences, ecology needs to establish an ecobank, in which investigators could share their best estimates of natural webs. An ecobank could serve as a depository for the many webs which may not merit independent publication but which are, nevertheless, highly useful as a basis for further analysis. Establishment of an ecobank with the advice and governance of active ecologists might encourage the development of common standards or language for presenting webs. Descriptions of a web could be refined as successive investigators contributed additional information about body sizes, population abundances or temporal variations, for example. Data could be distributed via telecommunications or standard computer media.

An ecobank is a long way from the entangled bank Darwin (1859) described. As he was an assiduous gatherer of data, I like to believe he would have approved the idea.

2. Empirical Laws

In the second panel of the cartoon, order emerges from the replication of chaos.

When I was a graduate student two decades ago, food webs were presented like (pre-Darwinian) butterflies, as works of natural beauty, to be admired for all their inscrutable complexity. The patent variation between one food web and the next web was taken as clear evidence (by my friends who were molecular biologists) that the structure of webs was not a fit subject of science, or (by my friends who were ecologists) that webs could only be understood one at a time, through detailed study of the population dynamics and natural histories of each kind of organism in the web.

The primitive theoretical concept behind all the work in this book is that it is worth looking at an ensemble (of webs, in this case) for order that may not be apparent in isolated individuals. Students of statistical mechanics could recognize this primitive theoretical concept as their own, but it is not uniquely their own; statistical mechanics owes it, I believe, to the social and biological sciences of the first two-thirds of the nineteenth century, and it probably goes further back.

Based on the first collection of food webs (Cohen 1978), I observed a few empirical regularities that had not been noticed before. The trophic niche overlap graphs were interval graphs surprisingly often (Chap. II.4); and the ratio of the number of kinds of prey to the number of kinds of predators seemed to be independent of the total number of kinds of organisms in the web (though the ratio was higher for sink webs as a group than for community webs as a group) (Chap. II.5).

The easy availability of data and the possibility of unsuspected order attracted attention from people who like to look for order. Based on the same data, MacDonald (1979) promptly observed that the ratio, which he called β, of the number L of trophic links to the number S of kinds of organisms in the 30 webs had a mean 1.88 and a fractional root mean square deviation of 0.27, with no notable difference between sink webs and community webs. For all 30 webs of Cohen (1978), the ratio lay between 1 and 3.

Simultaneously, Rejmánek and Starý (1979) plotted $L/[S(S-1)/2]$, a quantity they called the connectance C, as a function of S for 31 plant-insect-parasitoid webs, one data point for each web. It appears likely that their webs were constructed as source webs, in the terminology of Cohen (1978). The data points for nearly all 31 webs fell between the two hyperbolic curves $C = 2/S$ and $C = 6/S$. The curve $C = 4/S$ ran through the center of this band and through the center of the data, though Rejmánek and Starý (1979) preferred $C = 3/S$ as a description of central tendency. MacDonald (1979), in a note added in proof, pointed out that the hyperbolic relationship of Rejmánek and Starý (1979) between connectance and number of species is equivalent to the constancy of $\beta = L/S$ when $(S-1)/S$ approximates 1. If $C = 4/S$, then $\beta = 2$ approximately.

The hyperbolic relation between connectance and number of species was confirmed by Pimm (1982) and Auerbach (1984), with some additional data including Cohen's (1978).

Without reference to MacDonald (1979) or Rejmánek and Starý (1979), Briand (Chap. II.5) revised some of Cohen's (1978) community food webs and collected an additional 27. (Warning: Briand's connectance C is twice the connectance C of Rejmánek and Starý (1979), Pimm (1982), and Auerbach (1984).) Briand found that the number of trophic links "increases as a nearly linear function of S" and, based on a straight line fitted by the method of least squares to the logarithms of S and L, suggested that $L = 1.3S^{1.1}$.

Cohen and Briand (Chap. II.3) examined the relation of L to S with 62 community food webs. Without presuming to distinguish a power law relation with an exponent of 1.1 from simple linearity (corresponding to an exponent of 1.0), we found approximate proportionality between L and S. The coefficient of proportionality was roughly 1.9 with a standard deviation of 0.1. Still unaware of MacDonald's (1979) work, we at least noted the equivalence of our linear relation to the hyperbolic relation between connectance and number of species.

Cohen and Newman (Chap. III.2) gave the name "link-species scaling law" to the linear relation between links and species. Cohen, Briand and Newman (Chap. III.4) confirmed the relation with all 113 webs collected by Briand and estimated a coefficient of proportionality (the slope of the regression through the origin) of 2.0 with a standard deviation of 0.1.

Only in preparing this book did I realize that MacDonald (1979) was the first to remark that $\beta = L/S$ varies little from one web to another. Thus the "link-species scaling law" has been discovered independently three times, in slightly different but mathematically equivalent forms: first by MacDonald (1979), as the near-invariance of β; second by Rejmánek and Starý (1979) as the hyperbolic relation between connectance and the number of species; and third by Briand (Chap. II.5) and Cohen and Briand (Chap. II.3) as a linear relation between trophic links and species. One hundred of the 113 webs studied by Cohen, Briand and Newman (Chap. III.4) are independent of the webs used by MacDonald, and the webs studied by Rejmánek and Starý are independent of all the others. The link-species scaling law appears as a robust fact about food webs.

The story of the link-species scaling law is not the only case of simultaneous or independent discovery in the chapters that follow. Sugihara (1982) observed that it was natural to consider kinds of organisms that had identical predators and identical prey as a single unit. He called two kinds of organisms with the same diets and predators "trophically equivalent." At the same time, without giving an exact definition, Yodzis (1982) suggested that organisms with "similar" diets and predators could be considered as "trophic species". Unaware of these proposals, Briand and Cohen (Chap. II.2) introduced the "lumping" of trophically identical species, i.e., species that were trophically equivalent in the sense of Sugihara (1982), and began referring to the resulting equivalence classes as "trophic species" (Chap. II.3).

Based on this experience with the "link-species scaling law" and the concept of "trophic species", I would not be surprised if other empirical generalizations and concepts reported in this section of the book have also been anticipated in one form or another.

The empirical papers in this portion of the book do not pretend to be encyclopedic. Other students of food webs and community ecology have reported many other empirical generalizations. These generalizations concern food chain length (see Chaps. III.3–4), the shape of trophic pyramids, element recycling, niche overlaps (Cohen 1978) and niche packing (Sugihara 1982). Moreover, I would be surprised if no further empirical generalizations emerge from data such as those assembled in Chap. IV.

However, some caution is necessary about forming generalizations. May (1983) pointed out that webs from different classes of habitats, e.g., terrestrial vs. marine, may differ because of the difference in training and interests of the classes of people who study those habitats, rather than because of differences in nature. This valid caution does not apply to generalizations based on all food webs.

3. Facts or Artifacts?

In the third panel, the field ecologists react to the theoretical ecologists who muck about in their food webs.

Paine (1988) pointed out numerous dangers and pitfalls in the use of existing webs as a foundation for ecological generalizations and theory. Some of his cautions apply to generalizations based on all food webs, not just to generalizations about differences among classes of webs. For example, to explain the hyperbolic relation between connectivity and number of species, Paine (1988) proposed the hypothesis of "artistic convenience": "When S [the number of species] is small, more links can be portrayed; when S is substantially higher, only those deemed to be most meaningful are drawn and connectance is correspondingly reduced. Necessity for graphical clarity, then, results in the omission of some links."

The merit of this criticism of graphical food webs cannot be denied, but it does not apply to webs reported in matrix form. Here is a student research project: separate the 111 published webs in Chap. IV according to whether the original web was published as a picture or as a matrix or as both, and see whether the points (species, links) fall into separate clouds according to the format of reporting. Send me the results, please.

Put positively, the possible bias introduced by artistic convenience argues for the use of the venerable predation matrix in reporting webs. The predation matrix has advantages besides avoiding the siren of "artistic convenience." The elements of the matrix can indicate the magnitudes of flows. If an observer puts *every* observed feeding relation in the matrix, as Forbes did more than a century ago, then the analyst can experiment with different threshold levels in deciding which trophic links are important and which are not. The margins of the matrix can easily accommodate reports of observation effort (e.g., hours spent, traps set), sample sizes, means and variances of body weights, estimates of population abundance or biomass, and other useful descriptors that will bind food webs to the rest of population biology and ecology. Finally, matrices are machine readable, and therefore easily shared as data.

Unfortunately, large, detailed matrices are not easily read by people who, after all, are the final consumers of food webs. Hence, an artistically done graph of a web, even if incomplete, retains its value for conveying the ecologist's impression of a community.

In addition to the problem of communicating results, Paine (1988) raises fundamental problems of gathering the data. Some species are harder to see than others. Some species, while easily seen, are more mobile than others and therefore less readily incorporated in a web. Transient species, as is their custom, come and go. How should the ecologist deal with them? The trophic relations of individuals within a biological species sometimes (e.g., Hardy 1924) depend strongly on the individual's age, stage or size. Summaries at the species level should not overlook these ontogenetic differences.

All these problems (differential ease of obervation, transient species, and age-stage effects) also affect other studies of community ecology, such as studies of species-abundance distributions, energy flow, or island biogeography. Paine's criticisms should stimulate empirical webologists to be more explicit about how they deal with the problems he raised. In many papers that report food webs, the existence of these problems is less troubling than the absence of any mention in the report that the observer was aware of them and had an explicit, consistent procedure for dealing with them.

There are other problems with web data besides the ones that troubled Paine (1988). For example, "dimension" has different clear definitions in different contexts, but caution is required in each context. Schoener (1974), while not focusing on webs, used "dimension" to refer to any measurable variable useful in describing a species' niche, including feeding and distribution in time and space. Cohen (1978) considered only trophic, or feeding, dimensions. Cohen, Briand and Newman (Chap. III.4) and Briand and Cohen (Chap. II.6) consider the apparent flatness or solidity of the physical setting of the food web, at the scale of the human observer. L. Dyck, M.J. Sibbald and P.R. Sibbald (personal communication, December 1987) pointed out that a forest that looks three-dimensional on the scale of a human observer might look flat on the scale of an insect who lives on a leaf. Other organisms of intermediate size might perceive a mixture of apparent physical dimensions. They suggested that fractal "dimension" might be useful in characterizing habitats. The suggestion remains to be explored. The important point is that "dimension" is not a unitary concept in ecology, and alertness is needed in using it to characterize food webs.

Like "dimension", the term "variability" figures importantly in Chap. II.6. This term, too, is used in different senses in different contexts. Even within mathematical statistics, there are many measures of variability. In ecology, each formal measure can be multiplied by a variety of empirical interpretations. Although the classification of food web environments in Chap. II.6 as "constant" or "fluctuating" is subjective to some extent, it preceded our analysis of the influence of variability on food chain length, and therefore was not biased by that analysis. I hope that less subjective measures of variability, comparable among food webs, will be systematically used by future reporters of food webs.

The criticisms that the empirical webologists have leveled at theoretical analyses of their data return to them as a challenge. Their data are valuable, perhaps more valuable than they knew. The methods by which the data are gathered and reported need to be made more explicit, more consistent, and more trustworthy. The many imperfections of the data assembled in this book, and the likely imperfections of the theory based on those data, should provoke field ecologists to render the data obsolete by replacing them with more food webs more systematically observed and more carefully reported. At the same time, flawed data are not necessarily worthless data. Further analyses of the assembled data on community food webs are welcome.

All the preceding criticisms of generalizing from existing webs are reasonable, however valid they may be. I suspect that the energy that sometimes accompanies such criticisms may arise from an irrational source, rarely articulated. A field ecologist who has devoted years of life to the minute observation and analysis of one or a few ecological communities is aware, as no reader of a brief published report can be, of the special characteristics that distinguish the biota and physical habitat of his or her study sites, both from one another and from the study sites of others. A theoretically inclined reader who extracts a few simplified measurements from an already condensed publication in order to show how this lovingly observed community is like all other communities must indeed appear presumptuous and be a source of irritation to more than one field ecologist. There is an inescapable tension between the thrilling uniqueness of individual communities and general empirical ecological laws, whatever their explanation.

A helpful perspective comes from the study of the solar system. "Unique events are difficult to accommodate in most scientific disciplines. The solar system, however, is not uniform. All nine planets (even such apparent twins as the earth and Venus) and over 50 satellites are different in detail from one another ... [There are many examples.] All this diversity makes the occurrence of single events more probable in the early stages of the history of the solar system" (Taylor 1987, p. 477). Notwithstanding this diversity, there is no doubt among physicists that these single events and the dynamics of the solar system were, are and will be governed by uniform laws of physics and chemistry. Only through a thorough understanding of those laws has it become possible to collect appropriate data about the unique characteristics of planets and satellites from earth-based observations, lunar collections, and planetary missions. Only through an increasing theoretical understanding of those laws in combination with the best data has it become possible to make good inferences about unique early events.

Likewise, in the study of food webs, and of community ecology generally, ecologists require a good understanding of general laws to appreciate the common features and differences among food webs and ecological communities.

A. General Regularities

§1. Ratio of Prey to Predators in Community Food Webs

Joel E. Cohen

Whether the diversity of resources limits the diversity of consumers, and specifically, whether the number of kinds of prey limits the number of kinds of predators, has been of continuing interest in theoretical ecology and wildlife management (Haigh and Maynard Smith 1972; Levin and Paine 1974; Sullivan and Shaffer 1975). Food webs from the ecological literature were collected in machine readable form to study this question empirically. We report here that in community food webs, the ratio of the number of kinds of prey to the number of kinds of predators seems to be constant, near 3/4. This invariance has not been noticed in earlier studies of individual cases.

Before analysis, food webs were characterised as one of three types – community, sink and source. Community food webs describe all kinds of organisms (possibly restricted to some location, size or taxa) in a habitat, without reference to the eating relations among them. Sink food webs describe all the prey taken by a set of one or more selected predators, plus all the prey taken by the prey of those predators, and so on. Source food webs describe all the predators on a set of one more selected prey organisms, plus all the predators on those predators, and so on. Sink and source food webs, hypothetical or schematic constructions, and avowedly incomplete, partial or tentative food webs were excluded from further study. Fourteen community food webs were thus selected. The complete data and individual cases are discussed in Cohen (1978). When the report of a food web contained ambiguous or uncertain information about a feeding relation, the web was included in two versions, one based only on the unambiguous information and the other incorporating the additional uncertain or probable eating relations. The analysis here, based on all versions, makes no claim that the data points are statistically independent and attaches no probability values to the statistics calculated.

The food webs describe the diets or predators not of individual organisms but of kinds of organisms. A 'kind of organism' may be a stage in the life cycle or a

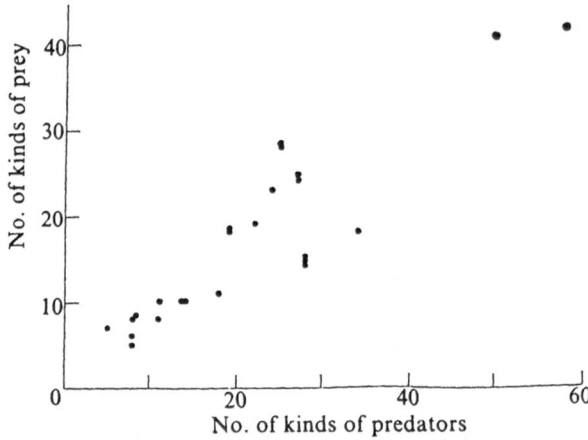

Fig. A.1.1. The number of kinds of prey and the number of kinds of predators in community food web versions

size class within a single species, or a collection of functionally or taxonomically related species, according to the practice of the original report. The numbers in the following analyses refer to these ecologically defined kinds of organisms, not necessarily to any conventional taxonomic unit. A predator is defined as a kind of organism that consumes at least one kind of organism in the food web. A prey is defined as a kind of organism that is consumed by at least one kind of organism in the food web. Some kinds of organisms may be both predators and prey.

In community food webs, the number m of prey is very nearly proportional to the number n of predators (Fig. A.1.1). A least squares regression of m against n gives

$$m = 1.79 + 0.71n . \tag{1}$$

The sample standard deviation of the regression coefficient is 0.07 and the linear correlation coefficient between m and n is 0.90. The standard error of estimate, or sample standard deviation from regression, is 4.62. As is obvious from Fig. A.1.1, the regression may be well approximated by a straight line through the origin. The least squares regression is

$$m = 0.77n . \tag{2}$$

The proportionality between the number of prey and the number of predators in Fig. A.1.1 is based on 24 versions of 14 food webs reported over a period of decades. When the food webs were collected and encoded it was not known that such a simplicity would emerge. It therefore seems likely that this invariance in the proportions of predators and prey represents a fact about nature, rather than an artefact of collusion or convention.

Given that the proportion of prey to predators is a scale-invariant feature of community food webs, the proportion can be predicted quantitatively from other facts. For a given food web with m prey and n predators, let A be the number of predator-prey couples. (If X eats Y and Y eats X, the couples (X, Y) and (Y, X) are counted as distinct. If X eats X, (X, X) also counts as a couple. In the conventional graphical representation of a food web, A is the number of directed arrows from prey to predator.) Then within any food web

$$A = \text{(average prey per predator)} \times n$$
$$= \text{(average predators per prey)} \times m. \tag{3}$$

The grand mean over all 24 community food web versions, weighting each food web equally, of the average prey per predator is 2.418; the grand mean of the average predators per prey is 3.199. If these means apply to each food web, then substitution into equation (3) predicts

$$m/n = 2.418/3.199 = 0.756 \tag{4}$$

which differs trivially from the least squares regression in equation (2).

The simplicity of the argument from the proportionality between m and n to equation (4) may raise a suspicion that its success depends on an arithmetical fact rather than on the observed invariance of proportions of predators and prey in nature. A numerical example disproves this suspicion. Suppose a sample of community food webs consisted of two food webs. Suppose the first food web matrix had $m_1 = 8$ prey, $n_1 = 6$ predators, and $A_1 = 19.2$ predator-prey couples (neglecting the requirement that A_1 be integer for the sake of argument). Then its (average predators per prey)$_1$ is 2.4 and its (average prey per predator)$_1$ is 3.2. Suppose the second food web matrix had $m_2 = 4$, $n_2 = 10$, and $A_2 = 16$. Then its (average predators per prey)$_2 = 4.0$ and (average prey per predator)$_2 = 1.6$. Then the grand mean over both food webs of the average predators per prey is 3.2 and the grand mean of the average prey per predator is 2.4, which are close enough to the observed. But the straight line through the pairs (n, m) satisfies $m = 14 - n$. Only because nature assures a constant proportion of prey to predators do the grand mean of the average predators per prey and the grand mean of the average prey per predator apply to all food webs.

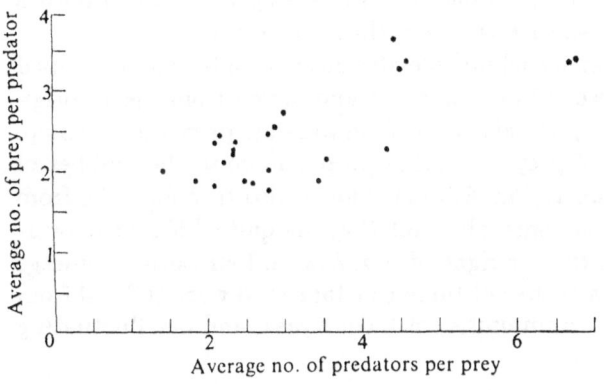

Fig. A.1.2. The average number of kinds of prey per kind of predator and the average number of kinds of predators per kind of prey in community food web versions

If the ratio of prey to predators in community food webs is a constant of the order of 3/4, then dividing equation (3) by n leads to the prediction that a regression (Fig. A.1.2) of average prey per predator against average predators per prey should be a straight line through the origin with slope 3/4. The regression coefficient of a straight line through the origin is 0.69, not far from 3/4.

In conclusion, in community food webs, the number of kinds of prey, as operationally defined by field ecologists, approximates 3/4 the number of kinds of predators. This results from the study only of an ensemble of food webs, rather than of individual cases.

§2. Community Food Webs Have Scale-Invariant Structure

Frédéric Briand and Joel E. Cohen

We have analysed 62 community food webs drawn from published studies and have found a remarkable regularity in ecosystem structure: in biological communities, the proportions of top, intermediate and basal species are, on average, independent of the total number of species. Hence, there is a direct proportionality between the numbers of prey and predators.

The finding (Chap. II.1) that, in community food webs, the ratio of prey to predators is 3:4 may be challenged on two grounds: first, it is based on a relatively small set of 14 webs, and second it may indicate that taxonomists have exercised greater taxonomic refinement in classifying organisms at higher than at lower trophic levels (Pimm 1982).

A community food web involves the feeding, that is, trophic, relations among all organisms found in a well-defined habitat by the original investigator. Organisms are separated into 'trophic species', which may be a single biological species, or a size class or stage in the life cycle of a single biological species, or a collection of functionally or taxonomically related biological species, according to the original report. Throughout this paper a 'species' refers to a 'trophic species', not necessarily to a single biological species. A 'top' species is a predator that has no predator. An 'intermediate' species is a species that is both a predator and a prey. A 'basal' species is a prey that has no prey.

The community food webs analysed include 40 webs assembled and described by Briand (Chap. II.5); of these, 13 are corrected and drawn from the 14 originally used by Cohen (Chap. II.1). Details of the food webs are presented in Chap. IV. We find that the number of prey is roughly proportional to the number of predators with a slope less than 1 (Fig. A.2.1a). This is also true for webs from constant and fluctuating environments, although they are quite different in overall structure (Chap. II.5). On the far right of Fig. A.2.1a, four outliers emerge from the general relationship: a cluster of three constant food webs (C), all from Fryer's study (1959) of littoral communities of Lake Nyasa, and one fluctuating

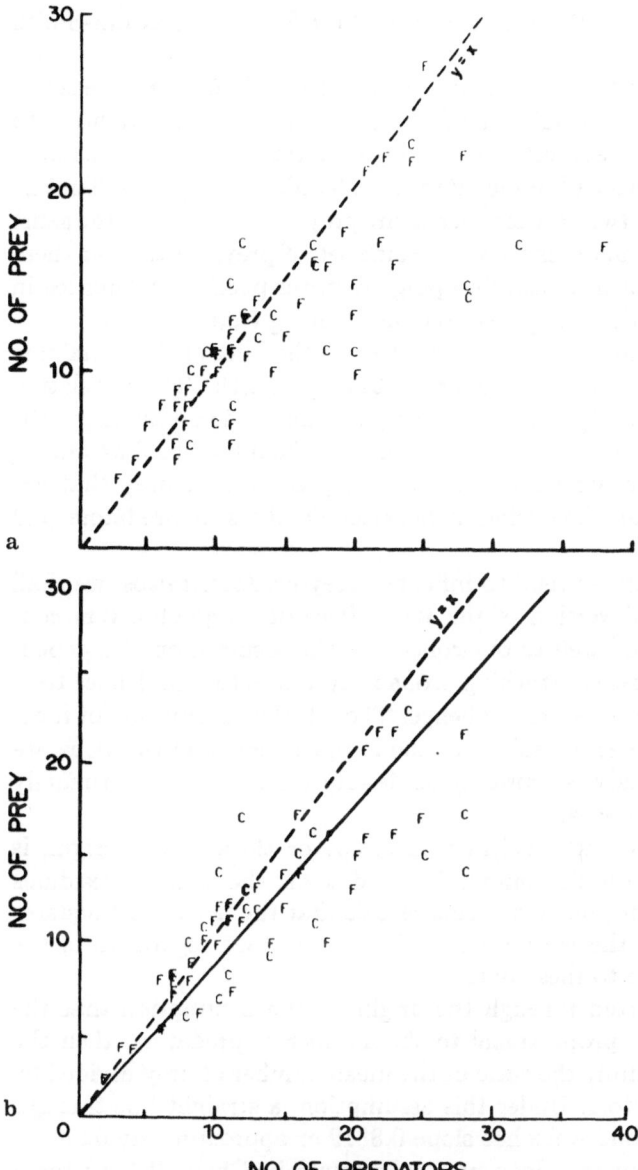

Fig. A.2.1a,b. Number of prey species as a function of number of predator species in 62 community food webs. F, fluctuating environment. C, constant environment. An environment is classified as 'fluctuating' if the original report indicates temporal variations of substantial magnitude in temperature, salinity, water availability, or any other major physical parameter. The magnitude, and not the predictability, of the fluctuations is the criterion of classification. The symbols F and C have been shifted from their exact locations by a small random amount to indicate when several food webs have exactly the same coordinates. (a) Original data; (b) after lumping. The solid line through the origin is fitted on the assumption that the variance of the residuals is proportional to the number of predators. The slope is 0.8819

food web (F) representing a salt meadow from New Zealand (Paviour-Smith 1956).

There can be no direct test of Pimm's conjecture (1982) of why the slope is less than 1, without repeating the original field studies with a uniform attention to taxonomic detail. As an indirect test, we examined the ratio of prey to predators in the 62 food webs, after we had 'lumped' trophically identical species. That is, in each food web, whenever two or more species are preyed on by exactly the same set of predators, and prey upon exactly the same set of prey, we treated them as one. This procedure, which we call 'lumping', removes possible differences in the propensity to split, both among observers and among trophic levels.

Lumping moves the outliers into or much closer to the bulk of the remaining data points, for both fluctuating and constant webs (Fig. A.2.1b). The correlation coefficient between numbers of predators and prey among the 43 fluctuating webs increases from 0.83 before lumping to 0.92 after, and from 0.58 to 0.64 among the 19 constant webs. In other words, eliminating predators or prey that are trophically identical tightens the relation between numbers of predators and prey.

Because individual observers tend to influence prey-predator ratios, we shall deal only with the lumped version of the webs. Because some observers contributed more than one food web to our collection, the assumption of independence that is required to justify attaching probability values to significance tests using the unlumped data is open to challenge. Though the assumption of independence is probably more acceptable with the lumped version of the webs, we shall base our statistical analysis primarily on descriptive statistics. Fortunately the patterns in the data are clear.

If a straight line, either with arbitrary intercept or through the origin, is fitted to a scatter plot of the 62 community food webs, the squared residuals increase with the number of predators. This reveals that the usual least-squares procedure, which assumes the variance of residuals constant regardless of the abscissa, is not appropriate to these data.

If a regression line is fitted through the origin on the assumption that the variance of the residuals is proportional to the number of predators, then the estimator of the slope is simply the ratio of the mean number of prey divided by the mean number of predators. Under this assumption, a straight line through the origin fitted to all 62 food webs has slope 0.8819 or approximately 0.9.

This slope is higher than the slope near 0.75 found in Chap. II.1, so there appears to be some merit in Pimm's suggestion (1982) that ecologists have exercised greater taxonomic refinement at high trophic levels than at low. This suggestion, however, is not quantitatively sufficient to account for the excess, that remains after lumping, in the number of predators over the number of prey.

Classical ecological theory views predators as generally limited by resources, and the diversity of predators in particular as being limited by the diversity of prey. From this perspective it would seem more natural to treat the number of prey as an independent variable and the number of predators as a dependent variable. However, when the number of predators is regressed against the number of prey, using a straight line through the origin with variance of residuals proportional to the abscissa, the standard error of estimate (with 61 degrees of freedom) increases from 3.1 to 3.5.

This observation means that the number of predators is a better predictor of the number of prey than the reverse. This raises the intriguing possibility that, in both constant and fluctuating environments, the number of predators is causally more important in controlling the number of prey than vice versa. Evidence and theory in favour of this suggestion have been independently reviewed by Jeffries and Lawton (1984).

That we find a linear relationship between number of prey and number of predators is not too surprising, since the x- and y-axes share a similar quantity, namely the intermediate species, which are both prey and predators. What is surprising is the tightness of the fit, considering the size and heterogeneity of the sample examined. This suggests two possibilities: either the redundant variable, that is, the number of intermediate species, is very large compared with the number of basal and top species in most communities, or the proportions of all species in a food web that fall into each of these three categories are, overall, independent of the total number of species.

Fig. A.2.2 illustrates the reality of the second alternative: in the 62 webs examined, the fractions of top, intermediate and basal species are, on average, independent of the total number of species, although there is a slight tendency for the fraction of top species to increase and for the fraction of basal species to decrease as the total number of species increases. To obtain a global estimate of the proportions of species in each of the three categories (top, intermediate, basal), we summed over all food webs the observed numbers in each category, and divided by the sum total of species over all food webs. The global proportions of top, intermediate and basal species correspond to the heights of the horizontal lines in Fig. A.2.2a-c.

The scatter of points about the horizontal lines in Fig. A.2.2, when constant and fluctuating food webs are considered together, agrees with the hypothesis that in each food web the top, intermediate and basal species are multinomially sampled from the total species in proportions that are constant for all webs. If the species counts are arranged in a 3×62 contingency table with rows for top, intermediate and basal species and one column for each web, a homogeneity test yields a χ^2 statistic of 138.9 with 122 df, which is not significant at the 0.1 level.

There is no evidence for a difference between constant and fluctuating food webs in the mean proportions of top, intermediate and basal species. A homogeneity test of a 3×2 contingency table with rows for top, intermediate and basal species and columns for constant and fluctuating food webs, and the summed species counts as cell entries, gives a χ^2 statistic of 1.4 with 2 df, which is not significant at the 0.1 level.

However, the proportions of top, intermediate and basal species in constant food webs, considered separately, are significantly more variable, and the proportions in fluctuating food webs, considered separately, are significantly less variable, than expected from multinomial sampling (using a 0.02 significance level). Separate homogeneity tests of the constant webs (in a 3×19 contingency table) and of the fluctuating webs (in a 3×43 contingency table) yield χ^2 of

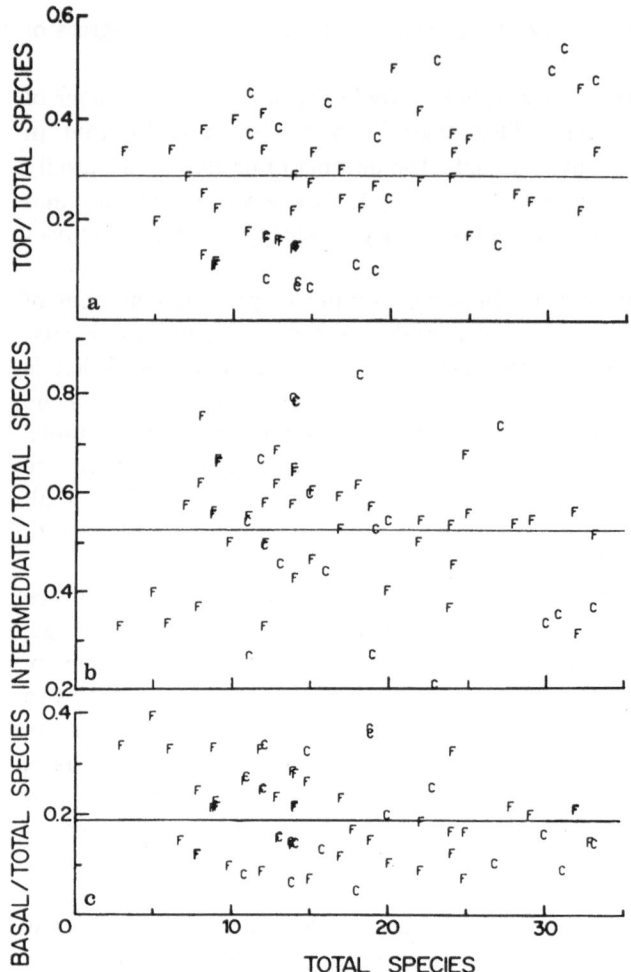

Fig. A.2.2a-c. Three ratios plotted as a function of the number of species. The fitted lines are constrained to be horizontal (slope = 0). (a) Top species/total species. The height of the line is 0.2853. (b) Intermediate species/total species. The height of the line is 0.5251. (c) Basal species/total species. The height of the line is 0.1896

82.8 with 36 df and 56 with 84 df, respectively. The visual counterpart of this statistical result is the appearance in each panel of Fig. A.2.2 of fluctuating food webs near the horizontal line and of constant food webs scattered above and below the band of fluctuating webs. Food web structure appears more constrained in fluctuating than in constant environments, as previously noted (Chap. II.5; Yodzis 1981).

We now show that the empirical regularities in Fig. A.2.1b can be derived from the approximate scale-invariance shown in Fig. A.2.2.

Let S be the total number of species in a single community food web, T the expected number of top species in that web, I the expected number of intermediate species, B the expected number of basal species, R the expected number of predators and Y the expected number of prey. By definition

$$S = T + I + B ,\tag{1}$$
$$R = T + I ,\tag{2}$$
$$Y = I + B .\tag{3}$$

By observation

$$T/S = p \quad \text{or} \quad T = pS, \quad \hat{p} = 0.2853 \text{ (Fig. A.2.2a)}\tag{4}$$
$$I/S = q \quad \text{or} \quad I = qS, \quad \hat{q} = 0.5251 \text{ (Fig. A.2.2b)}\tag{5}$$
$$B/S = r \quad \text{or} \quad B = rS, \quad \hat{r} = 0.1896 \text{ (Fig. A.2.2c) .}\tag{6}$$

Adding equations (5) and (6) and dividing by the sum of equations (4) and (5), we recover the observed regularity (Fig. A.2.1b)

$$Y/R = a \quad \text{or} \quad Y = aR, \quad a = (q + r)/(q + p) .\tag{7}$$

The predicted value $(\hat{q} + \hat{r})/(\hat{q} + \hat{p}) = 0.8819$ is identical to the observed $\hat{a} = 0.8819$ because of the formulas we used to estimate the slope a and the proportions p, q and r. However, the observation in Fig. A.2.1b that the average number of prey is a linear function of the number of predators is not a tautologous consequence of the estimation formulas. The proportionality of prey to predators follows from the scale-invariance we have discovered here. Were data available, it would be interesting to examine whether the distribution of biomass into top, intermediate and basal species is also scale-invariant.

We conclude that the values of any two of the three parameters p, q and r summarize succinctly a substantial amount of information about the empirical regularities found in community food webs and provide a factually grounded benchmark against which the deviations of particular food webs may be measured. Why these proportions take the values they do and why the proportions are scale-invariant remain open questions (see Chap. III).

§3. Trophic Links of Community Food Webs

Joel E. Cohen and Frédéric Briand

1. Problem and Hypotheses

How does the total number L of links in a web vary as the number S of species increases? At least three hypotheses are plausible. First, the number of potential links increases as S^2 because the maximal number of edges in a directed graph on S nodes is $S(S - 1)$. If there were a constant probability that any potential link were a real link, the mean $E(L)$ of L would be proportional to S^2. Second, if each species could eat or serve as food for only a finite number of other species, regardless of how many species were present in the community, the mean $E(L)$ would be proportional to S. Third, both of the preceding hypotheses might apply over different ranges of values of S. When the number of species in a community is small, L may be constrained only by the availability of potential links and hence vary as S^2. When the number of species in a community is large, L may be limited by the potential for interaction of each species and hence vary as S. The same relation between L and S might also arise because field ecologists might be more thorough in recording links when the total number of species in the community is small, but proportionally more prone to omission when the number of species is large.

According to these three hypotheses, plots against S, on the abscissa, of (a) the square root of L, (b) L, or (c) some power of L between 1/2 and 1, on the ordinate, should be approximately linear.

2. Definitions and Data

A community food web (henceforth abbreviated to "web") includes the feeding relations among all organisms found in a well-defined habitat by the original investigator. Organisms with identical sets of prey and identical sets of predators have been combined into a single "lumped" species (Chap. II.2). Throughout this chapter, "species" means trophic species, not necessarily a single biological species. A "top" species is a predator that has no predator. An "intermediate species" is a species that is both a predator and a prey. A "basal" species is a prey that has no prey. The number of basal, intermediate, top, and all species in a web will be denoted by B, I, T, and S.

A "trophic link" (hereafter, "link") is any reported feeding or trophic relation between two species in a web. Observers use various criteria to decide how much feeding justifies the reporting of a link and how much failure to observe feeding justifies reporting the absence of a link.

Webs are classified as arising in "fluctuating" or "constant" environments. The environment is considered to be fluctuating if the original report indicates temporal variations of substantial magnitude in temperature, salinity, water availability, or any other major physical parameter. This fluctuation may result from a pronounced seasonality, as in temperate terrestrial systems, from daily oscillation, as in intertidal systems, or from irregular perturbations, such as hurricanes. The magnitude, and not the predictability, of the fluctuations is the criterion of classification. Only 19 of 62 environments in our sample qualify as constant, including the deep sea and most, but not all, tropical systems. Since the classification of an environment as constant or fluctuating is to some extent subjective, we point out that this task was carried out before we had analyzed the data and uncovered any pattern.

The 62 webs analyzed here are drawn from published studies. Details are presented in Chap. IV.

3. Results

Fig. A.3.1 shows, plotted against S, $L^{1/2}$ (a), L (b), and $L^{3/4}$ (c). The slopes of the straight lines plotted through the origin are computed on the assumption that the variance of the (transformed) ordinates is proportional to the abscissa. Visual inspection of Fig. A.3.1a rejects the first hypothesis: the trend of the data points is distinctly concave compared to the fitted straight line. When the square roots of links vs. species are plotted separately (not shown here) for constant and fluctuating webs, both graphs show a concave trend like that of Fig. A.3.1a. Visual comparison of Fig. A.3.1b and c is less decisive. Plotting the 3/4 power of L (Fig. A.3.1c) brings the points closer to the fitted line at low values of S but, at high values of S, lets most of the points fall below the line.

We accept $E(L)$ as proportional to S. This approximation does no obvious violence to the data and simplifies further analysis.

If $E(L) = cS$ and the variance in L is proportional to S, then the estimate $c = 1.8559$ is the ratio of the total number of links, 1919, to the total number of species, 1034, in our 62 webs. The standard deviation of c is 0.0740.

The number L of links is the sum of the numbers L_{BI}, L_{BT}, L_{II}, and L_{IT} of links from basal to intermediate, from basal to top, from intermediate to intermediate, and from intermediate to top species, respectively. Fig. A.3.2 shows, plotted against S, the proportions of links in each category L_{BI}/L (a), L_{BT}/L (b), L_{II}/L (c), and L_{IT}/L (d). No increasing or decreasing trends are evident. Thus, the mean proportions of links of each kind are roughly invariant with respect to the total number of species in the web, though variability around the mean is evident. It follows that the average numbers of links of each kind, in addition to the average total number of links, increase in proportion to the total number of species, again with variability.

Table 1. Summary statistics of the numbers of species and links in 62 webs, by type of web, type of species, and category of link

Type of unit	Webs					
	Constant		Fluctuating		All	
	No.	Fraction	No.	Fraction	No.	Fraction
Webs	19		43		62	
All species	351	1.000	683	1.000	1034	1.000
B	66	0.188	130	0.190	196	0.190
I	177	0.504	366	0.536	543	0.525
T	108	0.308	187	0.274	295	0.285
All links	811	1.000	1108	1.000	1919	1.000
B-I	198	0.244	327	0.295	525	0.274
B-T	92	0.113	56	0.051	148	0.077
I-I	260	0.321	318	0.287	578	0.301
I-T	261	0.322	407	0.367	668	0.348

B, basal; I, intermediate; T, top

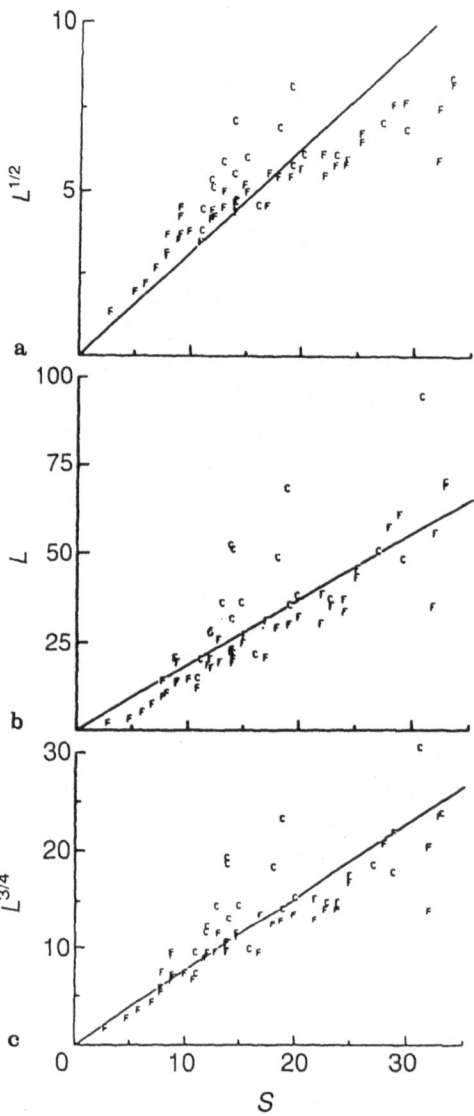

Fig. A.3.1a-c. Number of links (L) as a function of the number of trophic species (S) in 62 webs. Plotted against S are $L^{1/2}$ (a), L (b), and $L^{3/4}$ (c). C = constant environment; F = fluctuating environment. The symbols F and C have been perturbed from their exact locations by a small random amount to indicate when several food webs have exactly the same coordinates. In the straight lines through the origin plotted here and in Fig. A.3.3, the slopes are computed assuming that the variance of the ordinates (as transformed, in a and c) is proportional to the abscissa. In a, the trend of the data points is concave compared to the fitted straight line. In c, the points lie closer to the fitted line at low values of S but, at high values of S, most of the points fall below the line

Table 1 shows the numbers L_{BI}, L_{BT}, L_{II} and L_{IT} and proportions of links in each of the four categories, summed for constant, fluctuating, and all webs. The proportions of each category of links are highly variable among webs, compared

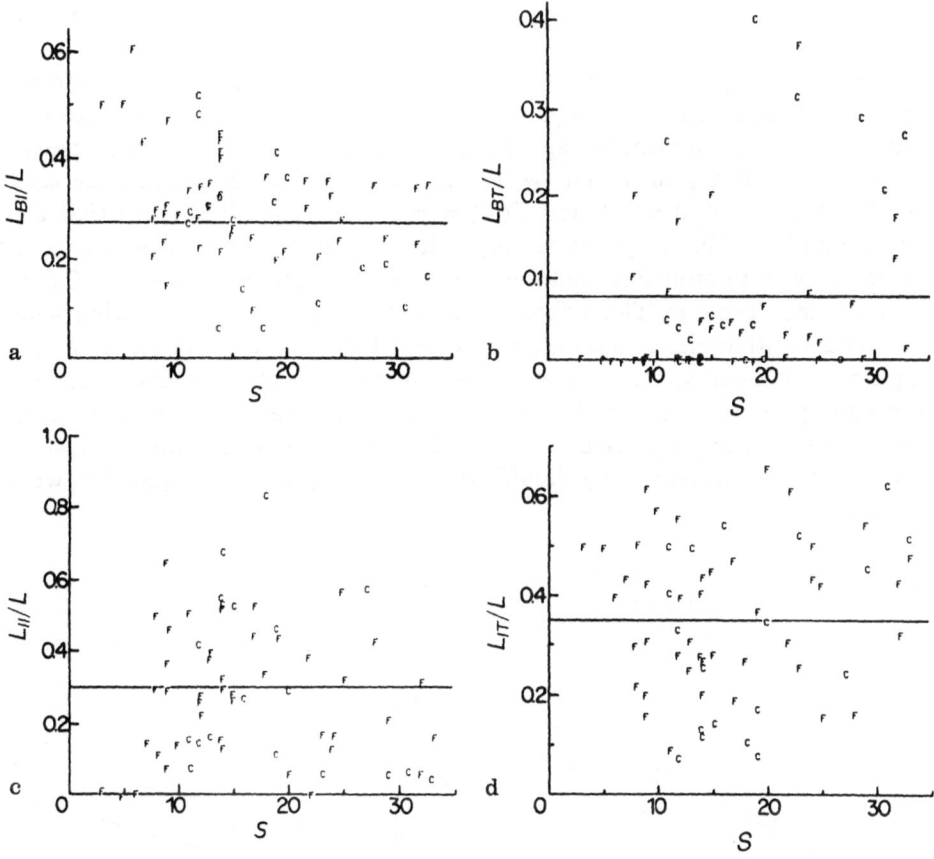

Fig. A.3.2a-d. The proportions of links in each category as a function of species. Plotted against S are L_{BI}/L (a), L_{BT}/L (b), L_{II}/L (c), and L_{IT}/L (d), where L_{BI}, L_{BT}, L_{II}, and L_{IT} are the numbers of links from basal to intermediate, from basal to top, from intermediate to intermediate, and from intermediate to top species, respectively. No increasing or decreasing trends are evident in the data. The points in the upper left corner of a are based on very few links. The heights of the fitted horizontal lines are the ratio of the links in the given category, summed over all webs, to the total links, summed over all webs

to the variation among webs that would be expected from multinomial sampling with proportions that are the same for all webs. If the counts of links are arranged in a 4×62 contingency table with one row for each type of link and one column for each web, a homogeneity test yields a χ^2 of 794.2 with 183 degrees of freedom (df). If the summed link counts of the constant webs are compared to the summed link counts of the fluctuating webs in a 4×2 contingency table (the counts are shown in Table 1), a homogeneity test yields a χ^2 of 33.0 with 3 df. A homogeneity test of the link counts for the 19 constant webs alone yields a χ^2 of 510.3 with 54 df, while the same test for the 43 fluctuating webs alone yields a χ^2 of 284.0 with 126 df. Under the assumption, which is open to doubt, that the observations of different webs are mutually independent, the astronomically low significance level of each of these values of χ^2 rejects the null hypothesis that the variation among webs in the proportions of links of each category is due to random sampling.

We now display the relation between number of links and number of species at a level of resolution finer than that of Figs. A.3.1 and A.3.2. Our previous analysis of community webs (Chap. II.2) established that there are fixed positive constants r, p, and q such that, within multinomial sampling error, for each web, $E(B) = rS$, $E(I) = qS$, and $E(T) = pS$. These equations mean that the average number of basal species is proportional to the total number of species and similarly for intermediate and top species. For all webs, $r = 0.190$, $q = 0.525$, and $p = 0.285$ (Table 1). The differences between constant and fluctuating webs are within multinomial sampling error (Chap. II.2). As a consequence of this simple proportionality, the geometric mean of any two of B, I, and T should be roughly proportional to S. Since, as Fig. A.3.2 implies, the number of links of each kind is also proportional to S (with substantial variability, in light of the above inhomogeneity), L_{BI} should be roughly proportional to $(BI)^{1/2}$, with variability.

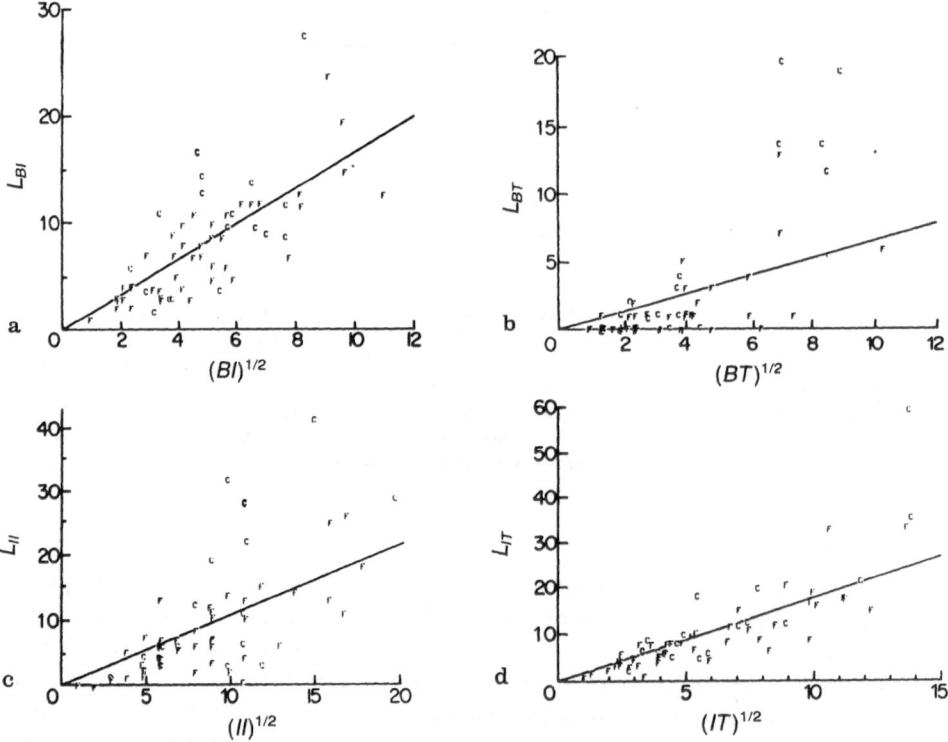

Fig. A.3.3. (a) Plot of the number L_{BI} of basal-intermediate links against the geometric mean $(BI)^{1/2}$ of basal and intermediate species and analogous plots for basal-top links (b), intermediate-intermediate links (c), and intermediate-top links (d). A straight line through the origin is most plausible as a description of the data in a and d, less so for c, and least so for b. In b, many webs, both constant and fluctuating, lack links from basal to top species

Fig. A.3.3a plots L_{BI} against $(BI)^{1/2}$. The remaining panels of Fig. A.3.3 give similar plots for basal-top links (b), intermediate-intermediate links (c), and intermediate-top links (d). The linear model in Fig. A.3.3b is least satisfactory because for many webs, both constant and fluctuating, there are no links from basal to top species.

The linearity or near-linearity in Fig. A.3.3 is not materially changed when the links of constant and fluctuating webs are plotted separately on the same axes or when the links of constant and fluctuating webs are plotted separately against the product of the corresponding species numbers – e.g., L_{BI} against BI rather than against $(BI)^{1/2}$ (not shown). For fluctuating webs, the apparent convexity of L_{BT} against $(BT)^{1/2}$ is somewhat diminished in the plot against BT.

Table 2. Regression coefficients of the number (y) of links in a specified category against the geometric mean number (x) of the corresponding types of species

Type of link	Webs					
	Constant		Fluctuating		All	
	Slope	SD	Slope	SD	Slope	SD
B-I	1.3824	0.0989	1.5817	0.1038	1.6802	0.0908
B-T	0.4437	0.1530	0.8317	0.1265	0.6580	0.0976
I-I	0.8258	0.0877	1.0907	0.1092	1.0645	0.0906
I-T	1.5464	0.0938	1.8053	0.1206	1.7891	0.0983

All regressions assume that the variance in the ordinate y is proportional to the abscissa x. SD = standard deviation of the estimated slope coefficient. B, basal; I, intermediate; T, top

Table 2 gives the regression coefficients, and their standard deviations, of the number of each kind of link against the geometric mean number of species in the source and sink class, for all webs (corresponding to the slopes of the lines plotted in Fig. A.3.3a-d) and for constant and fluctuating webs separately. Regressions (not reported here) that assume a line through the origin with the standard deviation of the residuals proportional to the abscissa give, in every case, a larger mean square residual and a visually poorer fit.

The regression coefficients in Table 2 and the values of r, q, and p in Table 1 can, in some cases, be combined to predict accurately the proportion of each kind of link shown in Table 1. For example, suppose that the proportion of basal-intermediate links is given by

$$p_{BI} = L_{BI}/L ,$$

that the regression in Fig. A.3.3a is summarized by

$$L_{BI} = a_{BI}(BI)^{1/2} ,$$

and that B, I, and T are all proportional to S. Then

$$p_{BI} = a_{BI}(BI)^{1/2}/[a_{BI}(BI)^{1/2} + a_{BT}(BT)^{1/2}$$
$$+ a_{II}(II)^{1/2} + a_{IT}(IT)^{1/2}]$$
$$= a_{BI}(rq)^{1/2}/[a_{BI}(rq)^{1/2} + a_{BT}(rp)^{1/2} + a_{II}q + a_{IT}(qp)^{1/2}] \, .$$

Table 3. Predicted fractions of links of each category, for constant and fluctuating webs separately and for all webs

Type of link	Webs		
	Constant	Fluctuating	All
B-I	0.2732	0.2561	0.2740
B-T	0.0685	0.0963	0.0791
I-I	0.2672	0.2966	0.2889
I-T	0.3911	0.3510	0.3579
χ^2	44.6144	31.2953	1.6049

The χ^2 statistic to measure goodness of fit between the observed fractions (given in Table 1) and the predicted fractions given here has 3 df. When computing the predictions for the constant webs, both the regression coefficients and the proportions of species of each type were derived from the constant webs only and similarly for the fluctuating webs. B, basal; I, intermediate; T, top

Table 3 shows the predicted proportions of links of each category, based on the regression coefficients a_{ij} from Table 2, the values of p, q, and r from Table 1, and a goodness-of-fit χ^2 (with 3 df) when the predicted proportions are compared with the observed proportions of links of each category. For all webs combined, the predicted proportions agree remarkably well with the observed; the discrepancy could be attributed entirely to sampling fluctuation.

However, this good agreement is not a strong confirmation that the number of links of each category scales according to the geometric mean rather than, say, according to the product. If, for example, in the above equations the regression coefficient of L_{BI} against BI is used and $(BI)^{1/2}$ is replaced by BI, and similarly for BT, II, and IT, then the predicted proportions also agree remarkably well, though not as well, with the observed ($\chi^2 = 3.95$). The excellent agreement between observed and predicted proportions of links of each category is rather robust with respect to the exact way in which the numbers of links scale with an increasing number of species.

For constant and fluctuating webs considered separately, the quantitative discrepancies between the observed and predicted proportions of each category of link are not large, but the χ^2 statistic indicates that the fit would be rejected at any conventional level of significance, under the assumption of independence among webs. Among constant webs, fewer basal-intermediate links and more basal-top links are observed than predicted. Among fluctuating webs, more basal-intermediate and fewer basal-top links are observed than predicted. These discrepancies precisely cancel when all webs are considered together. [When constant webs are compared with fluctuating webs, rather than each with the predictions of a model, the ratio of basal-intermediate links to total species is higher in constant webs ($198/351 = 0.564$) than in fluctuating webs ($327/683 = 0.479$), contrary to the comparison with the model. The ratio of basal-top links to total species is higher in constant webs ($92/351 = 0.262$) than in fluctuating webs ($56/683 = 0.082$), in parallel with the model comparison.]

When a web has intermediate species (as do all those in our sample), the presence of basal-top links gives the top species collectively a more flexible trophic strategy, in that some top predators prey on intermediate species and some (possibly the same) top predators prey on basal species. The deficit of basal-top links in fluctuating webs and the excess of basal-top links in constant webs, relative to the proportions expected from our simple model of scaling, suggests that fluctuating webs are trophically more constrained than constant webs.

Further evidence that fluctuating webs may be more severely constrained than constant webs is provided by comparing the standard deviations of characteristics of fluctuating and constant webs. Since the number of species in our sample of constant webs ranges from 11 to 33, while the number of species in our sample of fluctuating webs ranges from 3 to 33, we have, for the purposes of this comparison, removed from the sample of fluctuating webs those 13 webs with fewer than 11 species. Compared to the 19 constant webs, the remaining 30 fluctuating webs have smaller standard deviations of the number of: basal-intermediate links, basal-top links, intermediate-intermediate links, intermediate-top links, total number of links, basal species, intermediate species, top species, total species, predator species ($T + I$); and smaller standard deviations of the ratios: links per species, basal links ($L_{BI} + L_{BT}$) per basal species, intermediate links ($L_{BI} + L_{II} + L_{IT}$) per intermediate species, and top links ($L_{BT} + L_{IT}$) per top species. The 30 fluctuating webs were slightly more variable than the 19 constant webs only in the number of prey species ($B + I$).

4. Discussion

According to our newly assembled data, the mean number of links L in a web is approximately proportional to the total number of species S. The coefficient of proportionality is ≈ 1.8559 with a standard deviation of 0.0740.

The hypothesis that the mean of L is proportional to S may be derived from empirical observations that the connectance C varies approximately inversely as S. The observation was first made by Rejmánek and Starý (1979) in a collection of 31 plant-aphid-parasitoid webs and confirmed by Pimm (1980, 1982) in a sample of 18 miscellaneous webs, including those assembled by Cohen (1978). Since C is approximately proportional to LS^{-2}, if C varies approximately as $1/S$ then

L/S is approximately independent of S, or L is approximately proportional to S. (Conversely, if L is proportional to S, then C varies as $1/S$.) Pimm (1980; 1982, p. 89) interpreted his empirical generalization as a consequence of a behavioral supposition: "Suppose each species in a community feeds on a number of species of prey that is independent of the total number of species in the community."

The alternative hypothesis, that some power of L between $1/2$ and 1 is proportional to S, may be new and is also not ruled out by the data.

The only prior explicit examination of the relation between L and S appears to be Briand's (Chap. II.5 and Fig. B.5.2a) empirical finding, based on 40 "unlumped" webs, that $L = S^{1.1}$. He observed that this relation was "nearly linear."

Without presuming to discriminate between a power law exponent of 1.1 and one of 1.0, we find that Fig. A.3.1, which is based on 62 lumped webs, confirms the approximate correctness of Briand's finding and is consistent with the hypothesis of Pimm, at least within the range of S, 3–33, covered.

That the mean of L is approximately proportional to S indicates that the trend of SC or L/S is roughly independent of variation in S. This may appear to contradict Briand's (1983, p. 37) finding, based on a principal components analysis, that the product of species S times "upper connectance" is a major discriminator of variation among webs. However, upper connectance counts both links and "potential competitive links," and the latter increases as a nearly quadratic function of S (Chap. II.5 and Fig. B.5.2b). Therefore S times upper connectance is much more variable among webs than is SC, since connectance C, as used here and by Pimm (1980, 1982), counts only links.

Links are more subject to errors of omission than are species, because a feeding interaction between a predator and prey must be observed or inferred for a link to be recorded, whereas no special behavior need be observed for a species to be recorded. Consequently, future webs collected with more systematic attention to recording all links may yield larger estimates of C than that based on present data.

The likelihood of recording a link may vary more among observers than the likelihood of recording a species. Variability among observers in the probability of recording a link may explain why the above homogeneity tests for links, under the assumption of independence, reject the null hypothesis of multinomial sampling fluctuations with constant proportions.

The approximately linear relation in Fig. A.3.3 between the expected number of links of each category and the geometric mean number of species in the source and sink categories appears to be new.

Earlier observations have suggested that fluctuating webs are more severely constrained in trophic structure than constant webs (Chaps. II.2,5). The finding here that fluctuating webs have significantly fewer basal-top links, and constant webs have significantly more basal-top links, than expected from a simple model based on pooled proportions, may be interpreted to be consistent with the earlier observations. Similarly, the standard deviations of many characteristics of constant webs exceed those of fluctuating webs.

5. Conclusion

Together, this chapter and Chap. II.2 show that the main features of the structure of food webs – namely, the numbers of top, intermediate, and basal species and the numbers of links from each kind of predator to each kind of prey – all behave in quantitatively simple, interpretable ways as the number of species in webs ranges from 3 to 33. The data on which our quantitative generalizations are based are the most extensive and most carefully edited presently available. Nevertheless, because of variations among observers in field practices and definitions of concepts, the present generalizations will have to stand the test of more consistent and thorough field work in the future.

Our findings open at least three lines of further inquiry. First, how can these ecological generalizations be explained in terms of the behavior, genetics, and population dynamics of species, individually and in interaction? Second, do these ecological generalizations suffice to explain other significant features of food webs (Cohen 1978; Hutchinson 1959; Cohen 1983)? Third, what characteristics of individual communities account for their deviations from the overall trends?

§4. Food Webs and the Dimensionality of Trophic Niche Space

Joel E. Cohen

Ecological studies of where the organisms in communities are and what the organisms do (especially what they eat) frequently use the concept of niche space, the set of the environmental (including biotic) factors acting on an organism (Hutchinson 1944, 1965; Miller 1967; Vandermeer 1972; Pianka 1976). Studies of what organisms eat frequently also use the concept of a food web (Shelford 1913; Gallopín 1972).

Here is presented a new technique for using food webs to gain information about the minimum number of dimensions of a niche space necessary to represent, in a specific sense, the overlaps among observed trophic niches. Based on the application of this technique to data, it is inferred that, within habitats of limited physical and temporal heterogeneity, the overlaps among niches along their trophic (feeding) dimensions can be represented in a one-dimensional space far more often than expected by chance alone.

1. Materials and Methods

Classification and Selection of Food Webs

Prior to analysis, published or privately communicated food webs were characterized as describing a single habitat or as describing a composite of several habitats. Food webs were also characterized as attempting to describe all the kinds of organisms (possibly restricted to some location, size, or taxa) in a habitat, without reference to the eating relationships among them ("community food webs"); or as attempting to describe all the prey taken by a set of one or more predators, plus all the prey taken by the prey of those predators, and so on ("sink food webs"); or as attempting to describe all the predators on a set of one or more prey organisms, plus all the predators on those predators, and so on ("source food webs"). Source food webs were excluded from further study because they are uninformative about whether the community food webs of which they form a part are interval. Hypothetical or schematic constructions and avowedly incomplete, partial, or tentative food webs were also excluded. Fourteen community food webs and 16 sink food webs from 21 different papers were thus selected.

Units of Description

These food webs describe the diets or predators not of individual organisms but of kinds of organisms. A "kind of organism" may be a stage in the life cycle or a size class within a single species, or it may be a collection of functionally or taxonomically related species, according to the practice of the original report. This analysis assumes that a group of organisms qualifies as one "kind" of organism in a food web only if its niche, viewed as a region or set of points in niche space, is connected along the trophic dimensions – that is, only if it is possible to pass from any one point in the niche to any other without leaving the niche. For example, if two stages in the life cycle of a single species of insect were so different that the region in niche space corresponding to one stage were unconnected to the region corresponding to another, it is assumed that the two stages would have feeding habits sufficiently different that the stages would be distinguished as different "kinds" in a food web.

Machine Representation of Food Webs

Each food web selected for study was stored in a computer as a matrix with m rows and n columns. Each column corresponds to a predator or other kind of organism that consumes at least one of the kinds of organisms in the food web. Each row corresponds to a prey or other kind of organism eaten by at least one of the kinds of organisms in the food web. Some kinds of organisms are both predators and prey. Let w_{ij} be the entry in the i-th row and j-th column of a given food web matrix. Then $w_{ij} = 1$ if predator j eats prey i and $w_{ij} = 0$ if predator j does not eat prey i. Version A of a food web includes only eating relationships that could be unambiguously established from the original report; version B includes any additional eating relationships that were uncertain or probable.

The Overlap Matrix and the Number of Niche Overlaps

If two kinds of predators both eat some kind of prey, then along some trophic dimensions the niches of those two predators logically must overlap. The n by n overlap matrix which describes the overlaps among the trophic niches of the predators has 1 wherever the predator corresponding to the row and the predator corresponding to the column both eat some kind of prey in common, and 0 elsewhere. The overlap matrix is symmetric with respect to its main diagonal, which contains all 1s. The number of niche overlaps E is defined as the number of 1s above the main diagonal. Overlap matrices were constructed corresponding to version A and version B of each food web.

The Overlap Matrix and the Dimension of Trophic Niche Space

We say that a food web is interval, and that the trophic niche overlaps that it describes can be represented in a one-dimensional niche space, when its overlap matrix is the adjacency matrix of an interval graph (Klee 1969). An interval graph is the intersection graph of a set of intervals of the real line. More explicitly, a food web is interval if and only if, for each kind of predator i in the food web, there exists an interval i' of the real line such that for any two predators i and j, $\sum_{k=1}^{m} w_{ki}w_{kj} > 0$ when and only when the corresponding intervals i' and j' overlap. Not every food web with four or more predators is interval.

To test whether a food web is interval, a computer program implementing the algorithm of Fulkerson and Gross (1965) was written by Thomas Mueller. The performance of this algorithm was verified by hand for several hundred examples, and the same algorithm was used for both observed and artificially generated food webs (see below).

Monte Carlo Estimation of the Probability of an Interval Food Web

In order to compare the observed frequency of interval food webs with the frequency that would be expected if the food webs or niche overlaps were drawn by chance, it is necessary to estimate the frequency of interval food webs in a universe of possible food webs from which the observed food webs may be drawn. Two possible universes, or models of a random food web, are described here; the results of five other models are consistent with these.

Model 6 assumes that every predator in a given food web has a constant and independent probability p of preying on each prey. The probability p is estimated separately for each food web as $A/(mn)$ in which A is the sum of all elements in the food web matrix (that is, the observed number of feeding relationships) and mn is the maximum possible number of relationships in the food web. For each food web, 100 artificial food webs are generated by distributing a 1 with probability p and a 0 with probability $1 - p$ into each element of an m by n matrix, independently for each element.

Model 7 assumes that the number E of niche overlaps in a given food web is fixed but that the pairs of predators that have overlapping trophic niches are randomly determined. For each food web with E overlaps, 100 artificial overlap matrices are generated by distributing E 1s at random among the elements above the main diagonal.

Let f_{ij} be the proportion of the 100 artificial food webs that are interval according to model j using the parameter values (A, E, m, n) of food web i. For a set S of food webs (e.g., the set of version A community food webs) the mean μ and variance σ^2 of the number of interval food webs expected according to model j are $\mu = \sum_{i \in S} f_{ij}$ and $\sigma^2 = \sum_{i \in S} f_{ij}(1 - f_{ij})$, respectively. The probability of a discrepancy between an observed number of interval food webs in a set S and the expected μ is assessed by treating $z =$ (observed number of interval food webs $- \mu)/\sigma$ as a standardized normal random variable. Assuming the validity of the normal approximation, the probability that z exceeds 3.1 by chance alone is less than 0.001 (one-tailed test).

2. Results

Most food webs based on single habitats are interval (Table 1). The one sink food web and the two community food webs that are not interval are reviewed below. A higher proportion of food webs based on composite communities are noninterval. This finding does not conflict with the hypothesis that most or all single-habitat food webs are interval (see *Discussion*).

Table 1. Numbers of interval and noninterval food webs

Habitats	Community food webs		Sink food webs	
	Interval	Noninterval	Interval	Noninterval
Single	7	2	13	1
Composite	1 1/2[a]	3 1/2[a]	1	1

[a] One food web was interval in version A and noninterval in version B. There were no other discrepancies between versions A and B

Because the distinction between single and composite habitats is less clear-cut, both conceptually and in ecological reports, than that between community and sink food webs, the comparison between the observed number of interval food webs and the number expected by chance from two model universes of food webs retains only the distinction between community and sink food webs (Table 2). Community food webs are interval significantly more frequently than expected by chance, assuming either random eating relationships (model 6) or random niche overlaps (model 7). Sink food webs are interval significantly more frequently than expected by chance, assuming random niche overlaps (model 7) but not assuming random eating relationships (model 6), considering either version A (definite information only) or version B (additional uncertain information) food webs only. The significant excess of sink interval food webs when

Table 2. Comparison of observed frequencies of interval food webs with expectations assuming random predatory relations (model 6) or random niche overlap (model 7)

Set of food web versions	Versions in set, no.	Observed no. interval	Model 6			Model 7		
			Mean, μ	SD, σ	Normal deviate, z	Mean, μ	SD, σ	Normal deviate, z
All versions*								
Community food webs	24	14	4.83	1.13	8.11	2.95	1.26	8.73
Sink food webs	20	18	14.47	0.88	4.01	13.42	0.66	6.96
Version A								
Community food webs	14	9	3.27	0.93	6.16	2.19	0.96	7.09
Sink food webs	16	14	12.48	0.87	1.74	11.42	0.66	3.92
Version B								
Community food webs	14	8	2.89	0.82	6.21	1.82	0.86	7.19
Sink food webs	16	14	11.65	0.79	2.99	10.42	0.66	5.44

* Food webs for which versions A and B are identical are counted only once here

all versions are considered together ($z = 4.01$) is an artifact of the lack of independence between different versions of the same food web.

Individual Cases

One food web (Kohn 1959) reports prey organisms consumed by vermivorous species of the gastropod genus *Conus* in Hawaii at subtidal reef stations and at marine bench and deep water habitats. It is thus a sink food web describing a composite habitat, and it is not interval. The numbers of specimens examined of each predator range from 4 to 342. It seems plausible that, when only a few specimens of a predator are examined, some kinds of prey eaten on occasion might not be seen. The resulting omission of some trophic niche overlaps may cause a true underlying one-dimensional trophic niche space to appear to be more than one-dimensional. When only predators represented by more than 20 specimens (a threshold determined in advance) are included in a reanalysis, the food web is still not interval.

From this food web, the specimens taken at subtidal reef stations were selected to create the only sink, single-habitat food web which turned out to be noninterval (Table 1). However, if predators represented by 20 or fewer specimens taken at the subtidal reef stations are excluded, the resulting food web is interval. In this case, restricting attention to the adequately sampled predators is not enough to make the food web based on composite habitats interval but does yield an interval food web for a single habitat. Because the food webs of the single habitat and the composite community are reported by the same observer, the difference between them cannot be attributed to different definitions of "kind of organism."

The two single-habitat community food webs that are noninterval describe the sandy shore and Crocodile Creek of Lake Nyasa (Fryer 1959); a third food web describing the rocky shore is interval. The coded forms of these food webs incorporate extensive additions, based on the text, to the ambiguous food web graphs. The number of specimens of each predator examined is not reported, so it is impossible to exclude predators that were lightly sampled.

3. Discussion

Community Food Webs

The number of community food webs that are interval greatly (and significantly) exceeds the number expected assuming either random eating relations (model 6) or random trophic niche overlaps (model 7). The quantitative adequacy of two noninterval community food webs based on single habitats cannot be assessed. The finding that several composite-habitat community food webs are noninterval is consistent with the hypothesis that every niche space within a single habitat is one-dimensional. It is likely that the features that differentiate one habitat from another are multidimensional (Cody 1968; Schoener 1974) and different from the dimension of variation within a habitat.

Sink Food Webs

The only single-habitat sink food web that is noninterval becomes interval if lightly sampled predators are excluded. All single-habitat sink food webs based on sufficient sampling are interval. The number of sink food webs that are interval greatly (and significantly) exceeds the number expected, assuming random trophic niche overlaps. The parameters of the sink food webs evidently specify a region of the model universe 6, which assumes random eating relationships, in which the frequencies of interval food webs are nearly as high as those observed.

Because all of the adequately sampled sink food webs are consistent with a one-dimensional niche space in single habitats, the failure of the observed frequency of interval sink food webs to be significantly larger than expected from some models in no way weakens the conclusion that all or nearly all single-habitat community or sink food webs are interval.

Nonuniqueness of the One Dimension

If a one-dimensional niche space can represent trophic niche overlaps in a single habitat, the single dimension identified in one community may differ from that in another. In a single habitat, the one dimension may be chosen from a manifold of monotonically related dimensions such as predator size and prey size (Schoener 1967).

What Is the One Dimension?

A few food web studies provide enough information on feeding and distribution to suggest what the one dimension may be. For example, among Hawaiian snails (Kohn 1959), *Conus sponsalis, C. abbreviatus, C. ebraeus,* and *C. chaldaeus* have all possible pairwise overlaps of diet on marine benches and, in all four species, individuals between 27 and 28 mm long were found on the marine bench at station 5. If the dietary overlaps found from the pooled marine bench sample are faithfully reflected at station 5, the length of the snails is then a candidate for the single dimension of a space in which trophic niche overlaps can be represented. On the other hand, on reef platforms, the food web is again interval. There the diets of *C. ebraeus* and *C. sponsalis* overlap, but neither diet overlaps with that of *C. flavidus* or *C. lividus*, which do overlap with each other. Because all four species are found between 0% and 30% of the distance from the shore to the outer edge of the reef platforms at stations 3, 7, and 9, that distance measure can be excluded in this case as the one dimension along which trophic niche overlaps can be represented.

Operational Definitions of "Dimension"

Different kinds of studies of niche space, such as those of resource partitioning (Cody 1968; Schoener 1974) or those based on competition experiments, use different operational definitions of "dimension". Niche overlap inferred from food webs is a necessary but not a sufficient condition for exploitation competition when one common limited resource is food. Niche overlap is neither necessary nor sufficient for interference competition (Pianka 1976). Therefore, a low level of exploitation competition may be inferred when a low level of niche overlap is observed in food webs; but a high level of niche overlap implies only the possibility of a high level exploitation competition. A concordance among the results of the different kinds of studies of "dimensionality" would represent a major empirical discovery. If a concordance among the different operational definitions of "dimension" is taken for granted but turns out to be contrary to fact, the word will become a conceptual trap for the unwary.

Why One Dimension?

Several interpretations are possible of why the trophic niche space of single habitats appears to be representable in one dimension. If the finding were a tautology because we say that communities describe composite habitats when their niche spaces turn out not to be one-dimensional, then we would not have the embarrassment of the two single-habitat community food webs that are not interval. This interpretation cannot explain the excess frequency of interval food webs observed in comparison with expectations from random models. We dismiss the accusation of tautology.

It is plausible to expect a predator that can take prey at two different values of any natural continuous variable (such as prey size, seed hardness, altitude, or humidity) to be able to take prey at all intermediate values of the same variable. This argument implies only that a trophic niche should be convex, and hence (Klee 1969) that three independent dimensions are always sufficient to represent trophic niche overlap. The argument does not explain why *one* dimension suffices.

It may be shown that there is no necessary connection between the one dimensionality of a community's niche space and the qualitative stability (May 1973) of the dynamical system implied by its food web. The possibility of a statistical association between qualitatively stable and interval food webs remains uninvestigated.

The finding that single-habitat food webs are interval while trophic niches are commonly described in multidimensional terms may reflect the difference between community ecology and physiological ecology. Organisms may have more degrees of freedom in their physiological capacities to exist under varied circumstances than the biotic, especially trophic, interactions with other kinds of organisms in their community permit them to enjoy.

Extensions

When food webs are not interval, a combinatorial approach can reveal whether the niche overlaps could be represented by the overlaps of regions in a higher dimensional space (Roberts 1969a), but it is necessary to have quantitative information about the actual shape of niches before applying this theory. When a food web is not interval, it may also be worth examining how far it is from being interval (Kendall 1969).

Shortcomings of This Approach

These results suffer from at least four major shortcomings. First, the concepts in terms of which the data are reported and the results are framed are ambiguous (e.g., what constitutes a "single habitat"?). Second, statistical features of the data used, especially the sampling design and reporting, leave much to be desired. Third, even if the concepts were clear and the statistics of the data impeccable, the claimed results do not attempt to answer important quantitative questions. In particular, most available food webs record feeding relationships as either present or absent. It is impossible to determine whether the high frequency of interval food webs depends in some special way on replacing underlying continuous variables that describe the frequency of predation by a dichotomous representation. Finally, a derivation of the claimed results from a more fundamental dynamic theory is lacking. Each of these shortcomings opens opportunities for further empirical and theoretical investigation.

A review of these results, including examples of the technique of analysis, the complete food web data, a discussion of each food web, a fuller analysis of the consequences, interpretation, and limitations, and recommendations for further research, as well as a synthesis with related results, appear in Cohen (1978). An overview of results since 1978 is given in Chap. III.6, where the theme of this chapter is taken up again.

B. Differential Regularities

§5. Environmental Control of Food Web Structure

Frédéric Briand

The past decade has seen a surge of interest in food web structure and organization, largely under the impulse of theoretical ecologists concerned with the relation between complexity and stability (see review by May 1981). However, due to the small number of natural food webs generally known from the published record, most of these studies have remained confined to rather abstract models of randomly constructed communities. As a result we find that some very basic questions still elude us, for instance the extent to which, and even whether, habitat type and environmental variability affect the structure of food webs in nature.

To tackle this problem, I assembled and analyzed a collection, the largest to date, of 40 community food webs drawn from the published record and representative of a wide variety of environments. Community food webs are defined as those webs which attempt to include all the kinds of organisms found in a particular habitat. A "kind of organism" (interchangeable henceforth with the term "species") may be an individual species, or a stage in the life cycle or a size-class within a single species, or it may be a collection of functionally or taxonomically related species. In every case the segregation follows the practice of the original report.

The 40 food webs studied are listed in Table 1, along with their source of reference. They include 13 webs previously described by Cohen (1978), although 5 of those required corrections. Webs only partially defined, too schematic in representation, or else based on information drawn from different locations, were omitted. For each community a food web matrix indicating trophic interactions was built, based on the graph and in some cases on additional information contained in the text of the original report. These matrices are given as webs 1–40 in Chap. IV.

Table 1. Origin and main structural parameters of the food webs analyzed. Trophic structure indicates how many kinds of organisms occupy each trophic level from producers to top predators. Species feeding on more than one trophic level are recorded at their highest position in the web. S and C denote species richness and connectance, respectively

Case No.	Community	Trophic structure (producers → top predators)	S	$C(\%)$	Reference
A. Fluctuating environments					
1	Cochin estuary	2-4-1-1-1	9	69.4	Qazim (1970)
2	Knysna estuary	3-7-4-1	15	47.6	Day (1967)
3	Long Island estuary	4-10-6-4	24	21.4	Woodwell (1967)
4	California salt marsh	2-2-6-2-1	13	56.4	Johnston (1956)
5	Georgia salt marsh	3-2-2	7	33.3	Teal (1962)
6	California tidal flat	2-7-5-3-2-5-1	25	30.3	MacGinitie (1935)
7	Narragansett Bay	3-5-7-4-1	20	33.2	Kremer and Nixon (1978)
8	Bissel Cove marsh	4-4-2-4-1	15	41.9	Nixon and Oviatt (1973)
9	Lough Ine rapids	2-3-4-1	10	51.1	Kitching and Ebling (1967)
10	Exposed intertidal (New England)	2-2-1	5	70.0	Menge and Sutherland (1976)
11	Protected intertidal (New England)	3-4-1	8	42.9	Menge and Sutherland (1976)
12	Exposed intertidal (Washington)	3-7-1-2	13	46.2	Menge and Sutherland (1976)
13	Protected intertidal (Washington)	3-6-2-2	13	48.7	Menge and Sutherland (1976)
14	Mangrove swamp (station 1)	1-3-3-1	8	57.1	Walsh (1967)
15	Mangrove swamp (station 3)	1-5-2-1	9	58.3	Walsh (1967)
16	Pamlico River	4-4-5-1	14	36.3	Copeland et al. (1974)
17	Marshallese reefs	3-3-3-3-1-1	14	28.6	Hiatt and Strasburg (1960)
18	Kapingamarangi atoll	8-9-2-3-5	27	20.8	Niering (1963)
19	Moosehead Lake	2-3-8-3-1	17	42.6	Brooks and Deevey (1963)
20	Antarctic pack ice zone	3-3-5-3-4-1	19	29.8	Knox (1970)
21	Ross Sea	3-2-1-1-1-1-1	10	55.6	Patten and Finn (1979)
22	Bear Island	6-10-2-3-2-2-2-1	28	27.5	Summerhayes and Elton (1923)
23	Canadian prairie	1-5-4-4-1	15	59.1	Bird (1930)
24	Canadian willow forest	4-3-1-3-1	12	36.4	Bird (1930)
25	Canadian aspen communities	3-11-8-2-1	25	30.7	Bird (1930)
26	Aspen parkland	9-10-6-4-3-1-1	34	20.0	Bird (1930)
27	Wytham Wood	4-6-4-5-3	22	27.3	Varley (1970)
28	New Zealand salt-meadow	7-19-10-9	45	13.5	Paviour-Smith (1956)
B. "Constant" environments					
29	Arctic seas	2-3-6-6-3-2	22	31.2	Dunbar (1954)
30	Antarctic seas	1-2-3-2-3-2-1	14	52.7	Mackintosh (1964)
31	Black Sea epiplankton	2-3-5-1-1-1-1	14	83.5	Petipa et al. (1970)
32	Black Sea bathyplankton	2-3-5-1-1-1-1	14	84.6	Petipa et al. (1970)
33	Crocodile Creek	5-16-6-4-2	33	39.0	Fryer (1959)
34	River Clydach	4-4-1-2-1	12	56.1	Jones (1949)
35	Morgan's Creek	2-4-2-2-3	13	74.4	Minshall (1967)
36	Mangrove swamp (station 6)	8-7-4-2-1	22	34.8	Walsh (1967)
37	California sublittoral	6-10-3-5	24	26.1	Clarke et al. (1967)
38	Lake Nyasa rocky shore	3-10-9-9	31	67.1	Fryer (1959)
39	Lake Nyasa sandy shore	5-15-12-5	37	29.8	Fryer (1959)
40	Malaysian rain forest	3-3-4-1	11	52.7	Harrison (1962)

For purposes of comparison, I distinguished at the start between two broad categories of ecosystems: those exposed to high and those exposed to low temporal variability of the physical environment. By convention any system described in the original report as subjected to substantial variations in temperature, salinity, pH, water availability, or any other major parameter, was labelled "fluctuating." Accordingly, Table 1 lists 28 communities as representative of fluctuating environments and the remaining 12 as representative of "constant" environments. I emphasize that this distinction is based only on the amplitude of the changes, and not on their degree of predictability.

The striking disparity of food web structures encountered in nature can be appreciated from Table 1. Except for the closely related planktonic communities of the Black Sea (codes 31 and 32) that are similarly constructed, each network appears unique in design. If structural trends do exist, they are not readily apparent. For instance I find that the ratio of prey to predator species is far less constant than previously indicated by Cohen (1978). Nor is there any significant correlation (Student's t test; $P > .05$) between species richness and prey : predator ratio, percentage of specialized predators (those feeding only on one kind of prey), or food chain length. Taken singly, none of these variables can discriminate among fluctuating, constant, aquatic, terrestrial, tropical, or nontropical systems.

On the other hand, environmental variability is found to have a marked impact on the connectance, that is, the fraction of nonzero off-diagonal elements in the community matrix. Such a matrix indicates not only trophic interactions, as does the food web matrix, but also direct competitive interactions. Since interference competitors are not identified in the original reports, one must adopt simple and realistic criteria to that effect. I follow here the procedure used by Yodzis (1980), which yields a relatively high estimate of connectance but possesses such attributes. Whenever two predator species (say a and b) have at least one prey in common, I recognize them as potential interference competitors and so enter the elements A_{ab} and A_{ba} as nonzero in the community matrix. The connectance is calculated simply as

$$C = \frac{n}{S(S-1)} ,$$

where n denotes the number of nonzero interaction coefficients A_{ij} in the community matrix, and S the number of species in the system.

As shown in Fig. B.5.1a, for any given number of species, the connectance is significantly lower in fluctuating than in constant environments (t test, $P < .005$). This confirms a prediction tentatively advanced by May (1981), and is most interesting in light of the importance attached to connectance in stability theory. One recalls in particular the work of Gardner and Ashby (1970) and the proposition by May (1972) that communities in the neighborhood of equilibrium will tend to be stable if

$$i(SC)^{1/2} < 1$$

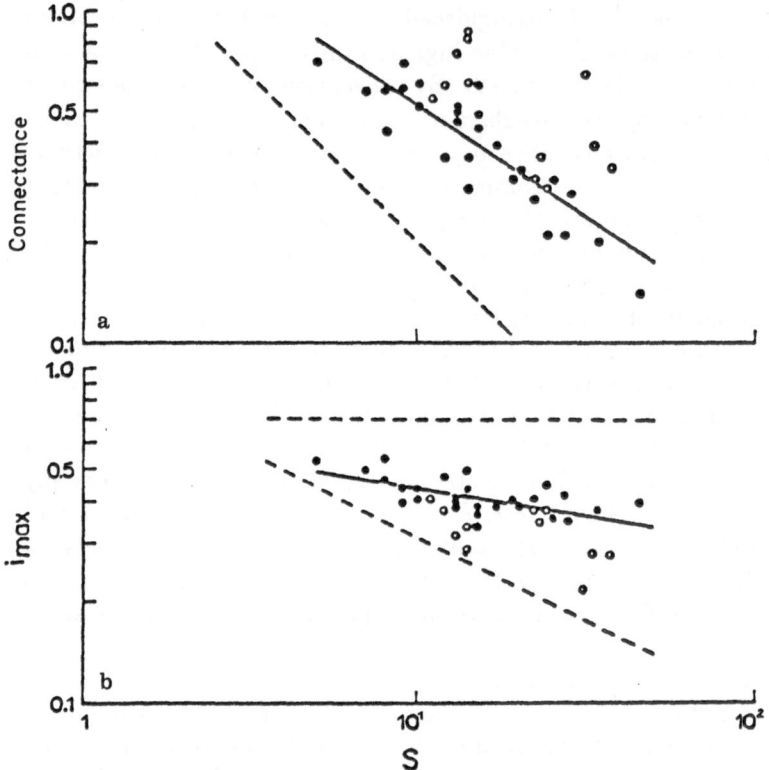

Fig. B.5.1. (a) Connectance of food webs as a function of species richness (S) in fluctuating (•) and constant (o) environments. The solid line represents the regression curve for fluctuating environments $y = 2.20x^{-0.65}$ ($r = -.83$, $P < .001$). The regression curve for constant environments $y = 2.71x^{-0.58}$ ($r = -.60$, $P < .05$) is not represented as it is based on a small sample size. The dashed line indicates the lower boundary for connectance, equal to $2S^{-1}$. (b) Maximum average interaction strength, calculated as $(SC)^{-1/2}$ (hence, according to May's equation, the upper bound on i which allows stability), as a function of species richness, where C = connectance. • = fluctuating environments; o = constant environments. The solid line represents the regression curve for fluctuating environments $y = 0.68x^{-0.17}$ ($r = -.62$, $P < .001$). For constant environments, the relation between the two variables is not significant ($P > .1$). The dashed lines indicate the upper and lower boundaries, equal to $2^{-1/2}$ and $S^{-1/2}$, respectively

and unstable otherwise. There S and C denote, respectively, species richness and connectance, while i represents the average strength of interaction among species.

Within the context of this relation, it is important to determine whether complexity (high S) is handled functionally or structurally by real systems. In other words, do complex systems retain their stability by reducing i, C, or both? The present study indicates that the answer will depend on the degree of environmental variability. In fluctuating systems the decrease in C associated with high S is so abrupt that only a slight reduction of i will be required to preserve stability (see Fig. B.5.1b). On the other hand, complex systems subjected to more

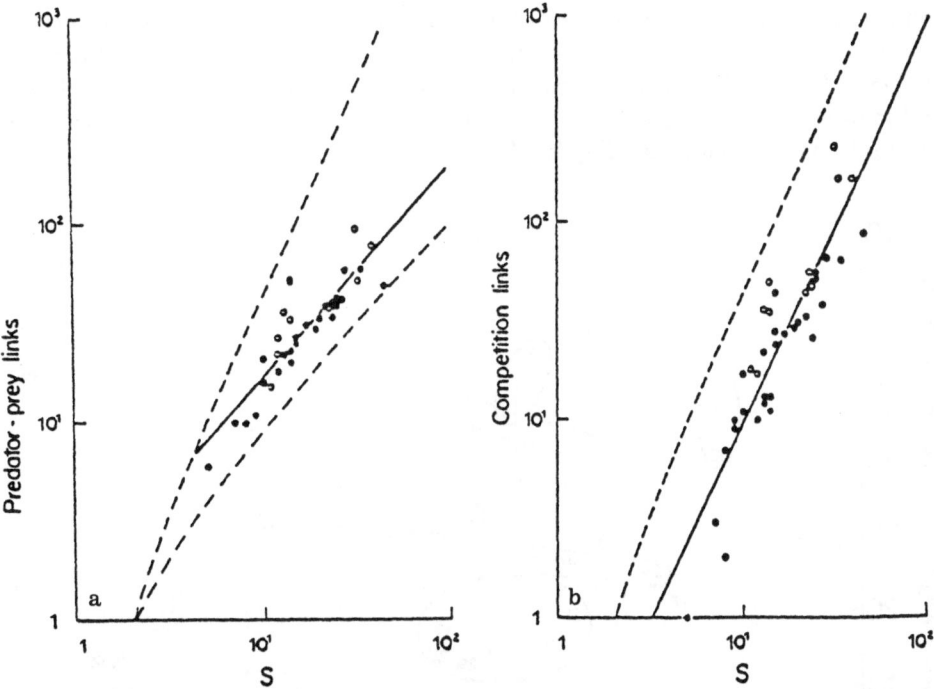

Fig. B.5.2. (a) Number of predator-prey links as a function of species richness (S) in fluctuating (•) and constant (o) environments. The dashed lines indicate the upper and lower possible boundaries, equal to $S(S-1)/2$ and $S-1$, respectively. (b) Number of potential competitive links as a function of species richness in fluctuating (•) and constant (o) environments. The dashed line indicates the upper possible boundary, equal to $S(S-1)/2$. There is no lower boundary

constant environmental conditions must depend on much weaker interactions, or else be more fragile.

I suggest that the difference in connectance patterns between the two environments results from the optimization of feeding, which imposes structural constraints in one case and functional adaptations in the other. In fluctuating systems, environmental perturbations do limit the time available for feeding. There, it would appear advantageous for the consumer species to rely on briefer but more intense periods of predation. If this is correct, i then is the factor that must be maximized, at the expense of C when necessary. By contrast, in constant environments the structure of complex food webs need not be constrained to accommodate as large an i as possible. In such environments, weaker interactions may be tolerated since they can be exploited on a more continuous and reliable basis. It is even conceivable that in such systems both C and i might be large. This would violate the conditions for stability, but the risk appears acceptable considering the low probability of environmental disruption.

It is perhaps worthy of note that the components of connectance relate quite distinctly to species richness: on one hand, the number of trophic links, that is, the total of nonzero entries in the food web matrix, increases as a nearly linear function of S ($y = 1.3x^{1.10}$); on the other hand, the number of potential competitive links, calculated by scoring one link for every pair of predator species sharing one or more prey species between them, increases as a quadratic function of S ($y = 0.07x^{2.09}$). Both regressions are highly significant ($r = .88$; $P < .001$), and in each case the dependent variable is markedly lower in fluctuating than in constant environments (see Fig. B.5.2).

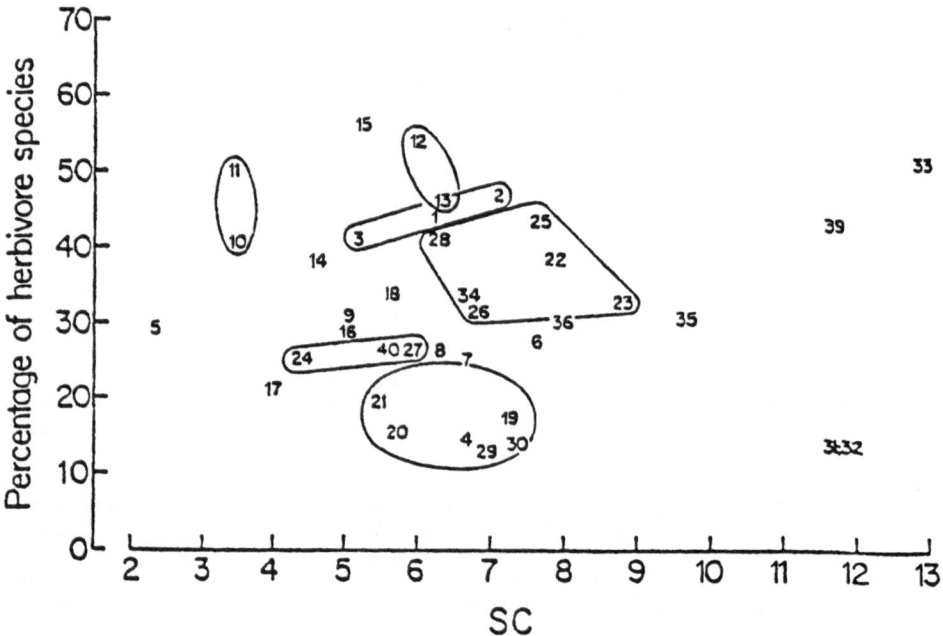

Fig. B.5.3. Segregation of ecosystems as a function of the percentage of strictly herbivorous species in the community and the product of species richness and connectance (SC). The code numbers identify the communities listed in Table 1. From left to right the groups represent Atlantic coast intertidal (10, 11), forest (24, 40, 27), estuarine (3, 1, 2), pelagic (21, 20, ...), Pacific coast intertidal (12, 13), and mixed terrestrial ecosystems (28, 25, ...). Although river and salt-marsh habitats are well represented, their respective communities do not appear distributed closely on this graph

Finally, some remarkable relations do emerge when one attempts to relate habitat type and food web structure. As shown in Fig. B.5.3, intertidal, forest, estuarine, pelagic, and mixed terrestrial communities appear as distinct groups in the space defined by connectance, species richness, and percentage of strictly herbivorous species; the last variable is chosen as indicative of food web shape. Clearly, then, ecological networks tend to be more similar within, than between, classes of ecosystems, and this is the case regardless of geographic location and taxonomic composition. I emphasize, however, that the limits delineating each group are only drawn tentatively and must be interpreted with caution. At the least, they imply that environmental constraints will impose a far greater rigidity of web shapes and a much smaller choice of trophic patterns than previously assumed.

§6. Environmental Correlates of Food Chain Length

Frédéric Briand and Joel E. Cohen

A community food web (Cohen 1978) describes the feeding relations in a community of organisms. A trophic species (Briand and Cohen 1984) (hereinafter species) in a web is a collection of organisms that feed on a common set of organisms and are fed on by a common set of organisms. Species x is linked to species y when energy flows from x to y, that is, when y feeds on x. A chain is an energy path or sequence of links that starts at a species that eats no other species in the web and ends at a species that is eaten by no other species in the web. The length of a chain is the number of links it comprises. The mean chain length of a web is the arithmetic average of the lengths of all chains in the web.

Two major hypotheses and one empirical generalization have been proposed to relate chain lengths to environmental conditions. The first hypothesis, known as the "energetic hypothesis" (Hutchinson 1959), proposes that chain length is limited by the inefficiency with which energy is transmitted by predation and by the minimal energy requirements of predators. Limited available energy may make it impossible to support enough individuals to maintain a population, may make it impossible for individuals to find enough prey to survive, or may constrain chain length through other mechanisms. In its simple form, this hypothesis predicts that chains should be longer in ecosystems with higher primary productivity. It has been tested experimentally (Pimm and Kitching 1987) and rejected for small artificial ecosystems, and it remains to be tested further experimentally. From a review of nine studies ranging from energetically impoverished to highly productive environments, Pimm (1982) concluded that there was no evidence for food chains being longer in more productive habitats.

The second hypothesis, known as the dynamical stability hypothesis (Pimm and Lawton 1977), is based on the finding in specific mathematical models of ecosystems that the longer the chains, the more severe the restrictions that must be imposed on the coefficients of the models for equilibrium to be feasible or stable. Further, in certain models, ecosystems with longer chains take longer to return to equilibrium once perturbed, so that webs with longer chains may be less likely to persist in nature. This hypothesis predicts that chains should be longer in ecosystems exempt from large perturbations. To our knowledge, there is no reported evidence for or against this hypothesis.

The empirical generalization (Briand 1983a), based on 34 webs, proposes that chains tend to be longer in three-dimensional than in two-dimensional environments. An environment is classified as having dimension 2 if it is essentially flat, like a grassland, the tundra, a sea or lake bottom, a stream bed, or the rocky intertidal zone. An environment is classified as having dimension 3 if it is solid, like the pelagic water column or a forest canopy. Webs from habitats integrating both flat and solid environments are considered as having "mixed" dimension.

To evaluate the relative influence on chain length of the primary productivity, the variability, and the dimensionality of the environment, we studied a collection of 113 webs, culled from 89 published and 2 unpublished studies, to cover as wide a diversity of natural environments as possible. Most of the world biomes are represented. There are 55 continental (23 terrestrial and 32 aquatic), 45 coastal, and 13 oceanic webs, ranging from arctic to antarctic regions.

Only webs partially defined, presented too sketchily, or based on information explicitly drawn from different locations were excluded from this collection. The webs were not screened by rejection of outliers or by any other statistical procedure based on the data. Only obvious biological errors were amended in editing the data. Although all webs were treated consistently in this collection, the practices of field ecologists in observing and reporting webs are not standardized. As the apparent characteristics of an individual web may reflect the idiosyncrasies of its observer, it is appropriate with these data to attend to broad trends and major differences among distributions.

The 113 webs studied are listed in Table 1 together with their sources and the following characteristics: mean chain length, maximal chain length, number of species, number of links, productivity, variability, dimensionality, and geographic origin. The details of all of these webs are fully documented (Chap. IV); the frequency distributions of chain length of all webs have been reported (Cohen, Briand and Newman 1986). This large collection allows comparisons to be made that are more sensitive than before to small differences in mean chain length.

The productivity of a web is classified as low if the net primary productivity of its ecosystem falls below 100 g of carbon per square meter per year and high if it exceeds 1000 g of carbon per square meter per year. Of 113 webs, 22 were classified as having low productivity, 10 as having high productivity, and 6 as having intermediate productivity. The remaining 75 webs were unclassified for want of information.

The variability of a web's habitat is classified as fluctuating or constant. The environment is fluctuating if the original report indicates temporal variations of substantial magnitude in temperature, salinity, water availability, or any other major physical parameter. The magnitude, not the predictability, of the variations is the criterion of classification. Of 113 webs, 64 were classified as fluctuating and 17 as constant. The remaining 32, previously (Cohen, Briand and Newman 1986) unclassified, are considered here as intermediate.

Of 113 webs, 40 were classified as having dimension 2 and 28 as having dimension 3. Forty-five webs previously (Cohen, Briand and Newman 1986) recorded as having neither dimension 2 nor dimension 3 are here considered as having mixed dimension.

Some subjective judgments are involved in classifying webs as fluctuating or constant and as two-dimensional or three-dimensional. For the first 40 webs in the series (Cohen 1978; Chap. II.5), the facts supporting these judgments are already documented.

All calculations were performed for both mean and maximal chain lengths. Maximal chain lengths varied in parallel with mean chain lengths throughout. We present the mean (within-web) chain lengths descriptively using box plots (Tukey 1977; McNeil 1977). We attempt no formal statistical tests of differences between

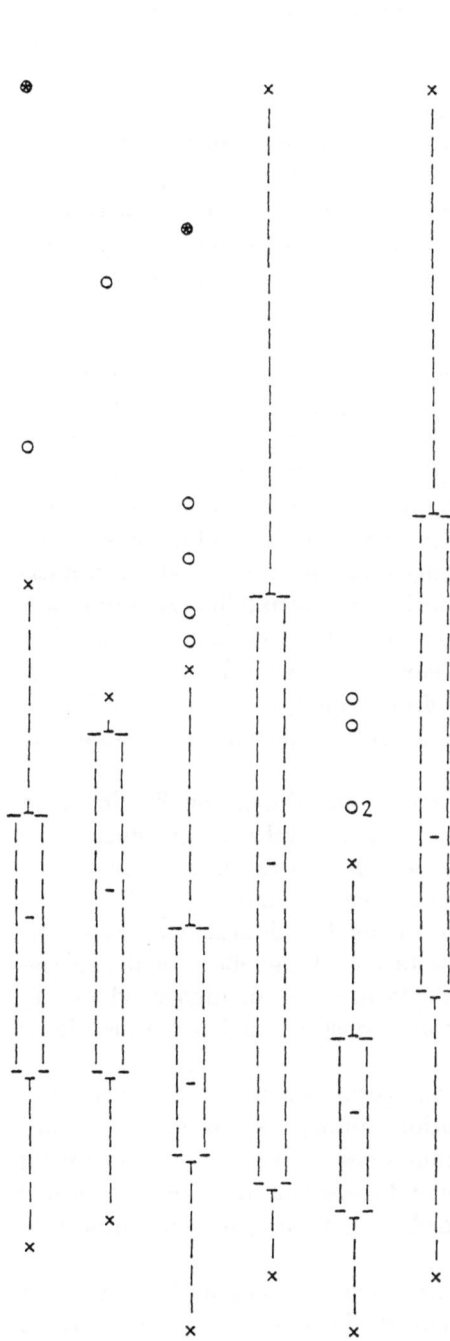

Fig. B.6.1. Box plots of the frequency distributions of mean (within-web) chain lengths in 113 webs classified according to productivity (low or high), environmental variability [fluctuating (Flu.) or constant (Con.)], and dimension (2 or 3). Some webs are omitted from each frequency distribution because they were intermediate. For each box, the upper edge corresponds to the upper quartile (75th percentile or $Q3$) of the distribution being plotted, and the lower edge corresponds to the lower quartile (25th percentile or $Q1$), and the dash in the middle corresponds to the median (50th percentile or $Q2$). The numerical values of these ordinates appear below each box. Vertical lines extend from the upper quartile $Q3$ up to the largest observation (marked by x) less than $Q3 + (Q3 - Q1)$, and from $Q1$ down to the lowest observation (also marked by x) greater than $Q1 - (Q3 - Q1)$. Webs more extreme than those represented by x are represented by one (o), unless a number next to the symbol indicates a larger number of webs coincident at this value. Outlying webs more than $1.5 \times (Q3 - Q1)$ distant from the nearest quartile are emphasized by (⊕); n, number of webs

	Productivity		Variability		Dimension	
	Low	High	Flu.	Con.	2	3
Q 3	3.3	3.6	3.0	4.0	2.6	4.3
Q 2	3.0	3.0	2.5	3.2	2.4	3.2
Q 1	2.5	2.4	2.2	2.1	2.0	2.7
n	22	10	64	17	40	28

distributions because it is doubtful that the webs in our collection form a random sample from a well-defined universe of webs (Cohen, Briand and Newman 1986).

Fig. B.6.1 shows that the distributions of mean chain lengths are similar, with virtually identical medians, in webs differing markedly in productivity. Contrary to the energetic hypothesis, high-energy systems do not support longer chains, on average or maximally, than energetically impoverished environments. The possibility remains that energy influences chain length but that highly productive systems attract a greater fraction of energetically less efficient consumers, which prevent the assembly of longer food chains. Lacking detailed data on the energetic efficiency of the web species, we cannot exclude this possibility (Yodzis 1983).

The distributions of mean chain length are relatively distinct in fluctuating compared to constant webs and quite distinct in webs having dimension 2 compared to those having dimension 3. The upper quartile of mean (within-web) chain length for the 40 webs of dimension 2 is 2.6 links, which falls below the lower quartile (2.7 links) of mean chain length for the 28 webs of dimension 3.

With a sufficiently large collection of fully described webs, it would be possible to cross-classify each web by its productivity, variability, and dimension and thereby to study the dependence of chain length on all three variables simultaneously. When the 113 webs are cross-classified by the variability and dimension of the environment only (and not by productivity, which is unknown for many webs), there are only two webs in constant environments of dimension 2. There are 27 webs in fluctuating environments of dimension 2, and this is the largest number in any cell of the cross-classification. Not enough webs are available to support further cross-classification.

It would be hasty to conclude that variability and dimensionality independently influence chain length. Of the two-dimensional webs, 27 are fluctuating and 2 are constant; of the three-dimensional webs, 13 are fluctuating and 7 are constant. Thus the proportion of constant webs is more than five times as high among three-dimensional webs as among two-dimensional webs. No such risk of confounding affects the interpretation of the effect of productivity in Fig. B.6.1, since webs from environments with low or high productivity include comparable fractions of fluctuating and constant, and two-dimensional and three-dimensional, habitats.

To assess the relative influence of environmental dimension and variability on chain lengths, we compared the distributions among webs of mean (within-web) chain lengths in fluctuating and constant webs having comparable, mixed dimension (Fig. B.6.2a) and in two- and three-dimensional webs of comparable variability in constant, fluctuating, or intermediate habitats (the last comparison being shown in Fig. B.6.2b).

If environmental variability alone markedly affects the length of chains, then the distributions in Fig. B.6.2a should be distinct. That is not the case: given a mixed dimension, constant environments do not support markedly longer chains than fluctuating environments, contrary to the dynamic stability hypothesis.

If environmental dimension alone markedly affects chain length, then the distributions for webs with intermediate variability in Fig. B.6.2b should be distinct.

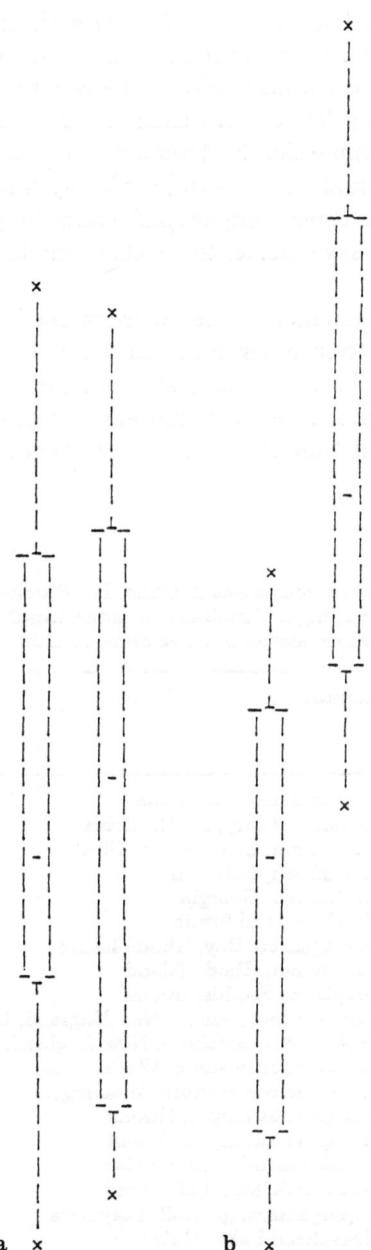

	Mixed dimension		Intermediate variability	
	Flu.	Con.	2-D	3-D
Q3	3.6	3.6	3.2	4.5
Q2	2.7	3.0	2.7	3.7
Q1	2.4	2.0	2.0	3.2
n	24	8	11	8

Fig. B.6.2a,b. Box plots of the frequency distributions of mean (within-web) chain length (a) in webs of mixed dimension, comparing fluctuating (Flu.) and constant (Con.) environments and (b) in webs of intermediate variability, comparing two- and three-dimensional environments. Symbols and other abbreviations are as in Fig. B.6.1

That is clearly the case. Further, in fluctuating habitats, the 27 webs with dimension 2 have a median 2.3 mean chain length, less than the median 2.8 mean chain length of the 13 webs with dimension 3. In constant habitats, the two webs with dimension 2 have a median 2.3 mean chain length, less than the median 4.0 mean chain length of the seven webs with dimension 3. Although there are too few webs in constant two- or three-dimensional habitats to justify any firm conclusion, the differences are consistent in the three comparisons: controlling for variability, webs in two-dimensional habitats have shorter mean chain lengths than those in three-dimensional habitats.

We conclude from our data that the dimensionality of the environment influences mean or maximal chain length more than environmental variability. Dimensionality is a major determinant of chain length in natural communities. Why this is so remains to be explained, although it is evident that environmental dimension may affect the probability per unit time of an encounter between predator and prey.

Table 1. Characteristics of 113 webs. Serial numbers and sources are as in Chap. IV. Productivity: 0, unclassified (unknown or intermediate); 1, low; 2, high. Variability: 0, intermediate; 1, fluctuating; 2, constant. Dimension: 0, mixed; 2, two dimensional; 3, three dimensional.

Web number	Mean chain length	Max. chain length	No. of trophic species	No. of links	Prod.	Var.	Dim.	Habitat
1	3.13	4	8	14	0	0	0	Cochin backwater, India
2	2.71	3	14	22	0	1	0	Knysna estuary, South Africa
3	2.30	3	24	34	0	1	2	Salt marsh, Long Island, USA
4	2.74	4	13	26	0	1	0	Salt marsh, California
5	2.00	2	6	5	2	0	0	Salt marsh, Georgia
6	3.82	6	25	43	0	1	0	Tidal flat, California
7	2.79	4	18	30	0	0	0	Narragansett Bay, Rhode Island
8	2.44	4	15	25	2	1	0	Salt marsh, Rhode Island
9	2.86	3	9	13	0	0	0	Lough Ine Rapids, Ireland
10	2.00	2	3	2	0	1	2	Exposed rocky shore, New England, USA
11	2.00	2	5	4	0	1	2	Protected rocky shore, New England, USA
12	2.25	3	9	13	0	1	2	Exposed rocky shore, Washington
13	2.50	3	9	14	0	1	2	Protected rocky shore, Washington
14	2.40	3	8	10	0	0	0	Mangrove swamp 1, Hawaii
15	2.33	3	7	7	0	1	0	Mangrove swamp 3, Hawaii
16	2.14	3	14	20	2	1	0	Pamlico estuary, North Carolina
17	3.56	5	14	23	0	0	3	Coral reefs, Marshall Islands
18	2.00	4	23	35	0	0	0	Kapingamarangi Atoll, Polynesia
19	3.00	4	17	32	1	1	3	Moosehead Lake, Maine
20	3.26	5	19	30	1	0	3	Antarctic pack ice zone
21	4.61	7	9	20	0	0	3	Ross Sea
22	3.69	7	28	58	0	1	0	Bear Island, Spitsbergen
23	2.40	4	15	27	0	1	2	Prairie, Manitoba
24	2.70	4	12	18	0	1	3	Willow forest, Manitoba
25	2.16	4	24	37	0	1	3	Aspen communities, Manitoba
26	2.93	6	32	56	0	1	0	Aspen forest, Manitoba
27	2.89	4	22	39	2	1	3	Wytham Wood, England
28	1.96	3	32	35	0	1	0	Salt meadow, New Zealand
29	3.14	5	16	22	1	0	3	Arctic seas
30	5.02	7	14	32	1	0	3	Antarctic seas

Table 1. Continued

Web number	Mean chain length	Max. chain length	No. of trophic species	No. of links	Prod.	Var.	Dim.	Habitat
31	3.90	6	14	51	0	0	3	Epiplankton communities, Black Sea
32	3.86	6	14	52	0	2	3	Bathyplankton communities, Black Sea
33	1.93	4	29	48	0	2	0	Crocodile Creek, Malawi
34	2.56	4	12	27	0	2	2	River Clydach, Wales
35	2.72	4	13	36	0	0	2	Morgan's Creek, Kentucky
36	2.07	4	19	35	0	0	0	Mangrove swamp 6, Hawaii
37	2.75	4	24	46	0	2	0	Marine sublittoral, southern California
38	2.13	3	31	95	0	2	0	Lake Nyasa, rocky shore, Malawi
39	1.80	3	33	70	0	2	0	Lake Nyasa, sandy shore, Malawi
40	1.88	3	11	15	0	2	3	Rain forest, Malaysia
41	5.92	8	18	49	1	2	3	Tropical seas, epipelagic zone
42	4.95	8	15	36	2	2	3	Upwelling areas, Pacific Ocean
43	3.13	5	20	38	0	2	3	Kelp bed community, south California
44	3.63	5	12	29	2	2	0	Marine coastal lagoons, Guerrero, Mexico
45	2.14	3	11	20	0	2	2	Cone Spring, Iowa
46	4.43	8	19	68	1	0	3	Lake Texoma, Texas
47	4.22	5	27	50	0	2	0	Swamps, south Florida
48	3.53	5	13	20	0	1	0	Nearshore marine 1, Aleutian Islands
49	2.56	4	12	20	0	1	0	Nearshore marine 2, Aleutian Islands
50	2.44	3	14	23	0	1	2	Sand beach, California
51	3.28	5	25	46	0	0	0	Shallow sublittoral, Cape Ann, Massachusetts
52	2.08	3	20	32	0	1	2	Rocky shore, Torch Bay, Alaska
53	1.95	2	22	31	0	1	2	Rocky shore, Cape Flattery, Washington
54	2.58	4	14	20	0	0	0	Western rocky shore, Barbados
55	2.46	3	12	18	2	1	2	Mudflat, Ythan Estuary, Scotland
56	2.22	3	10	14	0	1	2	Mussel bed, Ythan Estuary, Scotland
57	3.29	5	9	19	2	0	0	Brackish lagoons, Guerrero, Mexico
58	4.28	7	17	21	0	1	0	Sphagnum bog, Russia, USSR
59	2.37	4	29	61	0	1	3	Trelease Woods, Illinois
60	2.36	3	33	69	0	1	3	Montane forest, Arizona
61	2.00	3	8	10	1	1	2	Barren regions, Spitsbergen
62	3.00	4	11	12	1	1	2	Reindeer pasture, Spitsbergen
63	3.16	4	18	75	0	0	2	River Rheidol, Wales
64	1.67	2	19	28	0	0	2	Linesville Creek, Pennsylvania
65	1.85	2	13	25	0	0	2	Yoshino River rapids, Japan
66	2.93	4	10	18	0	0	2	River Thames, England
67	3.94	6	21	62	0	0	0	Mudflats, Mississippi River, Iowa
68	2.63	4	22	32	0	1	3	Loch Leven, Scotland
69	3.62	6	29	73	0	1	0	Tagus Estuary, Portugal
70	2.49	3	14	28	0	1	0	Crystal River Estuary, Florida
71	5.15	7	16	32	0	1	3	Lake Rybinsk, Russia, USSR
72	3.95	5	17	32	0	1	3	Heney Lake, pelagic zone, Quebec
73	2.38	3	10	15	0	1	3	Hafner Lake, Austria
74	2.38	4	21	36	0	1	2	Sand beach, South Africa
75	2.75	4	9	14	0	1	3	Vorderer Finstertaler Lake, Austria
76	2.67	4	14	17	1	1	0	Neusiedler Lake, Austria
77	3.63	5	13	24	0	2	0	Lake Abaya, Ethiopia
78	3.15	5	16	27	2	2	0	Lake George, Uganda
79	3.41	5	21	29	0	1	0	Lake Pääjärvi, offshore, Finland
80	3.35	5	27	70	0	1	0	Lake Pääjärvi, littoral zone, Finland
81	2.73	4	12	19	1	0	0	Sendai Bay, mesopelagic zone, Japan
82	3.71	5	10	14	0	1	0	Permanent freshwater rockpool, France
83	2.45	4	25	67	1	1	0	Lake Pyhäjärvi, littoral zone, Finland
84	3.61	5	12	23	0	1	0	Temporary pond, Michigan
85	3.61	5	27	49	2	1	0	Tasek Bera Swamp, Malaysia

Table 1. Continued

Web number	Mean chain length	Max. chain length	No. of trophic species	No. of links	Prod.	Var.	Dim.	Habitat
86	4.09	6	16	37	0	1	3	Suruga Bay, epipelagic zone, Japan
87	2.91	4	11	17	1	0	0	Ice edge community, High Arctic, Canada
88	1.95	2	16	42	0	0	2	Lestijoki River Rapids, Finland
89	2.89	4	18	32	0	0	3	River Cam, England
90	1.84	2	22	39	0	1	2	Old field, New Jersey
91	3.00	4	10	13	0	1	3	Shigayama coniferous forest, Japan
92	2.00	3	18	18	1	0	2	High Himalayas community, Tibet
93	2.12	3	26	70	1	1	2	Alpine tundra, Montana
94	3.35	5	12	19	1	1	2	Wet coastal tundra, Barrow, Alaska
95	2.50	4	10	12	1	1	2	Tundra, Prudhoe, Alaska
96	1.92	2	9	16	1	1	2	Tundra, Yamal Peninsula, Siberia
97	2.00	3	11	17	1	1	2	Tundra, South Yamal, Siberia
98	3.54	5	17	39	1	0	2	Sand dunes, Namib Desert, Namibia
99	2.51	4	48	138	1	0	2	Sonora Desert, Arizona
100	3.34	6	22	59	1	0	2	Rajasthan Desert, India
101	1.67	2	6	5	0	1	0	Temporary freshwater rockpool, France
102	3.97	7	9	27	1	2	3	Plankton, oligotrophic tropical Pacific
103	5.59	10	23	133	1	2	3	Tropical plankton community, Pacific
104	3.16	5	27	62	0	0	2	Rocky shore, Bay of Panama
105	3.67	5	10	22	0	1	2	Rocky shore, Gulf of Maine, USA
106	2.41	5	35	73	0	1	2	Rocky shore, Monterey Bay, California
107	2.50	3	10	14	0	1	2	Bay pilings community, New Jersey
108	2.27	3	14	20	0	1	2	Rocky shore, Cabrillo Point, California
109	2.88	4	21	57	0	1	2	Rocky shore, central Chile
110	2.13	3	13	23	0	1	2	Rocky shore, Cape Ann, Massachusetts
111	2.44	3	19	36	0	1	2	Mudflat, Cape Ann, Massachusetts
112	1.83	3	14	17	0	1	0	Low salt marsh, Cape Ann, Massachusetts
113	2.11	3	11	12	0	1	0	High salt marsh, Cape Ann, Massachusetts

Chapter III
A Stochastic Theory of Community Food Webs

§1. Theory: Circles of Complexity, Spherical Horses

Joel E. Cohen

The introduction to the empirical portion of this book describes a three-panel cartoon. In the middle panel, theoretical ecologists collected community food webs and discovered some entertaining empirical regularities. (That summarizes the preceding chapters of this book.) In the background of the last panel, mice were constructing a theoretical scaffolding to hold together the observed empirical regularities. This portion of the book presents that scaffolding.

The materials in a scaffolding are often used for more than one building. The theory in this portion of the book is no exception. This introduction describes where some of the previously used pieces of the theoretical scaffolding came from, and suggests some directions for future extensions.

There is a well-known joke about a theoretical physicist who decides to conquer biology using methods that proved so powerful in physics. After modestly taking a whole week to learn biology, he or she begins: "Consider a spherical horse." The history of the theory in this portion of the book might be viewed as a progression from spherical horses, to random spherical horses, to the neighs of a random spherical horse. To form a clearer image of a scaffolding built of spherical horses, read on.

1. Spherical Horses

Lotka, in his 1925 magnum opus *Elements of Physical Biology*, and in earlier papers, constructed an influential model of community ecology from the kinetic equations of chemistry. He proposed that the species (chemical or biological) of a community evolve according to a family of autonomous, generally nonlinear, first-order ordinary differential equations. As special cases, he considered a pair of equations that model interactions between a predator and a prey, and another pair of equations that model interactions between two competitors. The famous

"Lotka-Volterra equations" refer to one or another of these pairs of nonlinear, first-order ordinary differential equations. Lotka's perspective on ecological dynamics has influenced much ecological theory (e.g., there are 11 index references to Lotka or his equations in May 1981; 3 references in the subject index, but none in the name index, of Strong et al. 1984; and 7 references in the subject index of Diamond and Case 1986), including in particular models of food webs (e.g., Pimm 1982). It is ironic that Lotka apparently did not consider himself an ecologist, and hoped for recognition from physicists that was never forthcoming (Kingsland 1985).

The history that led Lotka to his view of ecological communities is perhaps equally ironic. Trained as a physical chemist, Lotka studied for the 1901–1902 academic year in Leipzig, where he was greatly influenced by the lectures of Ostwald, a Nobel Laureate in chemistry. Ostwald in turn was greatly influenced at that time by the views of Haeckel, who (in 1866) coined the word "ecology". Haeckel saw Darwinian evolution, and the survival of the fittest (interpreted as naively as possible), as the universal law of human groups as well as of nonhuman species. In 1906, not long after Lotka's studies in Leipzig, Ostwald helped found the Monist League. Its purpose was to advance the views of Haeckel. In the social and cultural setting of Germany, Haeckel's Monism provided an interpretation of Darwinism that eventually justified the systematic destruction of non-Aryans (Stein 1988).

Lotka, who would have been among those destroyed had he remained in Leipzig, carried away from Leipzig the purely scientific interpretation of Ostwald's Monistic views. Coming to America, Lotka began the program of research and publication that culminated in his 1925 book. His 1907 paper, "Studies on the mode of growth of material aggregates", followed a mathematical theory of stable age-structured populations with a model of isothermal monomolecular reactions (Cohen 1987). He viewed both models as "the study of the laws governing the distribution of matter among complexes of any specified kind, as determined by their general physical character."

Here is a spherical horse! Lotka, more than many who adopted his perspectives and used his models, knew a spherical horse when he saw one, even if he rode it. In 1932, Lotka ended a paper on what are now called the Lotka-Volterra equations for two competing species with the observation: "It is perhaps hardly to be expected that concrete examples of the law of growth for two populations here discussed shall be found in nature." Among the feature of nature these equations neglect are: the age structure of the competing species, genetic heterogeneity within each species, spatial and temporal heterogeneity in the environment and in the parameters of interaction, possible roles of learning, interference from other species that may be present, and exhaustion or resupply of nutrients. The list could be extended, but suffices to indicate why Lotka's abstract, general models are spherical horses.

Nevertheless, when Gause constructed microbiological "ecosystems" that were sufficiently simplified to be described by the Lotka-Volterra competition equations, the equations were enshrined as useful keys to vastly more complex nature.

Perhaps one reason for the influence of Lotka-Volterra equations on food web theory is that they provide a language that can formalize Darwin's theoretical intuitions about food webs. Darwin (1859, p. 72) described a region of Paraguay where no cattle, horses, or dogs run wild because a certain fly lays its eggs in the navels of these animals when the animals are born. He then hypothesized:

Hence, if certain insectivorous birds (whose numbers are probably regulated by hawks or beasts of prey) were to increase in Paraguay, the flies would decrease – then cattle and horses would become feral, and this would certainly greatly alter (as indeed I have observed in parts of South America) the vegetation: this again would largely affect the insects; and this, as we just have seen in Staffordshire, the insectivorous birds, and so onwards in ever-increasing circles of complexity.

In a second example, Darwin (1859, pp. 73–74) described the crucial role of British insects in pollinating British flowers, and the apparent influence of field-mice on insects. He again hypothesized:

Hence it is quite credible that the presence of a feline animal in large numbers in a district might determine, through the intervention first of mice and then of bees, the frequency of certain flowers in that district!

Hypotheses such as these ignore the age structure of the competing species, genetic heterogeneity within each species, spatial and temporal heterogeneity in the environment and in the parameters of interaction, possible roles of learning, interference from other species that may be present, and exhaustion or resupply of nutrients. Does this list sound familiar? Darwin's hypotheses are made to order for the stoichiometric equations of Lotka; in both, predation simply inhibits the population growth and population size of the prey and enhances the population growth and population size of the predator. Though their verbal formulation makes them appear to be horses of another color, Darwin's hypotheses about food chains and food webs are spherical horses as much as Lotka's.

Like Lotka, Darwin knew a spherical horse when he saw one. Immediately after the preceding example, he wrote (1859, p. 74):

In the case of every species, many different checks, acting at different periods of life, and during different seasons or years, probably come into play; some one check or some few being generally the most potent, but all concurring in determining the average number or even the existence of the species. In some cases it can be shown that widely-different checks act on the same species in different districts.

Darwin recognized age structure, seasonality, and spatial heterogeneity. This recognition did not make him afraid to build around his observations a scaffolding of spherical horses.

2. Random Spherical Horses

Meanwhile, back at the spherical ranch (where the spherical horses range), nuclear physicists were trying to understand the distribution of energy levels, or spectra, of nuclei. In a typical experiment, nuclei of some element are bombarded by neutrons with differing amounts of energy, and the number of particles flying out are counted as a function of the energy of the bombarding neutrons (Mehta

1986). The frequency histogram of counted particles as a function of energy, called the spectrum, displays discrete sharp peaks. The problem is to explain the location of these peaks.

In the 1950s, the physicist Eugene P. Wigner introduced the idea that the distribution of peaks could be understood as the distribution of the eigenvalues of a certain square symmetric matrix with random elements. By a happy coincidence, the set of eigenvalues of a matrix is called its spectrum, so Wigner proposed that energy spectra could be described by random matrix spectra. This weird idea gives an amazingly good quantitative description of observed energy spectra (Mehta 1986).

Wigner discovered that if the size (i.e., number of rows or columns) of his particular random matrix increases while the variance of each element falls like the reciprocal of the size of the matrix, then the empirical distribution function of the spectrum (which is just a statistical summary of where the eigenvalues are) approaches a specific limit, the so-called semicircle law. Moreover, there is a fixed number such that, with a probability that approaches one as the matrix gets arbitrarily large, the largest eigenvalue falls arbitrarily close to the fixed number. This means that the cumulative probability distribution function of the largest eigenvalue looks like a step function, jumping from nearly zero below the fixed number to nearly one above the fixed number. As the largest eigenvalue of a matrix determines the stability of a linear system described by that matrix, Wigner proved, in effect, that the stability of a linear system described by his random matrix obeys a similarly abrupt transition. Wigner's results (e.g., Wigner 1958) have been very extensively generalized and refined (e.g. Bai and Yin 1988).

In 1972, Robert M. May, another physicist by training, proposed that Wigner's mathematics could illuminate an important question of ecology: what is the relation between the complexity and the stability of communities? May's central, and powerful, idea was that if it is difficult to investigate the relation between stability and complexity in individual communities, it may be informative to investigate that relation in a hypothetical ensemble of random communities. In the absence of data, such communities could be constructed according to hypotheses chosen largely for analytical convenience, to take advantage of Wigner's mathematics.

May's model, truly a random spherical horse, might have been named Son of Lotka. His model describes the population dynamics of a set of interacting species by a set of nonlinear first-order differential equations, assumed to have a point of equilibrium. Around this equilibrium, the nonlinear dynamics are approximated by a linear equation $dx/dt = Ax$. Here x is a vector with as many elements as there are species in the community, and each element represents that species' deviation from its equilibrium. The matrix A is called the community matrix. The diagonal elements of A are fixed at -1, so that each species is stable by itself. The off-diagonal elements are set equal to zero with probability $1 - C$ and otherwise, with probability C, are chosen at random from a distribution with mean zero and some positive variance. The parameter C, $0 < C < 1$, is called the connectance. It is the probability of interaction between two species.

May (1972) found that, as the number of species became very large, if the product of the number of species times the connectance times the variance of the off-diagonal elements remained less than one, then the probability that the community would be stable approached one, i.e., certainty, while if the same product remained greater than one, then the probability that the community would be stable approached zero, i.e., impossibility. May interpreted his finding:

Applied in an ecological context, this ensemble of very general mathematical models of multi-species communities, in which the population of each species would by itself be stable, displays the property that too rich a web connectance ... or too large an average interaction strength ... leads to instability. The larger the number of species, the more pronounced the effect.

According to this conclusion, any increase in complexity, whether measured by number of species, connectance, or interaction strength, brings an initially stable community closer to instability. This conclusion contradicted the then-received wisdom among ecologists that the more complex a community is, the more stable it is. May's random spherical horse, amplified in his monograph (May 1973), kicked open the ecological barn door, and a stampede of theoretical and empirical studies thundered out. See the reviews and references of May (1981) and Pimm (1984).

It took a dozen years to notice that May's claims, in the generality with which they were originally stated, were mathematically false (Cohen and Newman 1984, 1985). Wigner had assumed symmetry, May had not. In a footnote, May (1972) had suggested that "the present results for the largest eigenvalue and its neighbourhood can be obtained by using Wigner's original style of argument on" the product of the community matrix times the transpose of that matrix, which is indeed symmetric. Attempts by others to fill in the details of May's sketch were doomed to failure.

Nevertheless, in a model in which the random community matrix changes randomly in time, conclusions very like May's hold under certain assumptions (Cohen and Newman 1984). For a model with a fixed community matrix, as May (1972) assumed, Geman (1986) discovered additional conditions sufficient to guarantee May's conclusions about stability. Additional conditions sufficient to guarantee May's conclusions about instability remain to be discovered for a model with a random community matrix fixed in time. This random spherical horse still limps in one leg.

But in an important sense, by the time the limp became obvious, it did not matter! Indeed, May's random spherical horse, limp and all, has arguably been much more useful than many a technically correct model in theoretical ecology.

First, May's model contributed importantly to ecological theory. The model raised the level of ecologists' thinking about the relation between complexity and stability. The model encouraged ecologists to think in terms of an ensemble of ecosystems composed according to some random process. The model strengthened the growing willingness among ecologists to think abstractly and exactly about ecological issues in general. It demonstrated that multi-species community models could, in principle, be analyzed using methods more recent than Lotka's general theory of equilibrium. It provoked a host of more detailed models.

Second, May's conclusions made quantitative predictions that could be tested against data. For example, for a fixed variance in the off-diagonal elements of the community matrix, the maximum connectance that is compatible with a stable community is predicted to fall as the reciprocal of the number of species in the community, asymptotically as the number of species becomes very large. Reinterpreted as a statement about communities with a finite number of species, this prediction prompted Rejmánek and Starý (1979) to collect food webs; as predicted, they found a hyperbolic relation between connectance and the number of species. Though other interpretations of this observation are now possible (Cohen and Newman 1988), it was May's model that prompted the observation in the first place.

A random spherical horse that promotes theory and observation is not a bad horse!

3. Neighs of a Random Spherical Horse

In another part of the spherical ranch, while Wigner was analyzing the asymptotic theory of the spectra of random matrices, two mathematicians, P. Erdös and A. Rényi, were developing a revolutionary theory of random graphs. A random graph consists of a set of vertices and some probabilistic rule for assigning edges between pairs of vertices. Think of each vertex as representing a species in an ecological community, and of each edge as some bidirectional relation between species, e.g., competition. In 1960, Erdös and Rényi showed that amazing things happen in a random graph if the number of vertices grows arbitrarily large while the probability of an edge between two vertices is a specified, declining function of the number of vertices. For example, if the probability of an edge between two vertices declines ever so slightly faster than the reciprocal of the number of vertices, then almost surely a large random graph will contain no cycles of any order; but if the probability of an edge between two vertices declines ever so slightly slower than the reciprocal of the number of vertices, then almost surely a large random graph will contain cycles of every order. Thus the reciprocal of the number of vertices is called the threshold function for cycles of all orders: when edges are asymptotically more (or alternatively less) probable than the reciprocal of the number of vertices, cycles of all orders are nearly sure to be present (or alternatively absent). Erdös and Rényi discovered threshold functions for a host of interesting graph properties, and many more have been discovered since (e.g. Bollobás 1985).

Given data about community food webs, these discoveries about random graphs are exciting because they reveal large-scale order in random mathematical objects, graphs, that are much closer to the form of most food web data than are the quantitative community matrices assumed in the models of Lotka and May; in this sense, because random graphs simplify quantitative relations into purely combinatorial ones, they are the neighs of random spherical horses. On the other hand, these discoveries are frustrating because random graphs are not random *directed* graphs, or digraphs, as food webs are; random graphs do not represent directed relations between species, such as one species eating another.

One possible response is to construct from food webs *undirected* graphs, such as trophic niche overlap graphs (Cohen 1978) or resource graphs or common enemy graphs (Sugihara 1982; Lundgren and Maybee 1985). If the probabilistic models of Erdös and Rényi (1960) are appropriate, then their methods and theorems give information about the asymptotic behavior of these undirected graphs (e.g., Cohen, Komlós, and Mueller 1979).

A second possible response, slower and more laborious, is to construct digraph models appropriate for food webs, taking inspiration from Erdös and Rényi (1960). Cohen (1978) considered six digraph models for food webs, all of which were unsuccessful; Cohen and Newman (Chap. III.2) consider three more, two of which clearly fail. Pimm, Yodzis (1984), Lawton, DeAngelis, Sugihara (1982) and others have considered many other models; see Pimm (1982) for references. The cascade model, which is the principal support of the theoretical scaffolding in the following chapters, is not the first idea that came to mind, though it incorporates elements of many of its predecessors.

The cascade model is an incomplete, cross-sectional description of dynamic ecological processes. In directing attention to a finite set of species (though they are trophic species rather than biological species), the cascade model reveals itself a descendant of the spherical horses of Darwin, Lotka, and May. (Where are the age structure and the genetic heterogeneity of the species, and where the temporal and spatial heterogeneity [e.g., Levin 1978] of the environment?) In supposing that the presence or absence of interaction between two species is randomly determined, the cascade model shows the genes of May's random spherical horse. In replacing the quantitative effects of one species preying on another by a simple all-or-none relation, i.e., in replacing the horse by its neigh, the cascade model shows the parentage of the random graph theory of Erdös and Rényi (1960).

Each of the following chapters points out limitations of the cascade model; but one limitation is perhaps not sufficiently emphasized. The cascade model, by construction, has no cycles, i.e., excludes the possibility that e.g. A eats B and B eats A, or A eats B, B eats C, and C eats A. The justification for this exclusion is empirical: published food webs rarely report such cycles. Nevertheless, in nature, energy flows not only uphill, in chains of grazing or browsing and carnivory, but also downhill, in chains of detritus, decomposition and decay (see e.g. Woodwell 1970; Cousins 1980). The cascade model neglects the trophic or chemical processes of recycling, but only because most of the available data do, too. As Kenneth Wachter suggested (personal communication, 16 May 1986), the cascade model really aims to describe the largest cycle-free portion of community food webs.

Stripping away dynamics (Darwin, Lotka), stability (Lotka, May), and that warm feeling that being able to tell a story about a model gives you, the cascade model concentrates on explaining the phenomenology of observed food web structure, using a minimum of hypotheses. This concentration on structural phenomenology does not deny the importance of other aspects of food webs;

it is merely a point of departure. The walls rise with the scaffolding, and the wallpaper comes later.

4. What Next?

The achievements of the cascade model have been summarized in the general introduction to this book, and are set out in detail in the following chapters. There are two main directions to explore next: seeing what else the cascade model can, or cannot, explain; and seeing what explains the structure and parameters of the cascade model.

The introduction to this book suggested that the cascade model, in conjunction with the standard species-area curve, might explain the very weak, or nonexistent, connection between the area of an island and the length of the longest chains in its food web.

Other areas where the cascade model might offer explanations are: the roles of predation and competition as a function of trophic position in a community; allometric relations between the body sizes of predators and prey; differences in mean or maximal food chain length between different classes of habitats; the pyramid of numbers and biomass in communities; and differential responses to environmental perturbations as a function of trophic position in a community.

In the other direction, if the cascade model continues to be useful, it becomes a challenge to explain the form and parameters of the model itself. I, and independently Warren and Lawton (1987), suggested that the upper triangular form of the predation matrix assumed by the cascade model might result from ordering the species in a community by size, if animals eat prey that are smaller than themselves. In one food web derived from laboratory experiment, rather than from field observation, Warren and Lawton (1987) found that when consumers were ordered by increasing size, the predation matrix was close to, but not quite, upper triangular. The role of size as the possible order in the cascade model deserves much further empirical study, using food webs derived from the field.

In addition to assuming an ordering of species, the cascade model assumes that the probability of a link from any species to any species above it in the ordering falls as the reciprocal of the number of species in the community. This hypothesis is mathematically equivalent to the link-species scaling law. Cohen and Newman (1988) use stability criteria inspired by May's (1972), in combination with a model of the incompleteness of ecological observations, to derive the link-species scaling law. Other possible derivations should be explored.

Thus serious first steps have already been taken to explain the two structural assumptions of the cascade model, namely, the ordering of species and the link-species scaling law. Still in need of explanation are the model's two parameters. One parameter is the coefficient of proportionality between the probability of a link and the reciprocal of the number of species. This coefficient is about 4 for large numbers of species. Why? The other is the number of species. Explaining the number of species occupied Darwin (1859) and has occupied every ecologist since.

§2. Models and Aggregated Data

Joel E. Cohen and Charles M. Newman

1. Introduction

A *food web* is a set of kinds of organisms and a relation that shows which kinds of organisms, if any, each kind of organism in the set eats. A *community food web* is a food web obtained by picking, within a habitat or set of habitats, a set of kinds of organisms on the basis of taxonomy, size, location, or other criteria, without prior regard to the eating relations among the organisms (Cohen 1978, pp. 20–21). In the past hundred years, ecologists have reported many community food webs. Briand (Chap. IV) collected and edited 62 of these, including 13 of those assembled by Cohen (1978). Several simple empirical generalizations describe the major features of these community food webs, viewed as an ensemble (Chaps. II.2, II.3).

The purpose of this chapter is to propose a simple explanation that accounts for these empirical generalizations in an economical way. The proposed explanation (the 'cascade' model of section 6) is one several attempted models. The unsuccessful models will also be reviewed to show why models that are simpler than the one we ultimately propose do not account for the major features of the data.

Section 2 introduces our terminology and summarizes the empirical generalizations that this work aims to explain. Sects. 3–6 describe successively more restricted stochastic models, based on random directed graphs, and their failures and successes in accounting for the observed generalizations. Section 7 reviews the results obtained, relates them to prior results, and points out some of their limitations.

The next chapter (Chap. III.3) tests further the most successful model proposed here, by using disaggregated data on individual community food webs.

2. Terminology and Empirical Generalizations

We shall follow the terminology and restate the major conclusions of Briand & Cohen (1984) and Cohen & Briand (1984) (Chaps. II.2–3).

By a *species*, we mean a class of organisms that prey on the same kinds of organisms and are preyed on by the same kinds of organisms. A species in this sense may result from lumping together kinds of organisms that were identified as separate by a reporting ecologist but that were recorded as having the same sets of prey and the same sets of predators (Briand & Cohen 1984). A species in this sense bears no necessary relationship to a biological species.

By a *link*, we mean any reported feeding or trophic relation between two species in a community food web. Observers use various criteria to decide how much feeding justifies the reporting of a link and how much failure to observe feeding justifies reporting the absence of a link (Cohen & Briand 1984).

A community food web graph represents a community food web as a directed graph or digraph. (The use of digraphs to represent food webs was proposed, apparently independently, by Harary (1961) and Gallopín (1972).) The vertices of the digraph correspond to the set of species in the community food web, and there is an arrow or directed edge from vertex i to vertex j in the digraph if and only if species j feeds on species i, that is, food flows from species i to species j. In the description of the theory of digraphs by Robinson & Foulds (1980), the possibility that $i = j$, that is, cannibalism, is excluded. As will be explained below, cannibalism was excluded from our data, independently of the theory of digraphs. Consequently the data are consistent with the assumptions of Robinson & Foulds (1980). Henceforth we shall use the single word *web* to mean a digraph that represents a community food web. We shall sometimes use the words species and vertex interchangeably.

A *predator* is a species that eats at least one species in the web. A *prey* is a species that is eaten by at least one species in the web. A *top* species is a species not eaten by any species in the web. Such a species is represented in the web by a vertex that is called a sink (Robinson & Foulds 1980, p. 20). An *intermediate* species is a species that has both at least one predator and at least one prey. A *basal* species is a species that eats no species. Such a species is represented in the web by a vertex that is called a source (Robinson & Foulds 1980, p. 20).

A species that neither eats nor is eaten by any species (an *isolated* species) is, according to the definitions just given, both a top and a basal species. However, either such species do not exist in reality or reports of webs, with rare exceptions, exclude them. In the whole collection of 62 webs that we shall analyse, only two or three isolated species in total were reported by the original sources, and these isolated species have been excluded in the editing of the data (F. Briand, personal communication).

We now distinguish special subsets of top and basal species. A *proper top* species is a top species that is also a predator, that is, a species that is eaten by none, but that eats at least one other species. A proper top species is represented by a vertex that is a proper sink in the terminology of Robinson & Foulds (1980, p. 20). A *proper basal* species is a basal species that is also a prey, that is, a species that eats none, but that is eaten by at least one other species. A proper basal species is represented by a vertex that is a proper source (Robinson & Foulds 1980, p. 20). Because isolated species are absent from our data, all reported top species are proper top species and all reported basal species are proper basal species. In the absence of isolated species, we can partition all species in a web into the sets of proper top, intermediate, and proper basal species.

A *basal-intermediate link* is a link from a (necessarily proper) basal species to an intermediate species; similarly for a *basal-top link*, an *intermediate-intermediate link*, and an *intermediate-top link*.

For a given reported web, let S denote the total number of species (vertices), T the number of (proper) top species, I the number of intermediate species, B the number of (proper) basal species, L the total number of links, L_{BI} the number of basal-intermediate links, L_{BT} the number of basal-top links, L_{II} the number of intermediate-intermediate links, and L_{IT} the number of intermediate-top links.

The adjacency matrix A of a web (or of any digraph) is an $S \times S$ matrix in which the element a_{ij} in row i and column j equals 1 if species i is eaten by species j, and equals 0 if species i is not eaten by species j. Thus species j is a basal species if and only if column j of A is 0, because column j of A is 0 if and only if species j eats no species in the web. Species j is a proper basal species if and only if column j of A is 0 and row j is not 0. Similarly species i is a top species if and only if row i of A is 0. Species i is a proper top species if and only if row i is 0 but column i is not 0. Species i is isolated if both row i and column i are 0.

As is conventional, let $E(.)$ denote the expectation or average of the random variable enclosed in parentheses. Let a bar denote the sample mean of the random variable it covers. Thus $\overline{B}$ is the sample mean number of basal species, while $E(B)$ is the expected number of basal species according to some model.

The three major findings of Briand & Cohen (1984) and Cohen & Briand (1984) may be stated as 'scaling laws', that is, as summaries of how the variables just defined change, or scale, as the total number of species in a web increases. Each of these scaling laws has two parts: (i) a qualitative part that states the approximate form of a scaling relationship, and (ii) a quantitative part that estimates the numerical value of the parameter or parameters in the scaling law. The scaling laws are cross-sectional, not longitudinal: they describe a comparison of many webs at the moment of observation, not the development of a single web over time resulting from the sequential addition of species.

Species Scaling (Briand & Cohen 1984)

(i) As S varies from 3 to 33 *lumped* species, $\overline{B}$, $\overline{I}$ and $\overline{T}$ are all approximately proportional to S. Equivalently, the proportions of species that are basal, intermediate and top show no pronounced trend, neither increasing nor decreasing, as S varies from 3 to 33.

(ii) Approximately, $\overline{B} = 0.19S$, $\overline{I} = 0.53S$, and $\overline{T} = 0.29S$ for all webs. (The sum $0.19 + 0.53 + 0.29$ exceeds 1 due to rounding. For more exact figures, see Table III.2.1.)

It seems plausible (Pimm 1982) that ecologists have been more interested in species at the top of webs than in species at the bottom, and that the coefficient 0.19 for the observed fraction of basal species is lower than the true fraction of basal species. When Briand & Cohen (1984) 'lumped' trophic species, they found that the ratio of basal species to top species increased relative to the ratio observed by Cohen (1977, 1978), as expected from Pimm's suspicions. Supposing that the number of top species in Table III.2.1 were correctly observed, and that the number of basal species were increased to equal the number of top species, as predicted by all of our models, the fraction of all species that are top species would decline to 0.26 and the fraction of all species that are basal would increase to 0.26. This number seems a reasonable estimate of the fractions of top and basal species, corrected for the possible undercount of basal species.

Table III.2.1. Summary statistics of the numbers of species and links in 62 community webs, by type of web, type of species, and category of link (from Cohen & Briand 1984; Chap. II.3)

	constant webs number	constant webs fraction	fluctuating webs[a] number	fluctuating webs[a] fraction	all webs number	all webs fraction
webs ...	19		43		62	
all species	351	1.000	683	1.000	1034	1.000
basal	66	0.188	130	0.190	196	0.190
intermediate	177	0.504	366	0.536	543	0.525
top	108	0.308	187	0.274	295	0.285
all links	811	1.000	1108	1.000	1919	1.000
basal-intermediate	198	0.244	327	0.295	525	0.274
basal-top	92	0.113	56	0.051	148	0.077
intermediate-intermediate	260	0.321	318	0.287	578	0.301
intermediate-top	261	0.322	407	0.367	668	0.348

[a] The environment of a web is considered to be 'fluctuating' if the original report indicates temporal variations of substantial magnitude in temperature, salinity, water availability or any other major physical parameter. Otherwise, the environment of the web is considered to be 'constant'

Link Scaling (Cohen & Briand 1984)

(i) As S varies from 3 to 33, $\overline{L}_{BI}$, $\overline{L}_{BT}$, $\overline{L}_{II}$ and $\overline{L}_{IT}$ are all approximately proportional to L. Equivalently, the proportions of links that are basal-intermediate, basal-top, intermediate-intermediate and intermediate-top show no pronounced trend, neither increasing nor decreasing, as S varies from 3 to 33.

(ii) Approximately, $\overline{L}_{BI} = 0.27L$, $\overline{L}_{BT} = 0.08L$, $\overline{L}_{II} = 0.30L$, and $\overline{L}_{IT} = 0.35L$, for all webs.

Link-Species Scaling (Cohen & Briand 1984)

(i) As S varies from 3 to 33, $\overline{L}$ is approximately proportional to S. Equivalently, the ratio of total links to total species in a web shows no pronounced trend, neither increasing nor decreasing, as S varies from 3 to 33.

(ii) Approximately, $\overline{L} = 1.86S$, for all webs. (More precisely, the coefficient of proportionality is 1.8559 with a standard deviation of 0.0740.) It will be convenient later to have a notation for the empirically observed ratio of links to species; we denote this quantity by d, to suggest 'density of links per species'. Thus in our data $d = 1.86$ approximately.

In stating these empirical generalizations, we have repeatedly emphasized that the range of variation in the total number of lumped species S among the webs collected by Briand is from 3 to 33. We cannot know whether these generalizations will continue to hold in webs with substantially larger S. The theory to be developed predicts that the scaling laws will continue to hold for larger S.

The scaling laws just stated are all first-order laws that describe trends only. They neglect entirely variability with respect to the trends. We shall discuss variability briefly in connection with the cascade model of §6.

A fourth empirical generalization plays a major role in attempts to explain the first three. Gallopín (1972, p. 266) observed that 'directed food webs are in general acyclic, although exceptions are possible'. Cohen (1978, p. 57) found one case of cannibalism, but no larger cycles, in four webs. In the 62 webs of Briand, cannibalism was reported by very few of the original sources, and then only for one species in each web. Because cannibalism is widespread in nature, particularly among invertebrates, the original investigators must have largely, but not consistently, ignored cannibalism. Consequently, Briand chose to exclude all of the few reported cases of cannibalism (F. Briand, personal communication).

To be precise in describing trophic cycles other than cannibalism, we now define (Robinson & Foulds 1980, pp. 24–25, 70) a *walk* in a digraph to be a finite sequence, consisting of vertices and edges alternately, beginning and ending with vertices, in which each edge goes from the vertex written on its left to the vertex written on its right. For example a, (a,b), b, (b,c), c is a walk from vertex a to vertex c through the edges (a,b) and (b,c). A walk in which the first vertex is the same as the last vertex is a closed walk. The *length* of a walk is the number of edges it contains, each counted as often as it occurs. A *cycle* is a closed walk in which all vertices are distinct except the first and last. If two cycles pass through the same set of vertices in the same order, differing only in the vertex that is written down first, the two cycles are considered to be identical. A *k-cycle* is a cycle of length $k > 0$. (If cannibalism is excluded, then 1-cycles are impossible. However, it will sometimes be convenient later to consider the possibility of a directed edge from a vertex to itself, that is, an 1-cycle or loop.) For $k > 0$, a digraph is *k-acyclic* if it contains no h-cycles, for $h = 1, \ldots, k$. A digraph is *acyclic* if it contains no k-cycles, for any $k > 0$.

If we use this language, all 62 of Briand's webs are acyclic except cases 21 and 30 (in the numbering of Chap. IV where the unlumped matrices are given). Webs 21 and 30 each contain a single cycle of length 2, and no longer cycles (F. Briand, personal communication). We summarize the distribution of cycles in the webs assembled by Briand with a fourth empirical generalization: acyclicity.

Acyclicity (Gallopín 1972)

Nearly all webs are acyclic.

It seems nearly certain that decomposers feed on what appear as top species and are food for what appear as basal species. The absence of cycles of length

greater than 2 implies that the reporters of webs ignore the decomposers. Therefore we cannot examine the ecological role of decomposers with these data.

3. Model 0: Anarchy

We shall say that a random variable Y is Bernoulli with parameter p, and shall
write $Y \sim B(p)$, for $0 \leq p \leq 1$, if $Y = 1$ with probability p and $Y = 0$ with
probability $q = 1 - p$. (In our notation, $\sim$ means 'has the distribution of' or
'distributed as' rather than 'asymptotically or approximately equals'.) We shall
say that a random matrix X is independently and identically distributed (i.i.d.)
Bernoulli with parameter p, and shall write $X \sim$ i.i.d. $B(p)$, if every element X_{ij}
of X is $\sim B(p)$, and all elements of X are i.i.d. (Since p is assumed constant for
all elements of the matrix, the additional requirement that they be *identically*
distributed is redundant, but is retained to accord with convention.)

Suppose A, the $S \times S$ adjacency matrix of a model web, were $\sim$ i.i.d. $B(p)$.
Then $E(L) = pS^2$. However, according to the link-species scaling law, $\overline{L} = dS$.
These two equations are simultaneously satisfied (with $E(L) = \overline{L}$) if $p = d/S$.
To avoid confusing the empirical estimate of density d with a model parameter,
we shall specify p in all of our models as c/S. The relation between the model
parameter c and the sample statistic d will vary from model to model.

Suppose that each species in a web of S species has an identical and independent chance p of eating any species, including itself, in the web, where, as
the number S of species increases, the probability p decreases according to c/S,
that is, let $A \sim$ i.i.d. $B(c/S)$ for $S \geq c$.

We now analyse the properties of model 0 and compare them with the empirical generalizations above, using $d = 1.86$ as an estimator for c.

Species Scaling

The probability that a species is a top species is q^S, and this is also the probability that a species is a basal species. Thus

$$E(T)/S = E(B)/S = q^S = (1 - c/S)^S \qquad (3.1)$$

is the expected fraction of species that are top species in a web of S species,
and also the expected fraction of species that are basal species. Thus model 0
predicts that the fractions of top and basal species should be equal.

Similarly, the probability that a species is a proper top species is $q^S(1 - q^{S-1})$,
and this is also the probability that a species is a proper basal species. Thus
model 0 predicts that the fractions of proper top and proper basal species should
be equal.

This prediction is only roughly consistent with the empirical observation that
0.19 of species are basal and 0.29 of species are top. However, like Pimm (1982),
we believe that ecologists often are more interested in species at the top of food
chains than in species at the bottom. If other predictions of the model turned
out to be correct, we would be prepared to accept the model's prediction that

in properly collected data, the expected fractions of top and of basal species are equal.

The right member of (3.1) increases monotonically and becomes close to the limiting value

$$\lim_{S\to\infty} E(T)/S = \lim_{S\to\infty} E(B)/S = e^{-c} \qquad (3.2)$$

even for moderate values of S. For example, $(1 - d/10)^{10} = 0.13$ and $(1 - d/20)^{20} = 0.14$ while $e^{-d} = 0.16$. Thus model 0 predicts that the fractions of top, basal and intermediate species should be very nearly independent of the number S of species in the web. The same conclusion applies to the asymptotic fractions of proper top and proper basal species, which are both equal to $e^{-d} - e^{-2d}$. The predicted change in these proportions for S between 3 and 33 would probably be undetectably small, given the variation among webs in the observed proportions (Chap. II.2).

While model 0 explains the qualitative part of the species scaling law, its predicted asymptotic fractions of top and basal species seem too low to explain the quantitative part of the species scaling law. The predicted asymptotic fraction 0.16 is substantially lower than the fraction 0.26 estimated above. The predicted asymptotic ratio of expected number of proper top or proper basal species to the expected number of non-isolated species, given by $[e^{-c} - e^{-2c}]/[1 - e^{-2c}]$, is 0.14, further still from the estimated fraction 0.26.

In going from the proportions of top or basal species to the proportions of proper top or proper basal species, the term involved in the corrections, $e^{-2c} = 0.024$, is small compared with the terms being corrected, given the observed ratio 1.86 of links to species, and appears in both numerator and denominator. When the proportions of top and basal species are corrected to the proportions of proper top and proper basal species, they decrease slightly. This slight decrease holds in the remaining models as well. For this reason, we shall not discuss proper top or proper basal species further until we come to model 3.

Link Scaling

We skip the analysis of link scaling because model 0 will be evaluated on other grounds.

Species-Link Scaling

The assumed behavior of p as a function of S is chosen to reproduce the observed species-link scaling.

Acyclicity

Although model 0 predicts that about 84% (that is, a fraction $1 - e^{-d}$) of webs will display cannibalism, model 0 should not be rejected on this basis because cannibalism has been suppressed from the data. However, according to theorem

1 below, model 0 also predicts that about 82% (that is, a fraction $1 - e^{-d^2/2}$) of webs will have one or more 2-cycles, which is grossly contrary to observation.

Effect of Lumping

According to model 0, it could happen that, for some $i < j$, column i is identical to column j and row i is identical to row j. In this case, if the simulated matrix were to be treated in the same way as the real data were treated, species i and j should be lumped. Our analysis so far has ignored the possible need to lump species in the simulated webs. We now show that the probability of needing to lump two non-isolated simulated species according to model 0 is so small that it is perfectly reasonable to ignore lumping, given the observed ratio 1.86 of links to (lumped) species.

Choose $i < j$. Define $P(\text{lump } i \text{ and } j)$ to be the probability that, in a matrix A with entries a_{hk} generated by model 0, column i equals column j and row i equals row j. Similarly, define $P(\text{lump non-isolated } i \text{ and } j)$ to be the probability that, in a matrix A generated by model 0, column i equals column j, row i equals row j, and column i or row i or both are not all zero. The $4S-4$ entries in the two columns and rows consist of $2(S-2)$ pairs of entries and one quartet of entries $(a_{ii}, a_{ij}, a_{ji}, a_{jj})$. To lump species i and j, we require that the two entries of each pair be equal and the four entries in the quartet be equal. Hence $P(\text{lump } i$ and $j) = (p^2 + q^2)^{2S-4}(p^4 + q^4)$, and $P(\text{lump non-isolated } i \text{ and } j) = P(\text{lump } i$ and $j) - P(i \text{ and } j \text{ are isolated}) = (p^2 + q^2)^{2S-4}(p^4 + q^4) - q^{4S-4}$. The expected fraction of species that are non-isolated but lost by lumping is then less than or equal to

$$(1/S) \sum_{j=2}^{S} \sum_{i=1}^{j} P(\text{lump non-isolated } i \text{ and } j)$$

$$= (1/2)(S-1)[(1 - 2pq)^{2S-4}(p^4 + q^4) - q^{4S-4}], \quad (3.3)$$

which, as S increases, approaches $c^2 e^{-4c} = 0.002$ when $c = 1.86$. Thus the expected fraction of non-isolated vertices of a random web generated according to model 0 that should be lumped is negligible, so we do not correct the previous calculations for lumping.

Effect of Disconnected Weak Components

All reported webs are *weakly connected* in the sense that the set of species cannot be divided into two non-empty subsets with no link between the two subsets. (The adjective 'weak' allows for the possibility that the linkage might be in one direction only.) A *weak component* is a maximal set of vertices (species) that is weakly connected. Thus all reported webs have only a single weak component. We now show that the expected fraction of non-isolated species that belong to a single weak component according to model 0 is asymptotically so close to 1 that it is reasonable to ignore the effect of disconnected weak components, given $d = 1.86$.

According to Erdös & Rényi (1960, p. 56, Theorem 9b), the fraction of all species (including isolated species) that belong to the largest weak component of a web is asymptotically

$$1 - (2c)^{-1} \sum_{k=1}^{\infty} k^{k-1} (2ce^{-2c})^k / k! \,. \tag{3.4}$$

Hence the fraction of all species that are not isolated and do not belong to the largest weak component is

$$(2c)^{-1} \sum_{k=1}^{\infty} k^{k-1} (2ce^{-2c})^k / k! - e^{-2c}$$

$$= (2c)^{-1} \sum_{k=2}^{\infty} k^{k-1} (2ce^{-2c})^k / k! \tag{3.5}$$

which is approximately 0.002 when $c = 1.86$. Thus 99.8% of the non-isolated species of a random web generated according to model 0 belong to a single weak component, so we do not correct the previous calculations for disconnected components.

In summary, model 0 can explain roughly the observed scale-invariance in the proportion of top, intermediate and basal species and the numerical similarity in the proportions of top and basal species. But it predicts fractions of top and basal species that are too low and fractions of food webs with cycles that are far too high.

4. Model 1: Finitely Acyclic Democracy

The most straightforward way to eliminate the problem of too many cycles is by assumption. We start with the weakest assumption that is *a priori* plausible.

Suppose there is a finite positive integer k and a finite positive real number c such that, for $S \geq c$, the adjacency matrix A of a web with S species is $\sim$ i.i.d. $B(c/S)$, conditional on A being k-acyclic.

Biologically, this model assumes that any species can eat any species with equal probability c/S provided that, in the resulting feeding relations, it never happens that species X eats species X (no 1-cycles), nor that species X eats species Y and species Y eats species X (no 2-cycles), nor that species X eats species Y, species Y eats species Z and species Z eats species X (no 3-cycles), nor that there are any cycles of length up to and including k, which is fixed and independent of S.

One way to simulate this model would be to generate Bernoulli matrices according to model 0 and then throw away those matrices A in which the trace (sum of the diagonal elements) of $A + A^2 + \ldots + A^k$ exceeds 0.

Before considering general k, we consider the special case of 1-acyclic democracy.

1-Acyclic Democracy

To generate an $S \times S$ Bernoulli matrix A with parameter c/S, conditional on no cannibalism (no 1-cycles), set the diagonal elements of A equal to 0 with probability 1. The off-diagonal elements of A are to be filled with independent random variables $\sim B(c/S)$ as before. Then $E(L) = (c/S)S(S-1)$. Since the species-link scaling law gives $\overline{L} = dS$, we can estimate c by $c = dS/(S-1)$, which approaches d for large S but is larger than d for finite S.

The probability that a species is a top species is q^{S-1}, where $q = 1 - c/S$, and this is also the probability that a species is a basal species. Thus

$$E(T)/S = E(B)/S = q^{S-1} = (1 - d/[S-1])^{S-1} \qquad (4.1)$$

is the expected fraction of species that are top species in a web of S species, and also the fraction of species that are basal species. This model predicts that the fractions of top and basal species should be equal. The asymptotic behaviour of $E(T)/S$ and $E(B)/S$ for large S is identical to that in (3.2) for model 0. The predicted asymptotic fractions of top and basal species are too low to accord well with observation.

A web will have a 2-cycle if there exist indices i, $j \neq i$ such that $a_{ij} = 1$ and $a_{ji} = 1$. For a given i and j, the probability that there is a 2-cycle through i and j is p^2, so the probability that there is no 2-cycle through i and j is $1 - p^2$. The probability that there is no 2-cycle in the entire web is

$$(1 - p^2)^{S(S-1)/2} = (1 - \{d/[S-1]\}^2)^{S(S-1)/2} \to e^{-d^2/2} = 0.18 \quad (4.2)$$

so that about 82% of such model webs would have at least one 2-cycle. This proportion is grossly too high and we are forced to abandon 1-acyclic democracy as unrealistic.

The calculated asymptotic fraction e^{-c} of webs under model 0 (anarchy) that have no 1-cycles may be multiplied by the calculated asymptotic fraction $e^{-c^2/2}$ of webs under 1-acyclic democracy that have no 2-cycles to give the predicted asymptotic fraction $e^{-c-c^2/2}$ of webs under model 0 that are 2-acyclic, that is, have neither 1-cycles nor 2-cycles, because under model 0 the diagonal elements of the adjacency matrix are independent of the off-diagonal elements.

k-Acyclic Democracy: The General Case

From the perhaps surprising finding that the predicted asymptotic fraction of top or basal species is e^{-c} under the anarchy model as under the model of 1-acyclic democracy, one might conjecture that the proportion is the same under the k-acyclic democracy model, for any finite $k > 0$. From the formula $e^{-c-c^2/2}$ for the asymptotic fraction of webs under model 0 that are 2-acyclic, and from the analogous formulas for undirected graphs of Erdős & Rényi (1960), one might conjecture that the asymptotic proportion of k-acyclic digraphs under model 0 is $\exp\left(-\sum_{h=1}^{k} c^h/h\right)$. The following theorem and corollary establish that both of these conjectures are correct.

Theorem 1. *Suppose that for some $c \geq 0$ and for $S \geq c$, the adjacency matrix A of a web with S species is $\sim$ i.i.d. $B(c/S)$. (This is model 0.) Let $M_k(S)$ be the number of distinct k-cycles in the web, $k = 1, 2, \ldots, S$, and let $Y(S)$ be the number of prey species of species 1, that is, the sum of column 1 of A. Let $M(S) = \sum_{h=1}^{S} M_h(S)$ be the total number of distinct cycles in A. Then for any $k \geq 0$, the random vector $(Y(S), M_1(S), \ldots, M_k(S))$ (which is interpreted as the scalar $Y(S)$ if $k = 0$) converges in distribution as $S \to \infty$ to a random vector with independent Poisson-distributed components with mean $(c, c, c^2/2, c^3/3, \ldots, c^k/k)$, that is, for any non-negative integers $y, m_1, \ldots, m_k$,*

$$\lim_{S \to \infty} P(Y(S) = y, M_1(S) = m_1, \ldots, M_k(S) = m_k)$$

$$= e^{-c}[c^y/y!] \prod_{h=1}^{k} \{e^{-(c^h/h)}[(c^h/h)^{m_h}]/m_h!\} . \tag{4.3}$$

For $0 \leq c < 1$, $(Y(S), M(S))$ converges in distribution as $S \to \infty$ to a bivariate random vector with independent Poisson-distributed components with mean $(c, -\ln(1-c))$.

Corollary. *Under the above assumptions, for any $c \geq 0$ and any $k \geq 1$, as $S \to \infty$*

$$P(Y(S) = y \mid M_1(S) = 0, \ldots, M_k(S) = 0) \to e^{-c}[c^y/y!] . \tag{4.4}$$

The left member of (4.4) is the probability that species 1 has y prey in the model of k-acyclic democracy. For $0 \leq c < 1$, the asymptotic probability that species 1 has y prey in an acyclic web is also Poisson, that is,

$$P(Y(S) = y \mid M(S) = 0) \to e^{-c}[c^y/y!] . \tag{4.5}$$

The corollary follows immediately from Theorem 1 and the definition of conditional probability. The proof of Theorem 1 is deferred to Appendix 1.

The corollary (with $y = 0$) implies that, in the model of k-acyclic democracy, the fraction of species with no predators, and the fraction of species with no prey, both approach e^{-c} as $S \to \infty$. The mean number of species on which a given species preys, and the mean number of species that prey on a given species, both approach c.

In summary, for fixed finite k, the model of k-acyclic democracy predicts that the expected fractions of top and basal species are equal and, asymptotically for large numbers S of species in a web, independent of S. These predictions are roughly consistent with the data. The model also predicts that the numerical value of this asymptotic fraction should be lower than that observed. However, in concluding that this discrepancy exists, we are assuming that it is appropriate to use the ratio $d = 1.86$ of links to species, observed in the *finite* range of S from 3 to 33, to estimate the *asymptotic* effective density of links c.

5. Model 2: Acyclic Democracy

Excluding cycles up to any fixed finite order k, as in model 1, might be qualitatively different, in the limit of large S, from excluding cycles of all lengths in
the limit of large S. To investigate this possibility, we have partly analysed the
next model.

Suppose there is a finite positive real number c such that, for $S \geq c$, the
adjacency matrix A of a web with S species is $\sim$ i.i.d. $B(c/S)$, conditional on A
being acyclic.

Biologically, this model assumes that any species can eat any species with
equal probability c/S, provided that, in the resulting feeding relations, it never
happens that species X eats species X (no 1-cycles), nor that species X eats
species Y and species Y eats species X (no 2-cycles), nor that species X eats
species Y, species Y eats species Z and species Z eats species X (no 3-cycles),
and so on, excluding all cycles of length up to and including S.

The theoretical results available to us so far require us to discuss separately
two cases: $0 \leq c < 1$, and $1 \leq c$.

In the first case, (4.5) implies that the fractions of top and basal species
are equal and, asymptotically for large S, independent of S. These predictions
are roughly consistent with the data. However, since the expected value of the
observed density d must be no larger than the model parameter c, and since
$d > 1$, this first case is not of empirical interest, given our data.

In the second case, $1 \leq c$, we have so far no exact results concerning the
asymptotic proportions of top and of basal species. By symmetry these proportions must be equal. The results of our numerical investigations, which we will
now describe, can be interpreted to be consistent with the conjecture that, for
$S \gg c$, the fraction of top species and the fraction of basal species both approach
e^{-d^*}, where d^* is the asymptotic (large S) effective density of links. We know
that this is also the case when $c < 1$ since then $d^* = c$. However, when $c \geq 1$,
we have no theory so far that permits us to compute c from d^* or vice versa.

To estimate the fractions of zero rows and of zero columns according to model
2, we have resorted to simulation, settling at last on the third of three approaches described in Appendix 2. This approach to simulation, which is actually a slight modification of model 2, guarantees that the model parameter c
equals the asymptotic ratio d^* of links to species. For $S = 10$ and $S = 20$, and
for each value of $c = 0.5(0.5)4.0$ (an abbreviation for the sequence of numbers
$0.5, 1.0, 1.5, \ldots, 4.0$), Table III.2.2 compares the simulated mean fractions of zero
rows and of zero columns in 100 acyclic matrices with the conjectured asymptotic fraction e^{-c}. For the lower values of c, the agreement between the sampled
fractions of zero rows or columns and e^{-c} is excellent. For the larger values of
c, e^{-c} falls more rapidly than the sampled fractions of zero rows or columns.
For large c, the difference between e^{-c} and the sampled fraction of zero rows or
columns is slightly smaller for $S = 20$ than for $S = 10$.

Table III.2.2. The simulated mean fractions of zero rows or zero columns in 100 acyclic $S \times S$ matrices with exactly Sc positive elements, generated by the third approach (Appendix 2) to simulating model 2, and the fractions predicted by the asymptotic function e^{-d^*} conjectured in (5.1) and by the function (6.2a) (with c replaced by $2c$) derived for model 3, the cascade model

c	$S = 10$		predictions		$S = 20$	
	rows	columns	$\exp(-c)$	model 3	rows	columns
0.5	0.5854	0.5974	0.6065	0.6321	0.5959	0.5962
1.0	0.3500	0.3494	0.3679	0.4323	0.3644	0.3600
1.5	0.2313	0.2445	0.2231	0.3167	0.2341	0.2470
2.0	0.1669	0.1769	0.1353	0.2454	0.1585	0.1642
2.5	0.1620	0.1232	0.0821	0.1987	0.1522	0.1332
3.0	0.1318	0.1042	0.0498	0.1663	0.1121	0.1142
3.5	0.1147	0.1056	0.0302	0.1427	0.0959	0.1133
4.0	0.1032	0.1000	0.0183	0.1250	0.0859	0.0820

For $S = 10$, the standard deviation (computed from the numerical simulation) of the proportion of zero rows (in a single matrix, not in the mean proportion) at first increases with increasing c and then declines slowly from a maximum of approximately 0.09 when $c = 1$ to a minimum of approximately 0.02 when $c = 4$. Since 100 matrices were generated, the standard deviation of the simulated mean proportions given in Table III.2.2 is one-tenth as large, that is, not exceeding 0.01. The standard deviations when $S = 20$ are similar, and the same conclusion applies. Thus the difference in Table III.2.2 between the sampled proportion of zero columns or rows and $e^{-c} = e^{-d^*}$ for the larger values of c appears to be real.

If this difference approaches 0 as $S \to \infty$, then we may conjecture, pending further theoretical progress, that in model 2,

$$\lim_{S \to \infty} E(T)/S = \lim_{S \to \infty} E(B)/S = e^{-d^*} . \qquad (5.1)$$

If this is so, then, like models 0 and 1, model 2 can explain roughly the observed scale-invariance in the proportion of top, intermediate and basal species and the numerical similarity in the proportions of top and basal species. But it predicts fractions of top and basal species that are too low according to the conjecture (5.1), and that are too low (according to our simulations) even for $S = 10$ (in Table III.2.2, $c = 2.0$ gives a fraction of 0 rows near 0.17, lower than the estimate from data of 0.26).

6. Model 3: Cascade

Many biologists might be reassured by the failure of the models considered so far because these models make the biologically implausible assumption that any species is capable, in principle, of eating any other species. These models assume that it is only a matter of chance that the grass does not eat the cow, nor the lamb

the wolf. Yet it is not absurd to consider such models. It is a healthy discipline
to require that they be rejected by quantitative data and not by 'intuitions' that
are often wrong.

Now that the previous models have been rejected for their quantitative failures, we must abandon the assumption that each species could potentially eat
any other, while imposing the least possible additional structure. We shall do so
by noticing an important feature of acyclic matrices.

An $S \times S$ matrix A is called *strictly upper triangular* if $a_{ij} = 0$ whenever
$i \geq j$. This means that the main diagonal and all matrix elements below the
main diagonal are zero; the non-zero elements of A, if any, lie strictly above the
main diagonal. For brevity, we shall henceforth call such a matrix *triangular*.

If the adjacency matrix of a web with S species is triangular, the species
labelled 1 can potentially be eaten by any species other than itself, but can eat
none. The second species can potentially be eaten by the species labelled 3 to
S, but can eat only species 1. And so on: the species labelled S can potentially
eat all the other species, but can be eaten by none of them. Thus a triangular
adjacency matrix describes a strict trophic hierarchy or cascade.

A digraph is acyclic if and only if its vertices can be numbered in such a
way that its adjacency matrix is triangular (for example, Robinson & Foulds
1980, p. 176). Thus the adjacency matrix A of a web is acyclic if and only if
some permutation, applied to both rows and columns of A, changes the matrix
to triangular form. Model 2 can be interpreted as saying that the luck of the
draw determines which species eat which others, provided that, when all is done,
the species can be arranged in a cascade. The order of species in the cascade is
determined (non-uniquely) after the trophic links are chosen.

We now suppose that the order of species in the cascade is determined before
the trophic links are chosen.

Suppose there is a finite positive real number c such that, for $S \geq c$, the
elements above the main diagonal of the adjacency matrix A are $\sim$ i.i.d. $B(c/S)$,
while the elements on or below the main diagonal are fixed with probability 1
at 0.

Theorem 2. *Suppose that for some $c \geq 0$ and for $S \geq c$, the adjacency matrix A
of a web with S species is triangular, with the elements above the main diagonal
$\sim$ i.i.d. $B(c/S)$. (This is model 3.) Let T be the number of zero rows (top species)
and B be the number of zero columns (basal species) in A. Then, with $p = c/S$,
$q = 1 - p$,*

$$E(T) = E(B) = [1 - q^S]/p \,, \tag{6.1a}$$

$$\mathrm{var}(T) = \mathrm{var}(B) = (1 - q^S)/p - (1 - q^{2S})/(1 - q^2) \,. \tag{6.1b}$$

Asymptotically,

$$\lim_{S \to \infty} E(T)/S = \lim_{S \to \infty} E(B)/S = (1/c)(1 - e^{-c}) \,, \tag{6.2a}$$

$$\lim_{S \to \infty} \mathrm{var}(T/S) = \lim_{S \to \infty} \mathrm{var}(B/S) = 0 \,. \tag{6.2b}$$

If T_P is the number of proper top species, B_P is the number of proper basal species, N is the number of not isolated species, and I is the number of intermediate species, then

$$E(T_P) = E(B_P) = S[(1 - q^S)/c - q^{S-1}]\,, \tag{6.3a}$$

$$E(I) = S[1 - 2(1 - q^S)/c + q^{S-1}]\,, \tag{6.3b}$$

$$E(N) = S(1 - q^{S-1})\,, \tag{6.3c}$$

and asymptotically

$$\lim_{S \to \infty} E(T_P)/E(N) = \lim_{S \to \infty} E(B_P)/E(N)$$

$$= \{[1 - e^{-c}]/c - e^{-c}\}/[1 - e^{-c}]\,, \tag{6.4a}$$

$$\lim_{S \to \infty} E(I)/E(N) = \{1 - (2/c)[1 - e^{-c}] + e^{-c}\}/[1 - e^{-c}]\,. \tag{6.4b}$$

For large c, e^{-c} is nearly zero so the asymptotic fraction of top or proper top or basal or proper basal species approaches $1/c$. Also, the total number L of trophic links is binomially distributed with mean and variance

$$\begin{aligned} E(L) &= pS(S-1)/2 = c(S-1)/2\,, \\ \mathrm{var}(L) &= pqS(S-1)/2 = c(S-c)(S-1)/(2S)\,, \end{aligned} \tag{6.5}$$

and the numbers of links of each kind have means

$$E(L_{BI}) = E(L_{IT}) = (S - 1)(1 + q^{S-1}) - (1 + q)(1 - q^{S-1})/p\,, \tag{6.6a}$$

$$E(L_{BT}) = (1 - q^{S-1})/p - (S - 1)q^{S-1}\,, \tag{6.6b}$$

$$\begin{aligned} E(L_{II}) &= pS(S-1)/2 - (S - 1)(2 + q^{S-1}) \\ &\quad + (1 - q^{S-1})(1 + 2q)/p\,. \end{aligned} \tag{6.6c}$$

Asymptotically, as $S \to \infty$,

$$E(L_{BI})/E(L), \ E(L_{IT})/E(L) \to 2[c(1 + e^{-c}) - 2(1 - e^{-c})]/c^2\,, \tag{6.7a}$$

$$E(L_{BT})/E(L) \to 2[1 - e^{-c} - ce^{-c}]/c^2\,, \tag{6.7b}$$

$$E(L_{II})/E(L) \to 1 - 2[c(2 + e^{-c}) - 3(1 - e^{-c})]/c^2\,. \tag{6.7c}$$

Proof. Only elementary calculations are required, noting that the probability that species i is basal is q^{i-1}, the probability that species i is top is q^{S-i}, the probability that species i is proper top is $q^{S-i} - q^{S-1}$, the probability that species i is proper basal is $q^{i-1} - q^{S-1}$, the probability that species i is intermediate is $1 - q^{i-1} - q^{S-i} + q^{S-1}$, and the probability that species i is not isolated is

$1 - q^{S-1}$. Also,

$$E(L_{BI}) = p \sum_{j=2}^{S} \sum_{i=1}^{j-1} q^{i-1}(1 - q^{S-j}) \,,$$

$$E(L_{BT}) = p \sum_{j=2}^{S} \sum_{i=1}^{j-1} q^{i-1} q^{S-j} \,,$$

$$E(L_{IT}) = p \sum_{j=2}^{S} \sum_{i=1}^{j-1} (1 - q^{i-1}) q^{S-j} \,,$$
(6.8)

$$E(L_{II}) = p \sum_{j=2}^{S} \sum_{i=1}^{j-1} (1 - q^{i-1})(1 - q^{S-j}) \,.$$

When (6.5) is solved for p and $E(L)$ is replaced by the observed number of links, it becomes apparent that p is what ecologists call the (lower) connectance (F. Briand, personal communication).

To compare the predictions of model 3 with observation requires an estimate of c. From (6.5),

$$c = 2E(L)/(S - 1) \,.$$
(6.9)

For a single finite S, replacing $E(L)$ by the total number of links, we estimate c as twice the total number of links divided by $S - 1$. However, for a single value of c common to all webs, we use an asymptotic estimate. Asymptotically, as $S \to \infty$, the link scaling law indicates that $\overline{L}$ is dS, and $S/(S-1) \downarrow 1$ as $S \to \infty$, so that c is estimated as $2d = 3.72$. We now examine the macroscopic predictions of model 3, using this single estimate of $c = 3.72$. We shall review the scaling laws stated in Sect. 2.

Species Scaling

Figure III.2.1 shows the predicted mean proportion of top species and a confidence interval of ± 2 standard deviations as a function of S, using (6.1) with a single value of $c = 3.72$, superimposed on the data of Briand & Cohen (1984). Figure III.2.2 shows the same for basal species. (Cf. Fig. A.2.2.)

The predicted mean proportion of top or of basal species changes so slowly in the observed range of S as to defy discrimination from constancy. According to (6.1) with $c = 3.72$, model 3 predicts the mean and variance in the proportion of top species to be (with identical results for basal species) as shown in Table III.2.3. Thus model 3 reproduces qualitatively the species scaling law.

Quantitatively, model 3 predicts asymptotic proportions of basal, intermediate and top species equal to 0.26, 0.48, and 0.26. (By using the remark after (6.4), we can easily see why the predicted proportion of top species is near one quarter. Because $e^{-3.72} = 0.024$, the fraction of top species is predicted to be

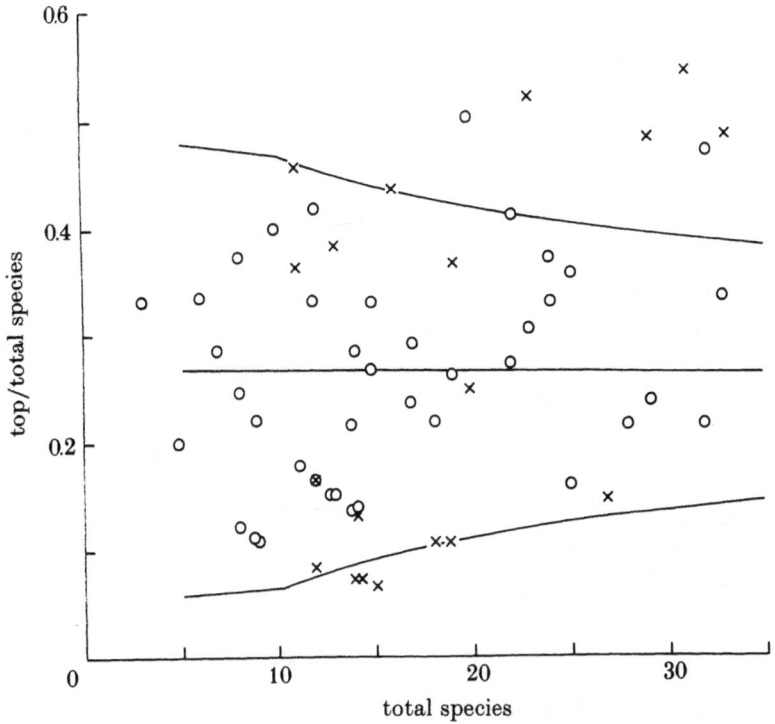

Figure III.2.1. The predicted mean proportion of top species (middle line) and a confidence interval of ±2 standard deviations (upper and lower lines) as a function of total species S, according to the cascade model. In this figure and Fig. III.2.2, ✕ is constant environment, o is fluctuating environment. The symbols ✕ and o have been perturbed from their exact locations by a small random amount to indicate when several food webs have exactly the same coordinate. The data are replotted from Briand & Cohen (1984)

slightly greater than one quarter.) The observed proportions are 0.19, 0.53, and 0.29. As we suggested above, if observer bias has lowered the fraction of basal species, a plausible estimate of the proportion of top and of basal species is 0.26, exactly as predicted by model 3. Thus the quantitative agreement between the predicted asymptotic mean and the observed mean is good. The model predicts a decrease in the standard deviation that is suggested by the data on basal species but that is not observed in the data on top species.

In summary, model 3 predicts the form and the parameter value of the species scaling law. It is only partly successful in explaining the variation with respect to the species scaling law.

We now show that models 0 and 1, and perhaps 2 (if conjecture (5.1) is valid), predict asymptotic fractions of top or basal species that are lower than those predicted by model 3. From (3.2), (4.4) and (6.2a), we must establish that for any non-negative c (for example, $c = 1.86$), $e^{-c} \leq (1 - e^{-2c})/(2c)$. We use $2c$ in place of c on the right of (6.2a) so that, asymptotically, models 0, 1 and 3 will all have the same effective density d^* of links. The inequality is equivalent to the inequality $c \leq (e^c - e^{-c})/2$, which is easily proved by noting that both sides

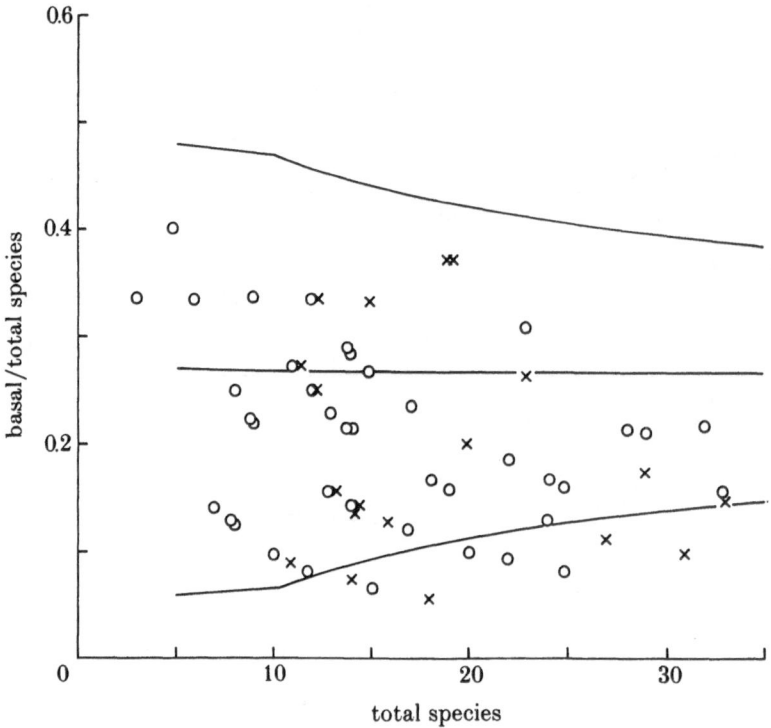

Figure III.2.2. The predicted mean proportion of basal species (middle line) and a confidence interval of ±2 standard deviations (upper and lower lines) as a function of total species S, according to the cascade model. The symbols and source of data are as in Fig. III.2.1

Table III.2.3. Predicted mean and variance in the proportion of top species, according to model 3

S	$E(T)/(S)$	$[\mathrm{var}\,(T/S)]^{1/2}$
5	0.269	0.104
15	0.265	0.086
25	0.264	0.069
35	0.264	0.059
∞	0.262	0

approach 0 when $c \downarrow 0$ and by comparing derivatives of both sides with respect to c.

This inequality raises a question. In Table III.2.2, the simulated fractions of top and basal species exceed e^{-c}. We have just shown that $(1 - e^{-2c})/(2c)$ exceeds e^{-c}. Might not $(1-e^{-2c})/(2c)$, shown in Table III.2.2 under the column headed 'predictions, model 3', be a better description of the simulated fractions of top and basal species in model 2 than e^{-d^*}? Table III.2.2 gives a weak hint that this may not be the case. Though, for $c = 4.0$, the simulated fractions of top and basal species are near those predicted by model 3, as S increases from 10 to 20 the simulated fractions move slightly away from $(1 - e^{-2c})/(2c)$ and towards e^{-c}.

Link Scaling

Figure III.2.3 shows the ratio of the expected number of links of each kind to the expected total number of links, based on (6.5) and (6.6) with $c = 3.72$, for S between 4 and 40. For $S > 10$, the ratios are effectively constant. For $S \leq 10$, the predicted curves for $E(L_{BI})/E(L)$ and $E(L_{IT})/E(L)$ reproduce the suggestion of a decline in the observed values of L_{BI}/L in Fig. A.3.2a and in the observed values of L_{IT}/L in Fig. A.3.2d. The predicted increase in $E(L_{IT})/E(L)$ might even be reflected in the data of Fig. A.3.2c. However, few of the real webs had 10 or fewer species, so these suggestions from the data are very weak. Overall, the qualitative predictions of model 3 are consistent with the qualitative link scaling law.

Quantitatively, model 3 predicts the asymptotic proportions of each kind of link shown in Table III.2.4. The principal discrepancy between the data and the model is that fewer basal-top links and more intermediate-top links are observed than predicted.

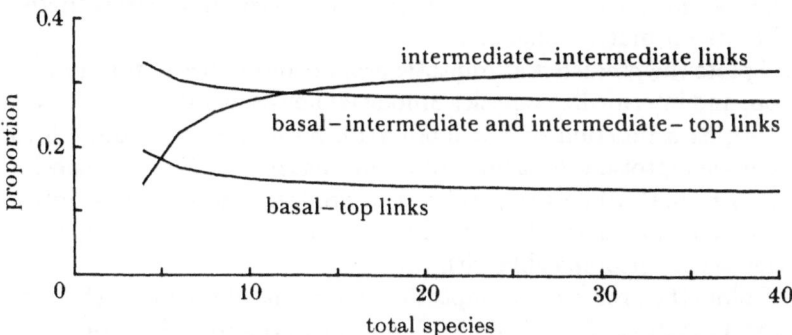

Figure III.2.3. The predicted ratio of the expected number of links of each kind to the expected total number of links, according to the cascade model with $c = 3.72$, for total numbers of species $S = 4(2)40$. For $S > 10$, the ratios change little

Table III.2.4. Observed proportions of each kind of link, and asymptotic predicted proportions according to model 3

type of link	observed proportion	predicted proportion from (6.7) with $c = 3.72$
basal-intermediate	0.27	0.27
basal-top	0.08	0.13
intermediate-intermediate	0.30	0.33
intermediate-top	0.35	0.27

Link-Species Scaling

That model 3 correctly predicts the qualitative relation between total links and total species follows from (6.5). Quantitative agreement is guaranteed by the choice of $c = 3.72$.

Acyclicity

Acyclicity is guaranteed by making the adjacency matrices triangular.

In summary, model 3 correctly predicts the qualitative species scaling and link scaling laws in webs with more than a handful of species. Quantitatively, model 3 also predicts, to a first approximation, the observed proportions of basal, intermediate and top species and the observed proportions of each kind of link.

Sensitivity Analysis

We are sceptical about the completeness of observation of trophic links, especially those that involve what are currently described as basal species. If moderately more trophic links were observed, would our quantitative predictions be radically altered? If so, the present quantitative estimates of model 3 are approximately right for the wrong reason, namely, that the effective density of links happened to be low. Thus it is important to know how the predicted asymptotic proportions of species and of links of each kind vary as c varies in the neighbourhood of its estimated value 3.72.

Figure III.2.4 plots the predicted asymptotic proportions of basal, top, proper basal, proper top, and intermediate species among all non-isolated species, based on (6.2) and (6.4), as a function of $c = 0.5(0.5)10$. As c increases from 3.5 to 4.5, the predicted asymptotic proportions of proper basal or proper top species declines from 0.25 to 0.21 while the predicted asymptotic proportion of intermediate species among non-isolated species increases from 0.49 to 0.58. Neither range of variation seems incompatible with the data.

Figure III.2.5 plots the predicted asymptotic proportions of links of each kind, based on (6.7), as a function of $c = 0.5(0.5)10$. As c increases from 3.5 to 4.5, the predicted asymptotic proportions of basal-intermediate or intermediate-top links declines from 0.27 to 0.25, the proportion of basal-top links declines from 0.14 to 0.09, and the proportion of intermediate-intermediate links increases from 0.31 to 0.40. Such changes improve the agreement between the observed and predicted proportions of basal-top links but worsen the agreement between the observed and the predicted proportions of the remaining classes of links. However, the changes in the predicted asymptotic proportions are not very radical in any case. In particular, the estimate of $c = 3.72$ happens to fall very near where the curve for basal-intermediate and intermediate-top links is flattest.

We conclude that the predicted asymptotic proportions of species and links of each kind are not so sensitive to the exact value of the observed ratio of links to species as to exclude the possibility of a somewhat greater effective density of links.

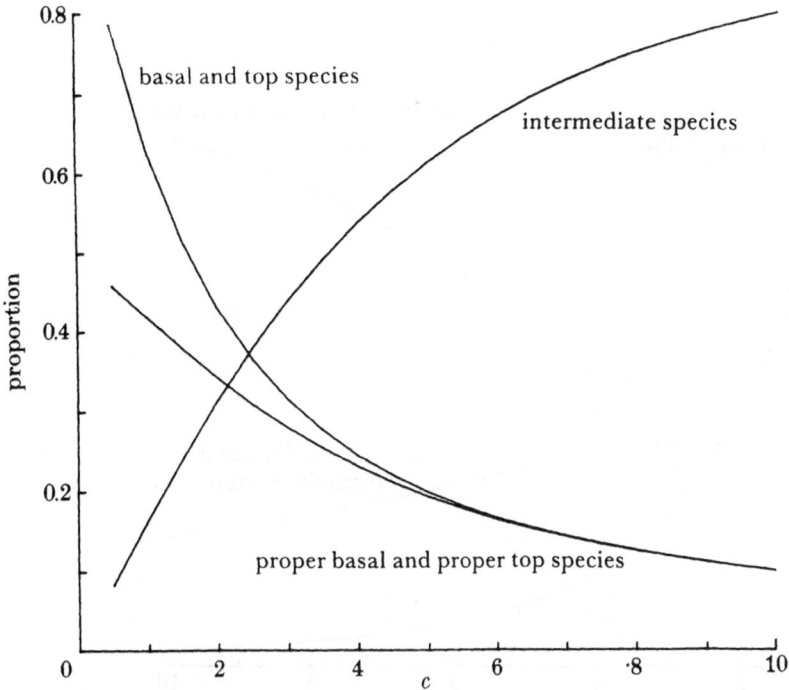

Figure III.2.4. The predicted asymptotic proportions of basal, proper basal, top, proper top, and intermediate species, as a fraction of non-isolated species, according to the cascade model, for $c = 0.5(0.5)10$. Because basal and top species are plotted here as a fraction of non-isolated species, the sum of proportions (basal + intermediate + top) exceeds 1. The excess over 1 is small once $c > 3$. The sum of proportions (proper basal + intermediate + proper top) equals 1

Lumping

Would lumping substantially alter the number of species and hence the proportions of interest in the cascade model? The same approach used to analyse lumping in model 0 shows that, for $i < j$, $P(\text{lump } i \text{ and } j) = (1-2pq)^{S+i-j-1}q^{2(j-i)-1}$ while $P(\text{lump non-isolated } i \text{ and } j) = (1-2pq)^{S+i-j-1}q^{2(j-i)-1} - q^{2(S-1)-1}$ The expected fraction of species that are not isolated and lost by lumping is then less than or equal to

$$(1/S)\sum_{j=2}^{S}\sum_{i=1}^{j}P(\text{lump non-isolated } i \text{ and } j)$$

$$= (q^{2S-3}/[Sr])[(S-1)(1+r)^{S-1}$$
$$- \{(1+r)^{S-1} - 1\}/r - S(S-1)r/2]\,, \qquad (6.10)$$

where $r = (p/q)^2$. As $S \to \infty$, (6.10) approaches $c^2 e^{-2c}/3 = 0.003$ when $c = 3.72$. In model 3 as in model 0, the effect of lumping non-isolated species is negligible.

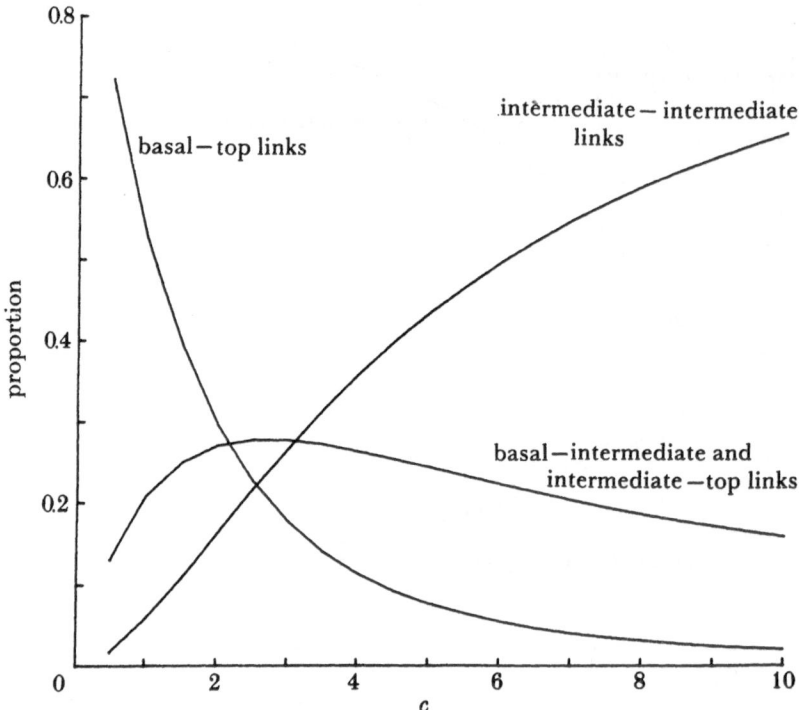

Figure III.2.5. The predicted asymptotic proportions of links of each kind, according to the cascade model, for $c = 0.5(0.5)10$

Effect of Disconnected Components

The effect of weak components is essentially identical in models 0 and 3. The calculation based on (3.4) remains the same, with the parameter c of (3.4) still estimated by $d = 1.86$ rather than by $2d$. As in model 0, asymptotically all but a negligible fraction of species belong to the largest weak component.

So far, we have taken c as exogenously determined, for example, by the feeding apparatus or behavioural flexibility of species, and have attempted to predict other structural features of webs from that parameter. Why might c assume a value in the vicinity of 3.72? Figure III.2.5 shows that $c = 3.72$ is in the range around 2.69 where the predicted asymptotic proportions of basal-intermediate and intermediate-top species are maximal. It is tempting to speculate, but without theoretical or additional empirical support at the moment, that the effective density of links is adjusted to maximize the proportions of links between basal and intermediate species, and between intermediate and top species.

7. Conclusions

In this section, we shall first summarize the conclusions we draw from the four models we have considered. We then relate our results to some earlier efforts to model webs. Finally, we mention two important limitations on our results.

Briand (1983; Chap. II.5), using 'unlumped' webs, first suggested, and Cohen & Briand (1984; Chap. II.3), using 'lumped' webs, demonstrated that the average total links of a web are nearly proportional to the total species of the web. Within the framework of the random digraph models considered here, this observation has the important implication that the probability of a given species eating or being eaten by another given species must vary as the reciprocal of the total number of species in the web. This has the further consequence that the number of predators or prey of a randomly chosen species is asymptotically independent of the total number of species in the web.

The exclusion of cycles of finite lengths or of all lengths as S increases is insufficient to reproduce quantitatively the species scaling law, although an open mathematical question remains in the analysis of model 2. That question is: when $c \geq 1$, what is the asymptotic mean fraction of zero columns or of zero rows in a random $S \times S$ matrix whose elements are independently and identically distributed Bernoulli random variables with mean c/S, conditional on the matrix being acyclic?

To explain the observed proportions of top and basal species, it appears to be necessary to suppose that there is an ordering, hierarchy, or cascade of species that constrains the possible predators and prey of each species. Under this assumption, it is possible to predict qualitatively, and to fair approximation quantitatively, the species scaling law and the link scaling law, by using a single parameter from the data, the ratio of total links to total species.

In evaluating the quantitative discrepancies between the observed and predicted proportions of each kind of species and each kind of link, it is important to recall that no fitting is involved in generating the predicted proportions. The only numerical parameter taken from the data is the observed ratio of the total number of links to the total number of species. In addition to its qualitatively correct predictions, model 3 gives seven numbers for the price of one. (Of these seven, only five are independent: two of the three proportions of kind of species, and three of the four proportions of kinds of links.)

The gross testing presented here demonstrates that the overall proportions of species or links are consistent with the predictions of model 3. The following chapter (Chap. III.3) examines how well model 3 describes individual webs.

Cohen (1978, pp. 58–61) considered six stochastic models of webs that are similar to those considered here. His model 6 models the adjacency matrix ('food web matrix') of a web with m prey and n predators by constructing an $m \times n$ matrix in which each element equals 1 with probability $L'/(mn)$ and equals 0 with probability $1 - L'/(mn)$, where L' is the observed number of links, independently for all elements. This model 6 is similar to model 0 here, but model 6 limits the number of prey to m and the number of predators to n. Model 0 here allows the adjacency matrix to be $S \times S$ so that the numbers of prey and predators are limited only by S. None of the models of Cohen (1978) rules out cycles (like our models 1 and 2) or imposes a cascade structure (like our model 3).

Lawlor (1978) observed that in randomly constructed matrix models of ecosystems, when the probability of a non-zero entry in the matrix is independent of the number of species, an overwhelming majority have 3-cycles if the number of species increases beyond 20 (contrary to his and others' informal observations that such cycles are rare in real webs). However, when the probability of a non-zero entry varies inversely as the number of species (as we suppose in this paper, on the basis of the link-species scaling law), Lawlor found (without giving the details of the calculations) that the proportion of random matrix models without 3-cycles increases with increasing numbers of species. He concluded that the usefulness of 'random' models of ecosystems depends critically on whether the models possess the specific structural patterns characteristic of real ecosystems. This conclusion we share.

We are aware of at least two major limitations of the scope of the models and data we have investigated here. First, we have dealt only with the combinatorial structure of webs, rather than with quantities of stocks and flows. Our approach is more like gross anatomy than like physiology. Second, we have dealt only with a static snapshot of webs, ignoring cyclical, successional, or other changes. That is, the gross anatomy is frozen, rather than in motion. In spite of these important limitations, we have provided, in the cascade model, a unifying perspective of simplicity and potential usefulness.

Appendix 1: Proof of Theorem 1

In this proof, we shall omit the explicit dependence on S where possible; for example, we replace $Y(S)$ by Y, $M(S)$ by M. Let $C^{(k)}$ be the set of possible distinct k-cycles. For $s \in C^{(k)}$, let $B_s^{(k)} = 1$ if cycle s occurs in (the web specified by) the random adjacency matrix A, $B_s^{(k)} = 0$ if s does not occur. Then the number of k-cycles in A is $M_k = M_k(S) = \sum_{s \in C^{(k)}} B_s^{(k)}$.

Since the random variables $\{a_{i1}\}_{i=1}^{S}$ are independent with $E(a_{i1}) = c/S$, it is a standard fact that Y converges in distribution to a Poisson variable with mean c.

Let #(.) denote the cardinality (number of elements) of the set in parentheses. Then

$$\#(C^{(k)}) = S!/[(S-k)!k], \quad E(B_s^{(k)}) = (c/S)^k$$

so that

$$\#(C^{(k)})E(B_s^{(k)}) \to c^k/k \quad \text{as} \quad S \to \infty .$$

The random variables $\{B_s^{(k)}\}_{s \in C^{(k)}}$ are non-decreasing functions of the independent elements $\{a_{ij}\}$ of A and hence are associated. (Recall that a finite family $\{X_1, \ldots, X_n\}$ of random variables is defined to be *associated* if $\text{cov}(f(X_1, \ldots, X_n), g(X_1, \ldots, X_n)) \geq 0$ for any real functions f and g that are coordinate-wise increasing.) A theorem independently discovered by Wood (1982) and Newman et al. (1984) (and stated as Theorem 11 by Newman (1984)) then implies that M_k converges in distribution to a Poisson variable with mean c^k/k, for $c \geq 0$, $k \geq 1$, provided that

$$\lim_{S \to \infty} \sum \mathrm{cov}(B_s^{(k)}, B_{s'}^{(k)}) = 0 \, , \tag{A1}$$

where the summation extends over pairs s, $s' \in C^{(k)}$ such that $s \neq s'$.

Similarly, according to Theorem 10 of Newman (1984), which is taken from Newman (1980), (4.3) holds if, in addition,

$$\lim_{S \to \infty} \mathrm{cov}(Y, M_h) = 0 \, , \tag{A2a}$$

$$\lim_{S \to \infty} \mathrm{cov}(M_h, M_j) = 0 \, , \tag{A2b}$$

for all $c \geq 0$ and all h, j such that $1 \leq h \neq j \leq k$.

So we must prove (A1) and (A2).

For $k = 1$, as noted in the text, each $B_s^{(k)}$ is just an a_{ii} so that $\mathrm{cov}\,(a_{ii}, a_{jj}) = 0$ for $i \neq j$ and (A1) holds. Also, for $k = 1$, $\mathrm{cov}\,(Y, M_1) = \mathrm{cov}\,(a_{11}, a_{11}) = (c/S)(1 - c/S)$ so that (A2a) holds for $h = 1$. Similarly $\mathrm{cov}\,(M_1, M_j) = 0$ for $j \neq 1$. We may henceforth assume $k, h, j \geq 2$.

Unless the two cycles, s and $s' \neq s$ share some directed edge, $B_s^{(k)}$ and $B_{s'}^{(k)}$ are independent. Similarly, a_{i1} and $B_s^{(k)}$ are independent unless the edge $(i, 1)$ is in s. Since, for $k > 1$, a_{11} and $B_s^{(k)}$ are independent,

$$\mathrm{cov}(Y, M_k) = \sum_{i=1}^{S} \sum_{s \in C^{(k)}} \mathrm{cov}(a_{i1}, B_s^{(k)})$$

$$= (S - 1) {\sum_{s \in C^{(k)}}}' \mathrm{cov}(a_{21}, B_s^{(k)}) + \sum_{s \in C^{(k)}} \mathrm{cov}(a_{11}, B_s^{(k)})$$

$$= (S - 1) {\sum_{s \in C^{(k)}}}' \mathrm{cov}(a_{21}, B_s^{(k)}) \, ,$$

where $\sum'$ is over those cycles s that include the edge $(2, 1)$. There are exactly $(S - 2)!/(S - 2 - (k - 2))!$ such k-cycles. If $i_1 = 2$, $i_2 = 1$, $i_3, \ldots, i_k$ are the vertices of such a cycle, with $i_{k+1} = i_1 = 2$, then

$$\mathrm{cov}(a_{21}, B_s^{(k)}) = \mathrm{cov}(a_{21}, a_{21} \prod_{j=2}^{k} a_{i_j i_{j+1}})$$

$$= (c/S)^{k-1} \mathrm{var}(a_{21}) = (c/S)^k (1 - c/S) \, . \tag{A3}$$

Thus

$$\operatorname{cov}(Y, M_k) = [(S-1)!/(S-k)!](c/S)^k(1-c/S) \to 0 \quad \text{as} \quad S \to \infty,$$

which proves (A2a).

Suppose s is an h-cycle and s' is a j-cycle. Let $\beta = \beta(s, s')$ denote the number of edges shared in both s and s'. Analogously to (A3), we have

$$\operatorname{cov}(B_s^{(h)}, B_{s'}^{(j)}) = (c/S)^{h-\beta}(c/S)^{j-\beta}\operatorname{var}(\prod_{g=1}^{\beta} a_{i_g i_{g+1}})$$

$$= (c/S)^{h+j-2\beta}[(c/S)^{\beta} - (c/S)^{2\beta}]$$

$$= (c/S)^{h+j-\beta}[1 - (c/S)^{\beta}] = O(S^{-h-j+\beta}). \quad (A4)$$

Then, for some fixed $s_0 \in C^{(k)}$, and for $C_1 = \{(s, s')|s \neq s'\}$, $C_2 = \{s \in C^{(k)}|s \neq s_0\}$,

$$\sum_{C_1} \operatorname{cov}(B_s^{(k)}, B_{s'}^{(k)}) = \#(C^{(k)})\sum_{C_2}\operatorname{cov}(B_{s_0}^{(k)}, B_s^{(k)})$$

$$= \#(C^{(k)})(\sum\nolimits^0 + \sum\nolimits^1 + \ldots + \sum\nolimits^{k-1})$$

$$= \#(C^{(k)})(\sum\nolimits^1 + \ldots + \sum\nolimits^{k-2}) \quad (A5)$$

where $\sum^{\beta}$ denotes $\sum \operatorname{cov}(B_{s_0}^{(k)}, B_s^{(k)})$ over those $s \in C^{(k)}$ such that $\beta(s_0, s) = \beta$. The last equality in (A5) holds because there can be no $s \in C^{(k)}$ with $\beta(s_0, s) = k - 1$ and because each term in the $\sum^0$ sum vanishes.

Now the removal of $k - \beta > 0$ edges from a k-cycle s_0 leaves some number $\eta \geq 1$ of disconnected walks. A cycle $c \in \sum^{\beta}$ must reconnect these walks into a cycle (in an order that may differ from the order in s_0). Thus s is specified by the order of the walks shared with s_0 and by η new walks leading from an end point of one shared walk to a starting point of another shared walk. If these η new walks have lengths $L_1, \ldots, L_\eta$, with $L_1 + \ldots + L_\eta = k - \beta$, then for a given ordering of the shared walks, the number of such new walks is bounded above by $S^{L_1-1}S^{L_2-1}\ldots S^{L_\eta-1} = S^{k-\beta-\eta}$ and thus by $S^{k-\beta-1}$. So for $\beta < k$ the number of terms in any $\sum^{\beta}$ is bounded by $S^{k-\beta-1}$ times a combinatorial coefficient that depends only on k and β but not on S. By using (A4) with $h = j = k$, we may bound (A5) above by

$$\#(C^{(k)}) \sum_{\beta=1}^{k-2} O(S^{-2k+\beta}S^{k-\beta-1}) = O(S^k S^{-k-1}) = O(S^{-1}), \quad (A6)$$

which proves (A1).

We now prove (A2b). As in (A5), for $h < j$,

$$\operatorname{cov}(M_h, M_j) = \#(C^{(h)}) \sum_{s \in C^{(j)}} \operatorname{cov}(B_{s_0}^{(h)}, B_s^{(j)})$$

$$= \#(C^{(h)})(\sum{}^0 + \sum{}^1 + \ldots + \sum{}^h)$$

$$= \#(C^{(h)})(\sum{}^1 + \ldots + \sum{}^{h-1}) . \tag{A7}$$

In the last equality, $\sum^h = 0$ because no j-cycle can share h edges with the h-cycle s_0 if $h < j$. As in the derivation of (A6) as an upper bound for (A5), we see that with $h - \beta > 0$, the number of terms in $\sum^\beta$ is $O(S^{j-\beta-1})$. Then (A4) implies that (A7) is bounded by

$$\#(C^{(h)}) \sum_{\beta=1}^{h-1} O(S^{-h-j+\beta} S^{j-\beta-1}) = O(S^h S^{-h-1}) = O(S^{-1}) ,$$

which proves (A2b).

The claimed limiting behaviour of $(Y(S), M(S))$ for $0 \le c < 1$ now follows from (4.3) by approximating $M(S)$ by $M_k^*(S) = \sum_{h=1}^k M_h(S)$ for large fixed k. For fixed k, (4.3) implies that $(Y(S), M_k^*(S))$ converges in distribution as $S \to \infty$ to a 2-vector with independent Poisson components and mean $(c, \sum_{h=1}^k c^h/h)$. Moreover,

$$E|M(S) - M_k^*(S)| = E(M(S) - M_k^*(S)) = \sum_{h=k+1}^\infty \#(C^{(h)}) E(B_{s_0}^{(h)})$$

$$\le \sum_{h=k+1}^\infty c^h/h$$

$$\to 0 \quad \text{as} \quad k \to \infty \quad \text{for} \quad c < 1 .$$

Now for any real numbers r and t

$$|E(\exp\{i(rY(S) + tM(S))\}) - E(\exp\{i(rY(S) + tM_k^*(S))\})|$$

$$\le E(|\exp\{it(M(S) - M_k^*(S))\} - 1|)$$

$$\le E|t(M(S) - M_k^*(S))| \to 0 \quad \text{as} \quad k \to \infty .$$

Therefore, the limiting distribution of $(Y(S), M(S))$ equals the limiting distribution, as $k \to \infty$, of the limiting distribution, for any fixed k, as $S \to \infty$, of $(Y(S), M_k^*(S))$. This proves the claimed results when $0 \le c < 1$.

Appendix 2: Numerical Simulation of Acyclic Random Digraphs

We have programmed three numerical approaches to investigating the fraction of zero rows or columns in a matrix that is $\sim$ i.i.d. $B(c/S)$, conditional on being acyclic.

The first, and most naive, approach is to generate matrices that are $\sim$ i.i.d. $B(c/S)$ and reject those that have a cycle of any length. There are two difficulties with this approach. First, given a value of c, this approach generates acyclic webs very inefficiently. For example, with an arbitrarily chosen $c = 2.1$, the number of Bernoulli matrices that had to be generated to find 100 acyclic matrices of each size in a sample calculation was as shown in Table III.2.A1.

Table III.2.A1. Matrices generated according to a first naive approach

size of matrix (S)	number of matrices generated to get 100 acyclic $S \times S$ matrices
3	34972
5	16113
10	28726
15	62825
20	279401

We lack theory for what the numbers on the right of the table should be, either for finite S or in the limit as $S \to \infty$. (These results show, incidentally, first that the fraction of acyclic matrices among Bernoulli matrices, for fixed c, need not be a monotone decreasing function of S, and second that the fraction of 10-acyclic Bernoulli matrices, asymptotically as $S \to \infty$, according to Theorem 1, bears no close relation to the fraction of 10-acyclic 10×10 Bernoulli matrices. According to (4.3), the former fraction is $\exp(-\sum_{k=1}^{10} c^k/k)$, which is less than 10^{-157} when $c = 2.1$, while according to the numerical results above the latter fraction is approximately $100/28726$.)

A second difficulty with this first approach is that, so far, we lack theory to guide the choice of c when we want to compare the computed fractions of zero rows or columns with data. By throwing away the matrices with cycles, we change the expected number of matrix elements that equal 1 from $pS^2 = cS$ to some (so far) unknown smaller function of c and S. For comparison with data, we want to choose c so that the 'effective density' of links, estimated as (average number of matrix elements equal to 1)$/S$, equals the observed $d = 1.86$. In the numerical simulations described above, with $c = 2.1$, the total number of elements equal to 1, summed over 100 acyclic matrices, and the average effective density per matrix, were as shown in Table III.2.A2.

Table III.2.A2. Number and effective density of
links in naively generated acyclic matrices

size of matrix (S)	number of 1s in 100 acyclic matrices	effective density
3	239	0.80
5	537	1.07
10	1293	1.29
15	2124	1.42
20	3094	1.55

Depending on the matrix size S, the effective density can be quite different from c in model 2. Again, we lack theory for what the numbers on the right should be, either for finite S or as $S \to \infty$.

A second approach, based on the ideas of Erdös & Rényi (1960), avoids both of these difficulties, but encounters a subtler third difficulty. In this approach, to obtain an effective density c, we construct a random acyclic matrix with the integer part of cS (denoted int (cS)) edges. This is impossible if $cS > S(S-1)/2$ (or more generally if cS exceeds the maximum number of links possible in an $S \times S$ acyclic matrix). Provided int (cS) is sufficiently small, we add one edge at a time. We choose a 0 element of the matrix, with probability equal to 1 divided by the number of 0 elements that could be changed to 1 without creating a cycle. To identify the 0 elements that are available to be changed to 1 without creating a cycle, we maintain in an auxiliary matrix the transitive closure of the adjacency matrix. We continue adding edges until int (c/S) edges have been added. If, because of the sequence of edges chosen, the required number of 1's cannot be added to the matrix, then the partly completed matrix is abandoned and a fresh start is made. This generates a random acyclic matrix with effective density close to c.

The virtue of this second approach is that it guarantees $L/S = c = d^* = d$ approximately (recall that c is the model parameter with c/S being the probability of a random link, d^* is the asymptotic (large S) effective ratio of links to species, and d is the observed ratio L/S of links to species in real webs). A drawback, which we overlooked at first, is that this approach does not generate all random digraphs with S vertices and, say, E (always directed) edges with equal probability. In the probability distribution over digraphs assumed by model 2, any two digraphs with S vertices and E edges occur with equal probability. However, in the numerical approach just described, suppose $S = 6$ and we wish to choose randomly $E = 3$ edges. There are $6 \times 5 = 30$ ways to choose the first edge without creating a loop. Suppose, without loss of generality, that the edges are labelled so that the first edge is $(1, 2)$, that is, the edge goes from vertex 1 to vertex 2. There are then 28 ways to choose the second edge (edge $(1, 2)$ has already been chosen and edge $(2, 1)$, which would create a cycle, is forbidden). If the second edge is, for example $(3, 4)$, then there are 26 ways to choose the third edge. But if the second edge is $(2, 3)$, then there are only 25 ways to choose the third edge because two edges have already been chosen and *three* edges are forbidden ($(2,1)$, $(3,2)$ and $(3,1)$).

Our third approach modifies the procedure just described to avoid this difficulty. As each randomly chosen edge is added to a digraph, the number of available edges that could have been chosen at that stage is noted. The product of all the numbers of available edges is assigned to the generated digraph as a weight. This weight is the inverse of the probability of choosing the edges in the particular random digraph *in the order in which the edges occurred*. The weight assigned to a given digraph may vary depending on the order in which the edges are chosen. All the statistics (such as the mean or variance of the fraction of species that are top or basal) computed from the random digraphs generated according to this third approach incorporate the weights, so that all digraphs with a given number of vertices and edges are represented with equal probability.

When the unweighted simulations based on the second approach are compared with the weighted simulations based on the third approach, the simulated mean fractions of 0 rows and columns were generally slightly larger when weighted, but usually not by more than 0.01 and never (for the range of parameters in Table III.2.2) by more than 0.04. A conjecture that for large S and for c small compared to S the two approaches give identical mean proportions of 0 rows and columns seems plausible.

The simulations based on the second and third approach are not identical to those based on the first, naive approach. There is no variation in the number of edges (links) per acyclic digraph generated according to the second or third approach, while there is variation in the number of edges per acyclic digraph generated naively by the first approach. As in the parallel case of undirected graphs considered Erdös & Rényi (1960), we expect (but have not proved) that this difference in approach to simulating model 2 has no effect in the limit of large S.

§3. Individual Webs

Joel E. Cohen, Charles M. Newman and Frédéric Briand

1. Introduction

A *food web* is a set of different kinds of organisms and a relation that shows the kinds of organisms, if any, that each kind of organism in the set eats. A *community food web* is a food web whose vertices are obtained by picking, within a habitat or set of habitats, a set of kinds of organisms (hereafter called *species*) on the basis of taxonomy, size, location, or other criteria, without prior regard to the eating relations (specified by trophic *links*) among the organisms (Cohen 1978, pp. 20–21).

In the previous chapter (Chap. III.2), several models were proposed to describe the structure of community food webs. When models were tested against data on 62 community food webs in Chap. III.2, a crucial parameter in all the models, namely the ratio of links to species, was estimated from the aggregated data on all webs taken together. One model, the cascade model, successfully described, to a first approximation, the proportions of all species that are top, basal and intermediate, and the proportions of all links of each kind.

The purpose of this chapter is to test how well the cascade model describes webs when the ratio of links to species is estimated separately for each web.

In section 2 we describe the cascade model, show how to estimate the parameters of the model, and verify the correctness of the estimation procedure. In section 3 we test the assumption, made in Chap. III.2, that the ratio of links to species is constant for all webs. We then test seven predictions of the cascade model, estimating this ratio separately for each web. In section 4 we evaluate the results of this chapter and relate them to the results of Chap. III.2.

We shall use a number of terms with special meanings that are given in section 2 of Chap. III.2. These terms include: web, species, link, predator, prey, top, proper top, intermediate, basal, proper basal, adjacency matrix, isolated, triangular. We shall not repeat the definitions here.

Webs are classified as arising in 'fluctuating' or 'constant' environments. The environment is considered to be 'fluctuating' if the original report indicates temporal variations of substantial magnitude in temperature, salinity, water availability or any other major physical parameter. The magnitude, and not the predictability, of the fluctuations is the criterion of classification. Since the classification of an environment as constant or fluctuating is to some extent subjective, we point out that this task was carried out before we had analysed the webs and uncovered any pattern.

The 62 webs analysed here are drawn from published studies. They include the 40 webs assembled and described by Briand (1983; Chap. II.5). Of these, 13 are drawn from the 14 originally used by Cohen (1978). Details of the webs appear in Chap. IV.

2. The Cascade Model and Parameter Estimation

The cascade model assumes that the S species of a web may be labelled from 1 to S so that, for some finite positive real number $c \leq S$, the probability that species j feeds on species i is 0 if $j \leq i$. If $i < j$, then j feeds on i with probability $p = c/S$ and does not feed on i with probability $q = 1 - c/S$, independently for all $1 \leq i < j \leq S$.

All numerical predictions of the cascade model depend on the values of the model's two parameters c and S. These two parameters, in turn, depend only on the observed numbers of links and of species.

In the data we shall use to test this model, only proper top species (that is, those that eat at least one other species) and only proper basal species (that is, those that are eaten by at least one other species) are reported. Thus the total number of *observed* species in a web is not S but the number of *not isolated* species. The true number S of species in a web is not directly counted.

The expected number $E(N)$ of not isolated species depends on both c and S according to (6.3c) in Chap. III.2. Similarly, the expected number $E(L)$ of links in a web depends on c and S according to (6.5) in Chap. III.2. To test the predictions of the cascade model with individual webs, we estimate c and S by the method of moments. That is, if S' is the observed number of species (that is, S' is the observed value of the random variable N, the number of not isolated species), and L' is the observed number of links (that is, L' is the observed value of the random variable L, the number of links in a web), we replace $E(N)$ on the left of (6.3c) in Chap. III.2 by S' and $E(L)$ on the left of (6.5) in Chap. III.2 by L'. The resulting equations are restated as (A1) in the appendix. We then solve this system of two nonlinear equations for the two unknowns c and S by using Newton's method, as described in the appendix, except for the one web with $S' = 3$. For this web, we take $S = 3$ and then compute c by solving (A1a).

As a check on the correctness of the numerical solutions c and S, we used the numerical values of c and S to compute $E(L)$ from (6.5) in Chap. III.2 and $E(N)$ from (6.3c) in Chap. III.2. In figures not shown, we plotted L' as a function of the calculated $E(L)$ and S' as a function of the calculated $E(N)$. A line of slope one through the origin passed through all the plotted characters except, as expected, the web with $S' = 3$, verifying that the computed numerical solutions for c and S in fact satisfy (A1) adequately.

The computed values of S are not in general integers. We could force them to be integers by replacing S with the integer closest to S and then solving (A1a) for a new value of c. A simpler alternative, which we adopt here, is to interpret the equations of the cascade model derived in Chap. III.2 as applying whether S is integral or positive real.

From (6.3c) in Chap. III.2 or (A1b), it follows that if $S \gg 1$ and $e^{-c} \ll 1$, then S is approximately S'. In the data plotted in Fig. III.3.1, S does not greatly exceed the observed values of S'. In fact $S - S' < 2.1$ for all webs but one. For the exceptional web (Paviour-Smith 1956), $S - S' = 5.5$, where $S = 37.5$. This exceptional case is visible as the outlying fluctuating web in the lower right corner of Fig. III.3.1. Briand & Cohen (1984) also noted that this web was an outlier on a plot of prey against predators based on unlumped data (Fig. A.2.1a). This web appears to be unusual in both the relation between links and total species and the relation between predators and prey.

3. Testing The Predictions of The Cascade Model

The tests of the cascade model in Chap. III.2 use a single value of c for all webs. If this procedure is correct, then a plot of c against S, estimated individually for each web, should display no increasing or decreasing trend. Substantial variability in c as a function of S is expected because the realized number of not isolated species need not exactly equal the mean $E(N)$ and the realized number of links need not exactly equal the mean $E(L)$.

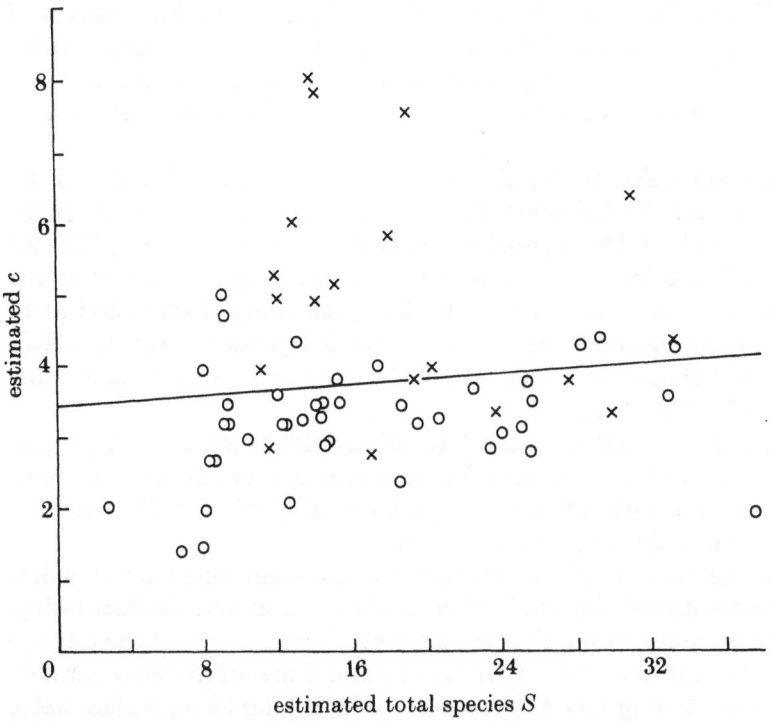

Figure III.3.1. The estimated value of c as a function of the total number of species S. The straight line $c = 3.438 + 0.017S$ is an ordinary least-squares regression line fitted under the assumption that the variance of the residuals is independent of S, and without constraints on the slope or intercept. The slope 0.017 has a standard deviation of 0.022. There is no evidence of a rise or fall in c with increasing S. In this and all subsequent figures, the plotted symbols have been perturbed by a small random amount from their exact positions to indicate when several symbols coincide. ×, constant web; o, fluctuating web. Only 'lumped' webs are used (Briand & Cohen 1984; Chap. II.2)

Figure III.3.1 shows that there is no evidence of a pronounced trend in the estimated c as a function of the estimated S. Because S and $S-1$ are close to the observed number S' of not isolated species, the observation that $c = 2L'/(S-1)$ has no significant trend as a function of S follows from the link-species scaling law (Chap. III.2) that L'/S' has no increasing or decreasing trend as a function of S'.

The observation of a slightly positive slope in Fig. III.3.1 is consistent with two earlier observations. First, by using multiple versions of the unlumped community webs of Cohen (1978), Yodzis (1980) observed that with increasing S', the observed (lower) connectance C', defined by $C' = 2L'/[S'(S'-1)]$, decreases nearly but not quite as fast as $1/S'$. Now $C' = c(1/S')[(S-1)/(S'-1)]$ and the last factor $[(S-1)/(S'-1)]$ approximates 1. Therefore if c has no trend as a function of S, C' would be expected to decline approximately as $1/S'$. Secondly, by using 40 unlumped webs, Briand (1983; Chap. II.5) observed that the number L' of links was proportional to $S'^{1.1}$ rather than to S'. Because of the overlaps among the sets of data used by Yodzis (1980), Briand (1983) and here, the findings of Yodzis (1980) and Briand (1983) are by no means independent of ours. Moreover, the webs analysed by Yodzis and Briand were unlumped while ours are lumped. Thus there is no persuasive evidence against the natural null hypothesis that for the lumped webs studied here, c is effectively constant as S increases.

Of the 19 constant webs, 14 fall above the regression line in Fig. III.3.1. Of the 43 fluctuating webs, 34 fall below the regression line. The difference in the proportions of webs above the regression line (74% for constant webs, 21% for fluctuating webs) is too large to be attributed to chance ($\chi^2 = 15.7$ with one degree of freedom, a value with extremely low probability if one chooses to believe the underlying but doubtful assumption of independence among webs). For a given number of species, constant webs have more links than fluctuating webs (Briand 1983; Chap. II.5).

This difference demonstrates at the level of individual webs the aggregate difference in the ratio of links to species between constant and fluctuating webs. For constant webs, the ratio of links to species is $811/351 = 2.31$, while for fluctuating webs, the ratio is $1108/683 = 1.62$.

The use of a single value for c in Chap. III.2 overlooks differences in the typical values of c of two distinguishable kinds of webs, the constant and the fluctuating, making it all the more surprising that the aggregated predictions of the cascade model in Chap. III.2 are not worse. Here, since c and S are estimated separately for each web, we are testing how the cascade model applies to individual webs, both constant and fluctuating.

We now test seven predictions of the cascade model. In Figs. III.3.2–8 the abscissa is the expected value of some feature of a web, according to the cascade model, and the ordinate is the observed value of that feature. If the estimated values of c and S corresponded exactly to the true values of c and S and if the observed value of each feature in each web corresponded to the expected value, then all data points would fall along a line of slope one through the origin. The cascade model is a stochastic model, however, so the data points are expected to deviate from such a line, but not systematically. Since the scales of the abscissa and ordinate vary, a line of slope one through the origin is drawn in Figs. III.3.2–8 for comparison.

There is no reason to assume that half of the data points should fall above, and half below, the line of slope one, because we have not proved that, according to the cascade model, the variables of interest are symmetrically distributed about their mean. However, as the number of species in a web increases, it seems reasonable to suppose that the distributions of these variables approach normality. In this limit of large S, it seems reasonable to anticipate roughly half of the data points above and half below the line of slope one if the cascade model is correct.

As might be expected, in Figs. III.3.2–8 the variance of the observed number, plotted on the ordinate, increases as the expected number, plotted on the abscissa, increases. Since all the abscissae are increasing functions of the number of species in a web, the variances of observed numbers also increase with increasing size of web.

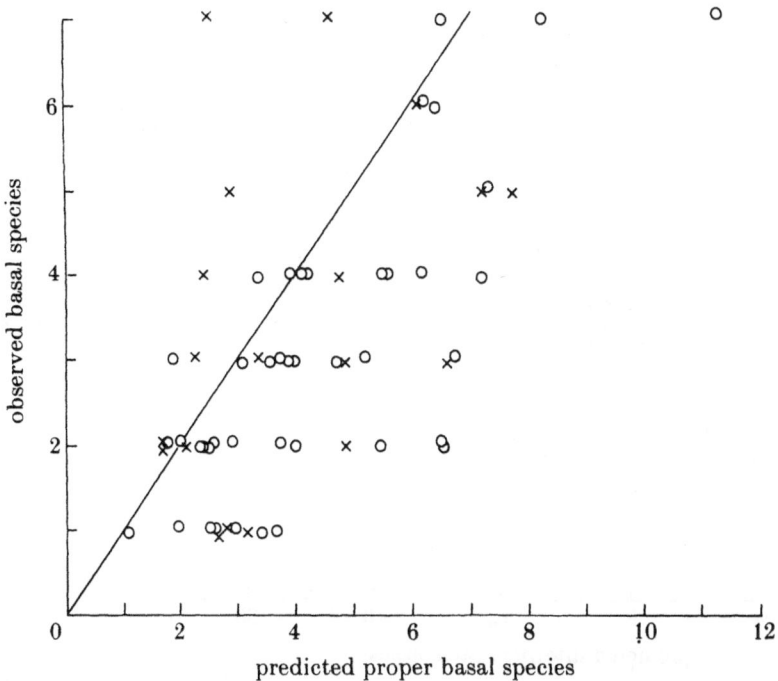

Figure III.3.2. The observed number of (proper) basal species as a function of the expected number of proper basal species according to the cascade model. In this and all the following figures, the solid straight line passes through the origin with slope 1. This is not a regression line, but should describe the trend of the data if the predictions of the cascade model are approximately correct

Figure III.3.2 plots the observed (proper) basal species against the expected proper basal species, computed from (6.3a) in Chap. III.2. There appear to be 'rows' of data points in Fig. III.3.2 because the observed numbers of basal species are constrained to be integers, while the expected numbers can vary continuously. The bulk of the data points, though by no means all, fall below the line of slope one. This finding is consistent with the fact that fewer basal than top species are observed and with the observation in Chap. III.2 that fewer basal species are observed than expected using an aggregate estimate of c. No difference between constant and fluctuating webs in the success of the cascade model is immediately evident from Fig. III.3.2. This absence of apparent difference is consistent with the finding of Briand & Cohen (1984; Chap. II.2) that the proportions of (proper) basal, intermediate, and (proper) top species are homogeneous between constant and fluctuating webs, within statistical fluctuations.

Figure III.3.3 plots the observed intermediate species against the expected intermediate species, computed from (6.3b) in Chap. III.2. The constant webs fall nearly evenly above and below the line of slope one (9 fall above, 10 fall below). The bulk of the fluctuating webs fall slightly above the line. This small difference is consistent with the insignificantly greater aggregate proportion of

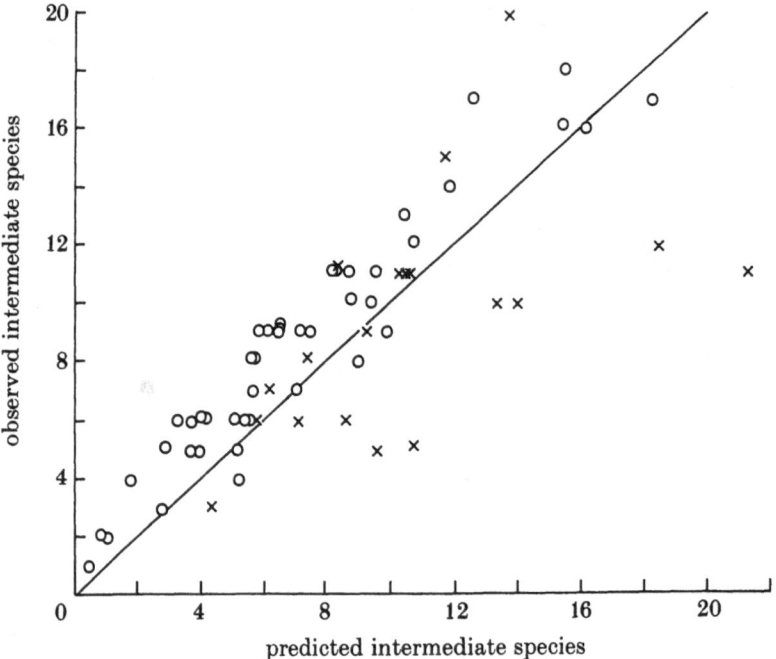

Figure III.3.3. The observed number of intermediate species as a function of the expected number of intermediate species according to the cascade model

intermediate species in fluctuating webs than in constant webs (54% versus 50%). Overall, the agreement between observed and expected species is good.

Figure III.3.4 plots the observed (proper) top species against the expected proper top species, computed from (6.3a) in Chap. III.2. The constant web with 17 top species appears to be an outlier. This same web, which describes the rocky shore of Lake Nyasa (Fryer 1959) also appeared as an outlier in a plot, with unlumped data, of prey against predators (Fig. A.2.1a). For both constant and fluctuating webs, there is a suggestion that the remaining points may rise convexly. At least in the middle range of expected values, however, the agreement between observation and expectation is good.

In summary, when the expected numbers of species of each kind are compared with the observed, the agreement is best for intermediate species and is fair for proper top and proper basal species. The cascade model describes the kinds of species in constant and fluctuating webs about equally well.

Figure III.3.5 plots the observed basal-intermediate links against the expected basal-intermediate links, computed from (6.6a) in Chap. III.2. There is no sign of systematic deviation between the points and the line of slope one, for the constant and fluctuating webs considered separately or together.

Figure III.3.6 plots the observed basal-top links against the expected basal-top links, computed from (6.6b) in Chap. III.2. Contrary to expectation, there are many webs with no basal-top links or one only. The line through the origin with slope one passes through the mass of the remaining points, but even for

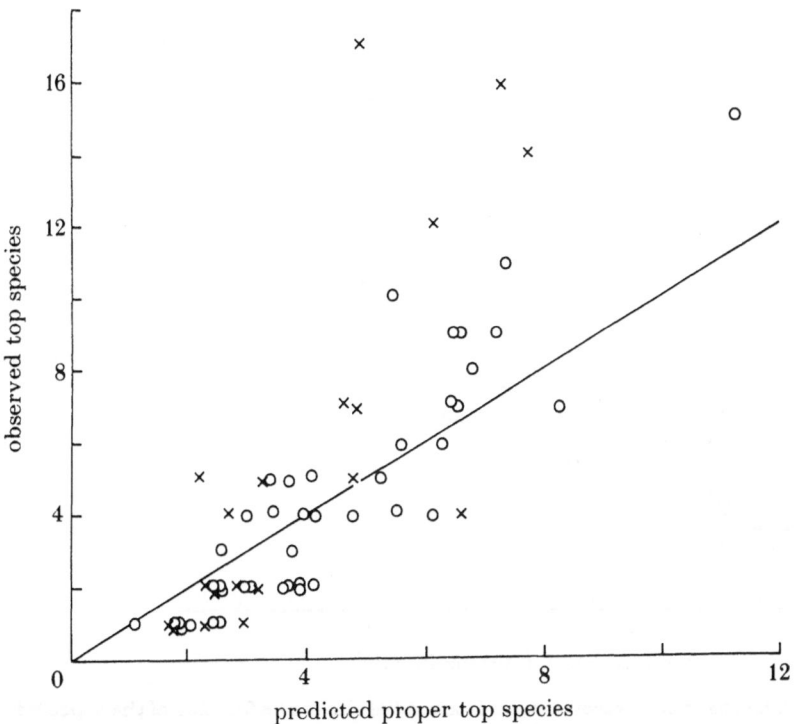

Figure III.3.4. The observed number of (proper) top species as a function of the expected number of proper top species according to the cascade model

these the scatter about the line is large, compared with that in Figs. III.3.5, 7 and 8.

Figure III.3.7 plots the observed intermediate-intermediate links against the expected intermediate-intermediate links, computed from (6.6c) in Chap. III.2. The apparent outlier with six observed intermediate-intermediate links in the lower right corner of Fig. III.3.7 is the same web that appears above as the potential outlier in Fig. III.3.4. This same web appears again as the outlier with 59 observed intermediate-top links in the top-right corner of Fig. III.3.8. Clearly this web is exceptional in several respects, when compared with other webs. Aside from this outlier, the remaining webs are scattered more or less symmetrically about the line of slope one, and no systematic deviations are evident.

Figure III.3.8 plots the observed intermediate-top links against the expected intermediate-top links, computed from (6.6a) in Chap. III.2. As in Fig. III.3.5 (apart from the single outlier), there is no sign of systematic deviation of the points from the line of slope one, for the constant and fluctuating webs separately or together.

In summary, the cascade model provides a good description of the numbers of basal-intermediate, intermediate-intermediate, and intermediate-top links, aside from one outlying constant web, and a rather poor description of the number of

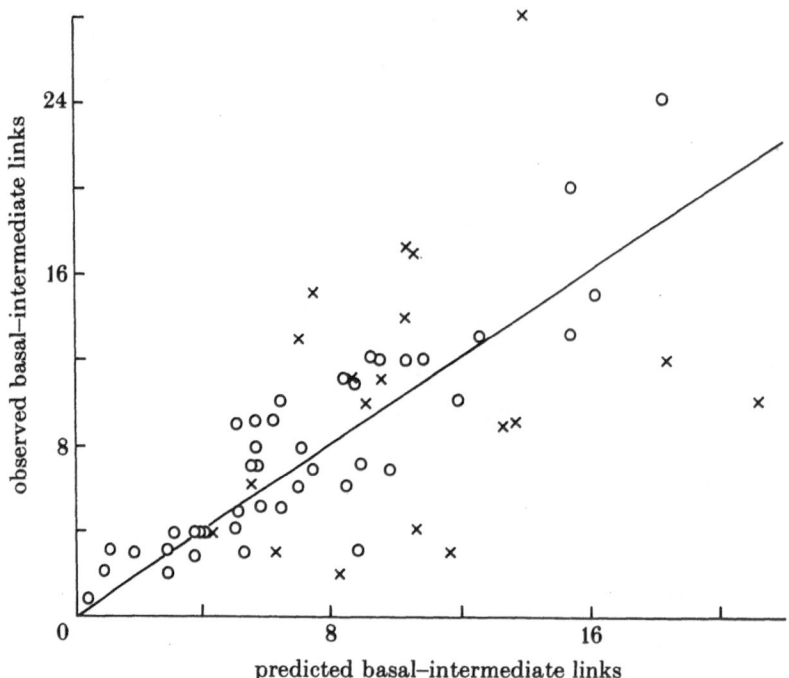

Figure III.3.5. The observed number of basal-intermediate links as a function of the expected number of basal-intermediate links according to the cascade model

basal-top links. The cascade model describes the links of constant and fluctuating webs about equally well.

4. Discussion and Conclusions

We have tested a model, called the cascade model, which assumes that species in a community are arranged in a hierarchy or cascade of potential feeding relations. This model assumes that whether a potential feeding relation becomes an actual feeding relation is determined randomly, independently of all other potential feeding relations. The probability that a potential feeding relation becomes actual is assumed to be the same for every potential feeding relation within a community, and to vary inversely as the number of species in the community.

Consequently, according to the model, for a randomly chosen species in a community, the mean number of other species that prey on it or that are prey to it is independent of the total number of species in the community. Thus the model is consistent with the hypothesis, suggested by Pimm (1982, p. 89), that 'each species in a community feeds on a number of species of prey that is independent of the total number of species in the community', provided that the term 'each species' is replaced by the term 'a randomly chosen species'.

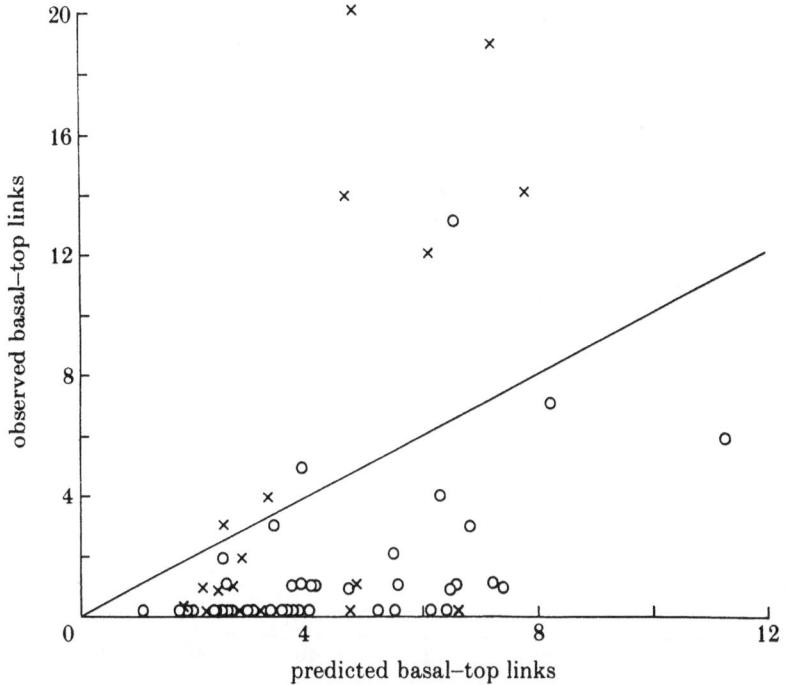

Figure III.3.6. The observed number of basal-top links as a function of the expected number of basal-top links according to the cascade model

In Chap. III.2 we showed that the cascade model describes several important properties of webs, to a first approximation, when a fixed probability parameter, estimated from aggregated data, is applied to all webs. There is no logical necessity for the cascade model to describe individual webs, given that it succeeds reasonably in the macroscopic analysis. Tests of the cascade model using data on individual webs are logically and empirically independent of tests using aggregated data. Indeed, it would be surprising to find that ecological 'assembly rules' as simple as the cascade model apply to communities that arise in diverse environments.

To test the cascade model's ability to describe individual webs, we used two numbers, the observed number of not isolated species and the observed number of links, to estimate the two parameters of the cascade model: the unknown true number S of species and the unknown constant c to which the probability of a feeding relation is proportional. We then computed the expected values of seven characteristics of webs and compared them with the observed.

In 62 webs, with exception of an occasional outlier, the cascade model describes well the numbers of intermediate species (Fig. III.3.3), basal-intermediate links (Fig. III.3.5), intermediate-intermediate links (Fig. III.3.7), and intermediate-top links (Fig. III.3.8). It describes fairly the numbers of proper basal (Fig. III.3.2) and proper top (Fig. III.3.4) species. It describes poorly the numbers of basal-top links (Fig. III.3.6).

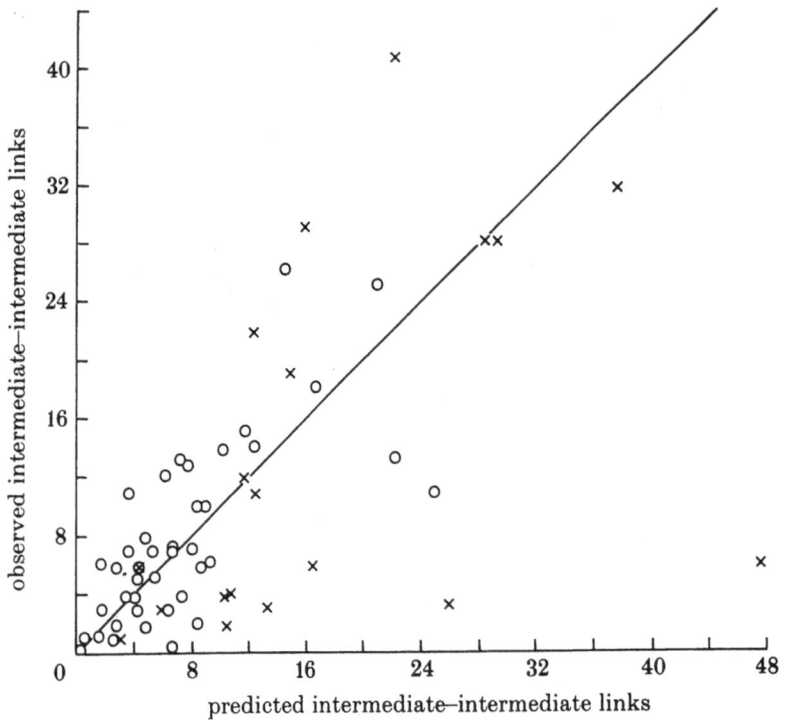

Figure III.3.7. The observed number of intermediate-intermediate links as a function of the expected number of intermediate-intermediate links according to the cascade model

For a given value of S, the probability c/S that a species will prey on another species, when their positions in the trophic hierarchy permit, is higher in constant webs than in fluctuating webs. Given c and S, the cascade model describes the numbers of kinds of species and kinds of links in constant and fluctuating webs about equally well, according to our examination of the data.

Cohen & Briand (1984; Chap. II.3) noted that the proportions of each kind of link appear to differ between constant and fluctuating webs. Since the cascade model describes the numbers of each kind of link about equally well in constant and fluctuating webs, the difference in proportions may be explained by the difference in the typical values of c for constant and fluctuating webs, rather than by some deeper structural difference between constant and fluctuating webs. The difference in the typical values of c between constant and fluctuating webs is not explained by the cascade model.

In testing the model, we present graphical comparisons of the observations and predictions so that the reader can make his or her own verbal summaries of how good or bad the fit is. We avoid formal statistical measures of goodness of fit because the data may not be independent and because we are interested in simultaneous inference about the model as a whole. The assumption of independence among webs appears doubtful, since some authors contributed more than one web and the proclivities of authors do appear to influence the structure of

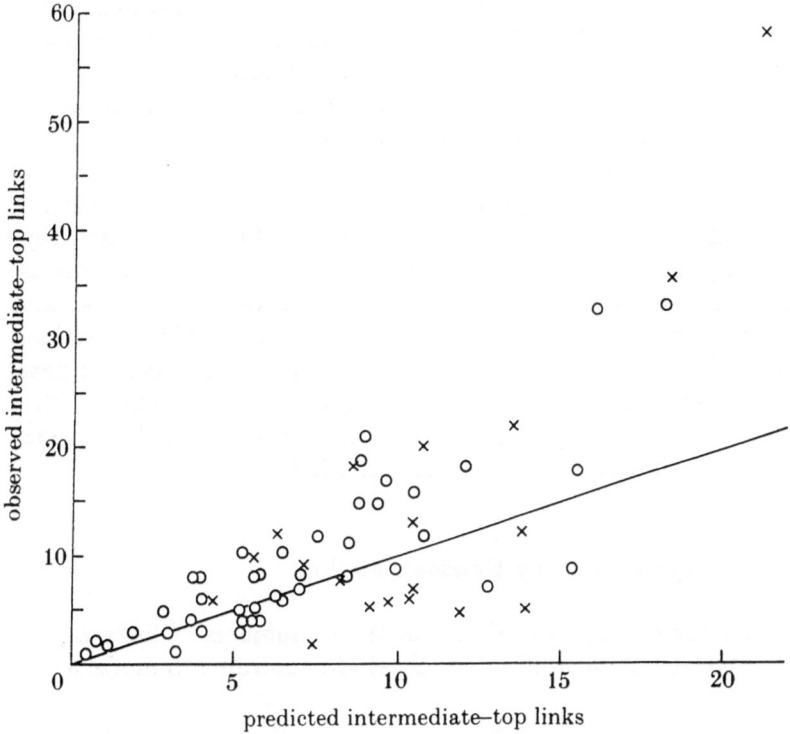

Figure III.3.8. The observed number of intermediate-top links as a function of the expected number of intermediate-top links according to the cascade model

webs. We are less concerned to test hypotheses about any portion of the cascade model than to see how well, on the whole, it describes simultaneously several major features of webs, some of which may not be independent. (For example, given the total number of not isolated species, the observed numbers of proper top, intermediate, and proper basal species are not independent.)

How should one evaluate the discrepancies between theory and observation most evident in Fig. III.3.6, and strongly suggested by Figs. III.3.2 and 4? One can be sceptical about the model, but not the data; or one can be sceptical about the data, but not the model; or one can be sceptical about both. We are sceptical about both.

As for the data, the earlier chapters by Briand and Cohen, jointly and separately, indicate that there is great variability among observers in the operational definitions of species and links and in the detail of published reports of field work. Often, these operational definitions are not even described in the published reports. A first step that field ecologists could make toward improving the data would be to describe in detail how the species and links are operationally defined. A second step would be to work toward some uniform definitions.

Nevertheless, the data analysed here are the best available at present. The regularities in these data merit theoretical attention.

As for the model, numerous assumptions underlying it are unrealistic. Is it plausible to assume that the species at the top of the hierarchy or cascade is equally likely to prey on all other species in the community? Is it plausible to assume that the prey species a predator eats are chosen independently of the abundance of the prey species and stochastically, once and for all, as the model implicitly assumes? We think not.

Nevertheless, the cascade model provides a very simple unifying perspective, quantitatively testable and open to improvement, that explains for the first time several empirical regularities in the structure of webs. The ecological generalizations explained by the cascade model still need to be derived from a persuasive and testable theory of behaviour, population dynamics, and trophic interactions.

The cascade model also needs to be tested further against macroscopic data. Can the cascade model explain the observed frequency of intervality (Cohen 1978) in food webs? Can the cascade model explain the observed frequency distributions of length of food chains? See Chaps. III. 4–6.

Appendix: Estimation of c and S by Newton's Method

Given observed numbers of species S' and observed numbers of links L', the parameter c and the true number of species S satisfy, according to model 3 (the cascade model):

$$0 = c(S-1)/2 - L' \equiv v_1 , \tag{A1a}$$

$$0 = (1 - [1 - c/S]^{S-1})S - S' \equiv v_2 . \tag{A1b}$$

We find c and S numerically by Newton's method (for example, Rektorys 1969, pp. 1180–1181).

Step 1. Let $c_0 = 2L'/(S'-1)$, $S_0 = S'$. (These are initial estimates.)

Step 2. Compute

$$J_{11} = (S_0 - 1)/2 ,$$
$$J_{12} = c_0/2 ,$$
$$J_{21} = (S_0 - 1)(1 - c_0/S_0)^{S_0 - 2} ,$$
$$J_{22} = 1 - (1 - c_0/S_0)^{S_0 - 1}[1 + c_0(S_0 - 1)/(S_0 - c_0) + S_0 \ln(1 - c_0/S_0)] .$$

(This is the Jacobian of the nonlinear system (A1), that is $J_{11} = \partial v_1/\partial c$, etc., evaluated at (c_0, S_0).)

Step 3. Compute $A = J^{-1}$, that is,

$$A = \begin{pmatrix} J_{22} & -J_{12} \\ -J_{21} & J_{11} \end{pmatrix} /(J_{11}J_{22} - J_{12}J_{21})$$

and v_1, v_2 from (A1) with c replaced by c_0, S by S_0, then

$$c_1 = c_0 - a_{11}v_1 - a_{12}v_2, \quad S_1 = S_0 - a_{21}v_1 - a_{22}v_2 \quad .$$

(The new estimates of c and S are c_1, S_1.)

Step 4. Stop if $|c_1 - c_0| + |S_0 - S_1| < \delta = 0.01$. (Stop when the procedure quasi-converges. The value of $\delta = 0.01$ was chosen to require the final estimates c_1, S_1 to be changing far less than the uncertainty in the data.)

Step 5. Otherwise, replace the value of c_0 by the value of c_1 and replace the value of S_0 by the value of S_1. Then go to step 2. (Iterate with improved estimates.)

When applied to the 62 pairs (S', L') from the webs assembled by Briand, this procedure stopped after at most 3, and generally 2, iterations, except for the pair $(S', L') = (3, 2)$ from one web (code number 10). For this web, the procedure diverged, and we used the initial estimate.

§4. Predicted and Observed Lengths of Food Chains

Joel E. Cohen, Frédéric Briand and Charles M. Newman

1. Introduction

The purpose of this chapter is to derive a quantitative theory of the length of food chains from a mathematical model of community food webs called the cascade model and to test this theory quantitatively against data from real food webs. The cascade model was developed and tested by Cohen & Newman (1985; Chap. III.2) and by Cohen et al. (1985; Chap. III.3). The predictions of the cascade model describe, to a first approximation, several major characteristics of a collection of 62 real webs: the proportions of all species that are top, basal and intermediate, and the proportions of all links from basal to intermediate species, from basal to top species, from intermediate to intermediate species, and from intermediate to top species.

This chapter determines what the cascade model implies for the frequency distribution of the length of food chains in webs with a finite number of species and compares the predictions with observations. The number of species in the observed webs ranges from 3 to 48. The theory of chain lengths is developed further for webs with a large number of species in the next chapter (Newman & Cohen 1986; Chap. III.5).

Section 2 reviews present biological theories of food chain lengths; section 3 presents terminology for chains and reviews the cascade model. Section 4 gives exact results about the frequency distribution of chain lengths for webs with a finite number of species and proposes a way to evaluate the goodness of fit of the cascade model's predictions to the observed frequency distribution of chain length in an individual web. A mathematical proof in section 4 (the only one in this chapter) is set off by *Proof* at the beginning and ■ at the end. Readers may defer or skip the proof with no loss of continuity.

In section 5, we find that the cascade model describes acceptably most, but not all, of the frequency distributions of chain length observed in 62 webs, other aspects of which were previously used to develop and test the model. Does the cascade model succeed in most of these webs because the model was selected to describe other aspects of the same data, since such selection might constrain the possible frequency distributions of chain length?

No, according to the results of section 6. There we examine the frequency distributions of chain lengths in a freshly assembled and edited collection of 51 webs that have not been previously related to the cascade model. The species-link scaling law (Cohen & Briand 1984; Chap. II.3), one of the central features of the cascade model, is not contradicted by these new data. The cascade model describes acceptably 46 of the 51 observed frequency distributions of chain lengths; this majority is even larger than the majority of its successes with the original 62 webs.

According to section 7, the mean and variance calculated from the expected numbers of chains of each length cannot validly be compared with the mean and variance of chain lengths in an observed web. If such a comparison is made, nevertheless, the mean chain lengths are described acceptably, but not the variances.

In section 8, we explain why we doubt the assumption that the 113 webs in our collection are a random sample from some statistical ensemble of webs. Under this dubious assumption, a Kolmogorov-Smirnov test rejects the null hypothesis that the cascade model's predictions describe the chain lengths in the ensemble of webs sampled by either the original 62 or the new 51 webs or all 113 combined.

Most of the 16 or 17 webs with chain lengths that the cascade model fits poorly have unusually large average chain lengths (greater than four links) or unusually small average chain lengths (fewer than two links).

Finally, in section 9, we review the accomplishments of the chapter, relate them to previous work, and propose several further studies. An appendix presents algorithms that were used to compute the frequency distribution of chain length and the length of the longest chain of a given web.

Chap. IV presents in detail the sources and full data on all 113 (62 + 51) webs.

2. The Length of Food Chains: Present Ecological Theory

Elton (1927 [1935], p. 56) justifies attention to food webs and food chains: 'The primary driving force of all animals is the necessity of finding the right kind of food and enough of it. Food is the burning question in animal society, and the whole structure and activities of the community are dependent upon questions of food-supply.'

To our knowledge, Elton (1927 [1935], p. 56) is the first to introduce the terminology 'food chains': 'There are, in fact, chains of animals linked together by food, and all dependent in the long run upon plants. We refer to these as "food chains", and to all the food-chains in a community as the "food-cycle".' Elton's 'food-cycle' has been generally replaced by 'food-web'.

In notes added to the second impression, Elton (1927 [1935], p. xxvii) remarks that 'the first food-cycle diagram was published by V.E. Shelford' in 1913. Elton does not remark that the community described by Shelford's diagram is hypothetical, but observes elsewhere (p. 57): 'Extremely little work has been done so far on food-cycles, and the number of examples which have been worked out in even the roughest way can be counted on the fingers of one hand'.

Systematic quantitative data about food chains have been assembled only in the last decade. To our knowledge, the first numerical data on the frequency distribution of chain lengths in real food webs are presented by Cohen (1978, pp. 56-59), who emphasizes the need for, but does not provide, a quantitative theory (see also Cohen 1983).

The most comprehensive, quantitative and empirically based modern presentation of theories about the length of food chains that we know of is Pimm's (1982, Chap. 6, pp. 99-130). He evaluates four hypotheses to explain why food chains rarely contain more than, roughly, five animal species (Hutchinson 1959, p. 147). Some recent perspectives on these hypotheses and their cousins are given by May (1983) and DeAngelis et al. (1983); see also Chap. II.6.

First, the *energetic* hypothesis suggested by Hutchinson (1959, p. 147) proposes that the length of food chains is limited by the inefficiency with which energy is transmitted along a chain and by the minimal energy requirements of predators at the top of a chain. This hypothesis could be interpreted to predict that food chains in ecosystems with higher primary productivity should be longer. Pimm's data do not confirm this prediction, though the ecosystems in Pimm's collection with extremely low primary productivity do have short chains. However, the energetic hypothesis could also be interpreted to predict that food chains in ecosystems with higher primary productivity can support energetically less efficient intermediate and top species without any change in chain length. Data on chain length alone, without detailed information on the energetic efficiency of the species in the chains, can neither establish nor disprove the energetic hypothesis. In a pioneering experimental study, Pimm & Kitching (1987) compared the chain lengths of artificial ecosystems with varying levels of energy input. They found no evidence of increasing chain lengths with increasing energy inputs.

Secondly, the *size* or *design* hypothesis predicts that chains should be limited in length by the requirement that a predator be larger than its prey. Pimm points out that parasites need not obey this requirement, and suggests that size or design requirements have no simple or easily testable effects on chain length.

Thirdly, the *optimal foraging* or *evolutionary shortening* hypothesis cites advantages in energetic efficiency that result from feeding low (near the primary producers) in food chains, and other energetic advantages that result from feeding high (near top predators), and suggests that the observed distributions of chain lengths result from an equilibrium of these opposing selective advantages. Although examples appear to illustrate one or another aspect of this hypothesis, precise quantitative predictions do not seem to follow from it.

Fourthly, the *dynamical stability* hypothesis argues first that, in several specific mathematical models of interacting populations, the longer the chains, the more severe the restrictions that must be imposed on the coefficients in the models for an equilibrium to be feasible or stable, and second that in certain models, those with longer food chains take longer to return to equilibrium once perturbed, so that systems with longer chains are less likely to persist in nature. The models (generally based on Lotka-Volterra equations) that support the dynamical stability hypothesis have not been independently verified. When these models are tested against data including data on chain length, it will be possible to decide what weight this hypothesis deserves as an explanation.

In addition to these four hypotheses, Kitching and Pimm (1985) describe seven environmental factors that may influence webs in phytotelmata. Phytotelmata are plant-held waters, such as occur in the axils of trees, bamboo internodal spaces, bromeliads, tree holes, and pitcher plants. The factors affecting webs include the size (surface area and volume) of the body of water, the latitude (hence climate), the size of the pool of species available to colonize the phytotelma, the evolutionary history of the host plants (see Beaver 1985), the particular host plant species, the successional stage, and altitude. Most of these factors influence webs in general. Kitching & Pimm give no quantitative predictions of the effects on chain length of changes in these factors.

Pimm (1982, appendix 6A) also presents a so-called 'null-hypothesis' about chain lengths. To our knowledge, his is the first simple quantitative model of web structure that is used to derive quantitative predictions about the frequency distribution of chain length. To describe Pimm's model, we repeat some definitions from Chaps. III.2–3. A *proper basal* species is a species that preys on no other species but is preyed on by at least one other. An *intermediate* species is a species that preys on at least one other species and is preyed on by at least one other species. A *proper top* species is a species that preys on at least one other species and is preyed on by no other species. If B_P, I and T_P are the numbers of proper basal, intermediate and proper top species in a community with L (trophic) links, Pimm constructs a predation matrix with $(B_P + I)$ rows and $(T_P + I)$ columns. All but L elements of the matrix are zero. The L elements that are equal to 1 are randomly assigned subject to three constraints: each proper top or intermediate species has at least one prey (at least one 1 in its column), each proper basal or intermediate species has at least one predator (at least one 1 in its row) and, to assure that the web is acyclic, the submatrix where intermediates prey on intermediates is strictly lower triangular. (The species are numbered from the top of the web to the bottom, contrary to the convention we adopt for the cascade model.)

For each of 13 real webs, Pimm computes the modal trophic level of each real top species (which, except for some minor details, is one greater than the modal length, defined below, of chains leading up to that species) and the modal trophic level of each (proper) top species in simulated webs generated as just described. He then adopts a conservative procedure for deciding when the vector of simulated trophic levels of (proper) top species is smaller than the vector of real trophic levels of top species. He concludes that the simulated trophic levels of top species are smaller than the real levels in a proportion P of simulations whose mean (over different real webs) is 'significantly less' (Pimm 1982, p. 104) than 0.5, though he gives no significance level, and therefore that real chains are shorter than would be expected 'at random' according to the null hypothesis.

This conclusion seems liable to two criticisms. First, assuming with Pimm that the observed webs are independent observations (we shall return to this assumption), we believe that Pimm's null hypothesis that the expected $P = 0.5$ should be replaced by the null hypothesis that P is approximately uniformly distributed between 0 and 1. P will not be exactly uniformly distributed under the null hypothesis because the number of trophic levels is a discrete, not a continuous, random variable. When we perform a one-sample Kolmogorov-Smirnov test of the null hypothesis that Pimm's 13 P values are drawn from a uniform distribution, we obtain a D_{13}-statistic of 0.389. The probability that a value that large or larger would occur by chance alone is between 0.02 and 0.05. We conclude that the data do not overwhelmingly reject Pimm's null hypothesis.

Secondly, Pimm's test of the hypothesis that the expected $P = 0.5$ is based on adding χ^2 values for each of the 13 webs; this is equivalent to treating the webs as independent. The webs are chosen from ten papers; Paine is the author or a co-author of two of these. We doubt that different webs reported by the same observer are independent in structure because the observer brings the same, usually unstated, biases to all his observations (Chaps. II.4–5, III.2–3). Under the worst dependence, Pimm's χ^2 value could be based on as few as nine independent observations. The probability that a D_9-statistic of 0.389 or larger would occur by chance alone is between 0.05 and 0.1 according to the Kolmogorov-Smirnov test.

We are less persuaded than Pimm that his null hypothesis is a bad idea. Pimm's model is in the same family, though perhaps not in the same genus, as the cascade model that we now review.

3. Terminology; The Cascade Model

This section reviews and introduces terminology, then describes the cascade model (as in Chaps. III.2–3).

A *food web* is a set of kinds of organisms and a relation that shows which, if any, kinds of organisms each kind of organism in the set eats. A *community food web* is a food web whose vertices are obtained by picking, within a habitat or set of habitats, a set of kinds of organisms (hereafter called *species*) on the basis of taxonomy, size, location or other criteria, without prior regard to the eating relations (specified by trophic *links*) among the organisms (Cohen 1978, pp. 20–21). Hereafter 'web' means 'community food web'. A *basal* species is a species that eats no other species, and a *top* species is a species that is eaten by no other species.

In the representation of a web by a directed graph or digraph (see Chap. III.2), each vertex corresponds to a (lumped trophic) species. An edge (always directed) (a, b) from vertex a to vertex b corresponds to a link from species a to species b, meaning that species b eats species a. An example of a walk in a digraph is the sequence a, (a, b), b, (b, c), c of alternating vertices and edges. The *length* of a walk is the number of edges in it. An n-walk is a walk of length n. The digraph of any web generated by the cascade model is acyclic, so no vertex (or species) can figure more than once in a walk in such a web. A *chain* is a walk from a basal species to a top species. A chain in this sense is identical to a 'maximal food chain' as defined by Cohen (1978, p. 56). An n-chain is a chain of length n, i.e. a chain with n links. The length of a chain is one less than the number of species involved in that chain.

Let S be the number of species in a web, and let C_n be the number of n-chains in an acyclic web, $n = 1, 2, \ldots, S - 1$. Algorithms for computing C_n for a given web are presented in the appendix. Chains of length greater than $S - 1$ are impossible. The *frequency distribution of chain length* is the vector $(C_1, \ldots, C_{S-1}) \equiv C$. The total number of chains in the web will be denoted

$$ C \equiv \sum_{n=1}^{S-1} C_n \,. $$

The *cascade model* assumes that species in a community web are arranged in a hierarchy, pecking order or cascade of potential feeding relations. Whether a potential feeding relation becomes an actual feeding relation is determined randomly, independently of all other potential feeding relations. The probability that a potential feeding relation becomes actual is the same for every potential feeding relation within a community, and varies inversely as the number of species in the community.

More formally, the cascade model assumes that the $S \geq 2$ species of a web may be labelled from 1 (at the bottom, subject to predation by all other species) to S (at the top, subject to predation by no other species). (In graph theory, this labelling is called a *topological sorting* (Gibbons 1985, p. 122) because for every edge (i, j) we have $i < j$.) The probability that species j feeds on species i is 0 if $j \leq i$. If $i < j$, then j feeds on i with propabiliby $p = p(S)$, i.e. with a probability between 0 and 1 that depends on S, and does not feed on i with probability $q = 1 - p$, independently for all $1 \leq i < j \leq S$. Unless a contrary assumption is explicitly given, it will be assumed that for some finite positive real number $c \leq S$, $p = c/S$, where c is a constant independent of S.

All numerical predictions of the cascade model depend on the values of the model's two parameters c and S. These two parameters, in turn, may be estimated from only two observations: the observed number, L', of links and the observed number, S', of species.

4. Frequency Distribution of Chain Length in Finite Webs; Testing Fit

As usual, $E(.)$ denotes the expectation (or mean) of the random variable enclosed in parentheses. According to the cascade model, with probability p of a random link, the expected number of n-chains in a web with S species is

$$E(C_n) = p^n q^{S-1} \sum_{k=n}^{S-1} (S-k) \binom{k-1}{n-1} q^{-k}, \quad n = 1, 2, \ldots, S-1 .$$

Proof. There is an n-chain going upward from vertex (species) i to vertex j if and only if: (a), $1 \le i \le S - n$; (b), $i + n \le j \le S$; (c), all n links on one of the $\binom{j-i-1}{n-1}$ possible walks of length n from i to j are present; (d), i is basal, i.e. no link is present from one of the $i - 1$ vertices below i to i; and (e), j is top, i.e. no link is present from j to one of the $S - j$ vertices above j. Therefore

$$E(C_n) = \sum_{i=1}^{S-n} \sum_{j=i+n}^{S} \binom{j-i-1}{n-1} p^n q^{i-1} q^{S-j} .$$

Now

$$\sum_{i=1}^{S-n} \sum_{j=i+n}^{S} = \sum_{k=n}^{S-1} \sum_{i=1}^{S-k} \quad \text{if} \quad k = j - i ;$$

therefore

$$E(C_n) = p^n q^{S-1} \sum_{k=n}^{S-1} \sum_{i=1}^{S-k} \binom{k-1}{n-1} q^{-k}$$

$$= p^n q^{S-1} \sum_{k=n}^{S-1} (S-k) \binom{k-1}{n-1} q^{-k} . \qquad \blacksquare$$

Figure III.4.1, which we discuss in more detail below, plots $E(C_n)$ as a function of n for parameter values that are typical of the webs in the sample of 62 webs analysed in Chaps. III.2–3.

This analysis leaves open a question concerning dependence, which we will answer roughly by numerical simulations of the cascade model. For typical webs, is there enough dependence between the number of chains of one length and the number of chains of another length to affect what statistical test we use to evaluate the goodness of fit between the observed and the predicted frequency distributions of chain length? In the cascade model of a web with S species, for any two different positive integers m and n, $1 \le m \ne n \le S - 1$, if C_m and C_n, the (random) numbers of chains of length m and n, were independent, then we might measure the goodness of fit of the observed to the expected frequency distributions of chain lengths by Pearson's χ^2 statistic. However, if C_m and C_n,

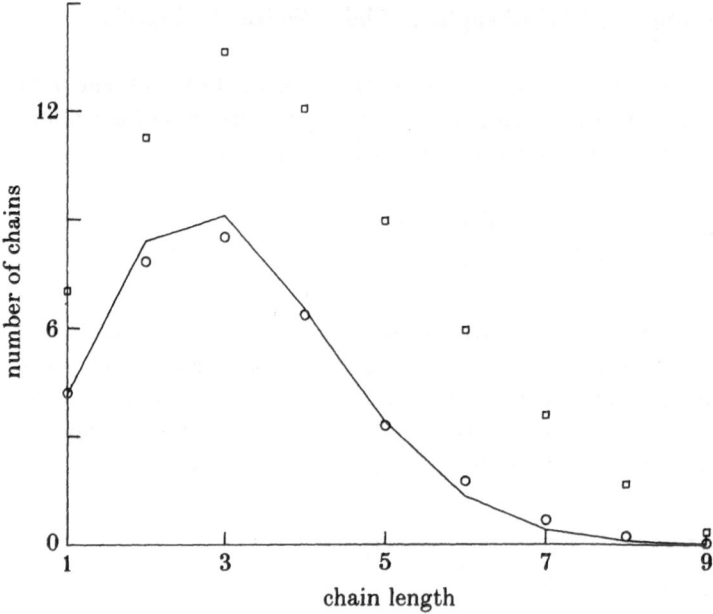

Figure III.4.1. Theoretically expected number (—) of chains of length 1 to 9 in a web of $S = 17$ species, according to the cascade model with $c = 3.75$, sample mean number (o) of chains of each length in 100 simulations of the cascade model, and sample mean plus one sample standard deviation (□) in the number of chains of each length. No chains with more than nine links occurred in the simulations; the expected total number of such chains per simulation is 0.003

$m \neq n$, were not independent, then the tabulated probability distribution of χ^2 would bear no relation to the actual probability distribution of the computed χ^2 statistic. In the case of dependence, it would be necessary to compute the correct probability distribution or find another way to measure goodness of fit.

To answer this question, we chose $S = 17$ as a typical number of species, because the mean number of species per web in the 62 webs analysed in Chaps. III.2–3 is 16.7. We chose $c = 3.75$, near the observed estimate of 3.71, so that the expected number of links per web would be 30, near the observed mean in the 62 webs of 30.95 links per web. Given these two parameters, we generated 100 random webs according to the cascade model and recorded various statistics.

The mean number, averaged over the 100 simulated webs, of chains of each length is plotted in Fig. III.4.1 along with the theoretically expected number derived above. The excellent agreement serves as a check both on the simulation and on the theoretical derivation. Also plotted in Fig. III.4.1 is the mean number plus one sample standard deviation in the number of chains of each length.

To investigate dependence among the numbers of chains of each length, we computed the dispersion matrix or variance-covariance matrix of the simulated random variables $\{C_n, n = 1, 2, \ldots, 9\}$. (No chains of length greater than nine occurred.)

Table III.4.1. Dispersion or variance-covariance matrix of the numbers of chains of each length $1, 2, \ldots, 9$ in 100 simulations of the cascade model with $S = 17$ and $c = 3.75$. (For example, the sample covariance of C_3 and C_4 was 19.85. No chains of length greater than 9 occurred)

chain length	chain length 1	2	3	4	5	6	7	8	9
1	8.08	2.42	−5.74	−8.08	−5.83	−3.85	−1.61	−0.58	−0.10
2	2.42	11.87	1.87	−1.90	−3.07	−3.39	−2.42	−1.14	−0.28
3	−5.74	1.87	26.82	19.85	16.19	7.96	1.28	−0.26	−0.14
4	−8.08	−1.90	19.85	32.64	28.02	16.16	5.29	1.12	0.06
5	−5.83	−3.07	16.19	28.02	32.13	20.00	8.31	2.49	0.37
6	−3.85	−3.39	7.96	16.16	20.00	17.70	10.26	4.14	0.72
7	−1.61	−2.42	1.28	5.29	8.31	10.26	8.52	3.98	0.80
8	−0.58	−1.14	−0.26	1.12	2.49	4.14	3.98	2.12	0.44
9	−0.10	−0.28	−0.14	0.06	0.37	0.72	0.80	0.44	0.10

Table III.4.1 gives the dispersion matrix. In general, the numbers of chains of similar length appear to be positively correlated, while the numbers of very short chains are negatively correlated with the numbers of very long chains.

To test whether $\{C_n, n = 1, 2, \ldots, 9\}$ could be treated as independent, we applied a test for independence given by Kendall & Stuart (1968, p. 271). If the $p \times p$ dispersion matrix D (for p random variables) has diagonal elements d_{ii} and determinant $\det D$ and is based on a sample of N observations, then the test statistic

$$-2(1 - [2p + 11]/[6n]) \ln \left(\det D / \prod_{i=1}^{p} d_{ii} \right)^{\frac{1}{2}N}$$

has approximately the distribution of χ^2 with $p(p-1)/2$ degrees of freedom. For the dispersion matrix in Table III.4.1, $p = 9$, $N = 100$, and we obtain a test statistic of nearly 1050 with 36 degrees of freedom. The test statistic is so large that it decisively rejects the null hypothesis that $\{C_n, n = 1, 2, \ldots, 9\}$ are independent.

We therefore measure the goodness of fit of the predicted frequencies $E(C_n)$ to the observed frequencies, for each web separately, by a Monte Carlo procedure. For brevity, let $E_n = E(C_n)$ be the expected number of chains of length n according to the cascade model and D_n the observed number in a given web. (We reserve C_n for the random variable that denotes the number of n-chains in the cascade model.) If M (for maximum) is the length of the longest chain observed in the given web, we take as data the vector

$$\boldsymbol{D} = (D_1, \ldots, D_M, 0),$$

where the final 0 is the total observed frequency of chains of all lengths greater than M (namely, none). We take as our theoretical predictions the vector of expectations computed using the values of S and c estimated by the iterative procedure in the appendix of Chap. III.3:

$$E = \left(E_1, \ldots, E_M, \sum_{h=M+1}^{S-1} E_h \right).$$

Table III.4.2 gives D and E for all 113 webs analysed here; and shows that the sum of the expected number of chains of each length, i.e. the expected total number of chains, does not, in general, equal the sum of the observed number of chains of each length, i.e. the observed total number of chains. The values of the parameters c and S used to compute E match the expected with the observed numbers of links, but these links can be arranged to yield widely varying numbers of chains.

Table III.4.2. Species, links, and numbers of chains of each length observed in 113 webs, and the cascade model's estimated parameters S, c, and expected numbers of chains of each length. Web numbers are identified in Chap. IV.

(Under 'S', the upper number for each web is the observed number of species, the lower number the estimated value of the parameter S. Under 'L', the upper number is the observed number of links, the lower number the estimated value of the parameter c. Under the number of chains of each length, the upper number is the observed number, while the lower number is the predicted number. The last positive predicted number is the number predicted for all chains of that length and longer)

web number	S	L	number of chains of length 1	2	3	4	5	6	7	8	9	10	>10
1	8	14	0	2	3	3	0	0	0	0	0	0	0
1	8.1	4.0	1.9	4.0	3.8	2.0	0.7	0.0	0.0	0.0	0.0	0.0	0.0
2	14	22	0	4	10	0	0	0	0	0	0	0	0
2	14.5	3.3	3.7	6.1	5.3	4.8	0.0	0.0	0.0	0.0	0.0	0.0	0.0
3	24	34	1	19	10	0	0	0	0	0	0	0	0
3	25.5	2.8	7.1	9.4	7.4	6.5	0.0	0.0	0.0	0.0	0.0	0.0	0.0
4	13	26	0	7	10	2	0	0	0	0	0	0	0
4	13.1	4.3	2.9	7.1	8.7	6.7	5.5	0.0	0.0	0.0	0.0	0.0	0.0
5	6	5	0	3	0	0	0	0	0	0	0	0	0
5	8.1	1.4	2.4	1.1	0.3	0.0	0.0	0.0	0.0	0.0	0.0	0.0	0.0
6	25	43	1	12	8	18	18	3	0	0	0	0	0
6	25.7	3.5	6.3	11.7	12.1	8.7	4.7	2.0	0.9	0.0	0.0	0.0	0.0
7	18	30	1	5	16	2	0	0	0	0	0	0	0
7	18.5	3.4	4.6	8.2	8.0	5.3	3.8	0.0	0.0	0.0	0.0	0.0	0.0
8	15	25	5	6	12	2	0	0	0	0	0	0	0
8	15.4	3.5	3.9	6.9	6.7	4.2	2.7	0.0	0.0	0.0	0.0	0.0	0.0
9	9	13	0	1	6	0	0	0	0	0	0	0	0
9	9.3	3.1	2.5	3.7	2.7	1.5	0.0	0.0	0.0	0.0	0.0	0.0	0.0
10	3	2	0	1	0	0	0	0	0	0	0	0	0
10	3.0	2.0	1.1	0.4	0.0	0.0	0.0	0.0	0.0	0.0	0.0	0.0	0.0
11	5	4	0	2	0	0	0	0	0	0	0	0	0
11	6.9	1.4	1.8	0.7	0.2	0.0	0.0	0.0	0.0	0.0	0.0	0.0	0.0
12	9	13	0	6	2	0	0	0	0	0	0	0	0
12	9.3	3.1	2.5	3.7	2.7	1.5	0.0	0.0	0.0	0.0	0.0	0.0	0
13	9	14	0	4	4	0	0	0	0	0	0	0	0
13	9.2	3.4	2.4	4.0	3.3	2.1	0.0	0.0	0.0	0.0	0.0	0.0	0.0
14	8	10	1	1	3	0	0	0	0	0	0	0	0
14	8.5	2.7	2.4	2.7	1.5	0.6	0.0	0.0	0.0	0.0	0.0	0.0	0.0

Table III.4.2. (Continued)

web number	S	L	\multicolumn{13}{}{number of chains of length}										
			1	2	3	4	5	6	7	8	9	10	>10
15	7	7	0	2	1	0	0	0	0	0	0	0	0
15	8.1	2.0	2.5	1.8	0.7	0.2	0.0	0.0	0.0	0.0	0.0	0.0	0.0
16	14	20	1	10	3	0	0	0	0	0	0	0	0
16	14.7	2.9	3.9	5.4	4.1	3.1	0.0	0.0	0.0	0.0	0.0	0.0	0.0
17	14	23	0	2	9	9	3	0	0	0	0	0	0
17	14.4	3.4	3.6	6.4	6.0	3.7	1.6	0.6	0.0	0.0	0.0	0.0	0.0
18	23	35	13	10	5	4	0	0	0	0	0	0	0
18	24.1	3.0	6.5	9.9	8.6	5.2	3.6	0.0	0.0	0.0	0.0	0.0	0.0
19	17	32	0	4	17	4	0	0	0	0	0	0	0
19	17.3	3.9	4.0	8.7	10.1	7.7	6.7	0.0	0.0	0.0	0.0	0.0	0.0
20	19	30	0	5	9	7	2	0	0	0	0	0	0
20	19.7	3.2	5.1	8.2	7.4	4.6	2.1	1.0	0.0	0.0	0.0	0.0	0.0
21	9	20	0	2	8	15	16	10	3	0	0	0	0
21	9.0	5.0	1.8	5.4	7.4	5.7	2.7	0.8	0.1	0.0	0.0	0.0	0.0
22	28	58	4	13	34	36	19	6	2	0	0	0	0
22	28.3	4.2	6.2	15.3	20.6	18.9	13.0	7.0	3.1	1.6	0.0	0.0	0.0
23	15	27	1	11	7	1	0	0	0	0	0	0	0
23	15.3	3.8	3.7	7.4	8.0	5.6	4.2	0.0	0.0	0.0	0.0	0.0	0.0
24	12	18	3	5	12	4	0	0	0	0	0	0	0
24	12.4	3.1	3.3	5.0	4.0	2.1	1.0	0.0	0.0	0.0	0.0	0.0	0.0
25	24	37	3	16	5	1	0	0	0	0	0	0	0
25	25.1	3.1	6.8	10.4	9.3	5.8	4.1	0.0	0.0	0.0	0.0	0.0	0.0
26	32	56	7	16	16	10	5	2	0	0	0	0	0
26	32.9	3.5	8.0	15.1	16.2	12.2	6.9	3.2	1.7	0.0	0.0	0.0	0.0
27	22	39	0	12	28	7	0	0	0	0	0	0	0
27	22.5	3.6	5.5	10.6	11.4	8.4	7.4	0.0	0.0	0.0	0.0	0.0	0.0
28	32	35	6	15	5	0	0	0	0	0	0	0	0
28	37.5	1.9	11.2	9.2	4.9	2.6	0.0	0.0	0.0	0.0	0.0	0.0	0.0
29	16	22	1	5	8	6	2	0	0	0	0	0	0
29	17.0	2.7	4.9	6.3	4.6	2.3	0.8	0.3	0.0	0.0	0.0	0.0	0.0
30	14	32	0	0	5	21	39	25	4	0	0	0	0
30	14.1	4.9	2.8	8.4	12.4	11.6	7.5	3.5	1.2	0.4	0.0	0.0	0.0
31	14	51	0	9	39	51	29	7	0	0	0	0	0
31	14.0	7.8	1.8	10.4	28.4	47.2	53.5	43.4	41.8	0.0	0.0	0.0	0.0
32	14	52	0	11	40	51	29	7	0	0	0	0	0
32	14.0	8.0	1.7	10.5	29.2	49.9	57.8	48.0	47.8	0.0	0.0	0.0	0.0
33	29	48	14	20	7	2	0	0	0	0	0	0	0
33	30.0	3.3	7.8	13.4	13.3	9.3	7.9	0.0	0.0	0.0	0.0	0.0	0.0
34	12	27	1	22	18	4	0	0	0	0	0	0	0
34	12.0	4.9	2.4	7.1	10.2	9.0	8.3	0.0	0.0	0.0	0.0	0.0	0.0
35	13	36	1	33	36	12	0	0	0	0	0	0	0
35	13.0	6.0	2.2	8.7	16.5	19.2	28.5	0.0	0.0	0.0	0.0	0.0	0.0
36	19	35	14	13	11	3	0	0	0	0	0	0	0
36	19.3	3.8	4.6	9.5	10.7	8.1	7.2	0.0	0.0	0.0	0.0	0.0	0.0
37	23	38	0	21	23	8	0	0	0	0	0	0	0
37	23.8	3.3	6.0	10.3	10.1	6.8	5.3	0.0	0.0	0.0	0.0	0.0	0.0
38	31	95	20	55	34	0	0	0	0	0	0	0	0
38	31.0	6.3	4.9	21.2	47.6	314.0	0.0	0.0	0.0	0.0	0.0	0.0	0.0
39	33	70	19	34	7	0	0	0	0	0	0	0	0
39	33.4	4.3	7.2	18.3	25.5	59.9	0.0	0.0	0.0	0.0	0.0	0.0	0.0
40	11	15	4	10	2	0	0	0	0	0	0	0	0
40	11.6	2.8	3.2	4.1	2.8	1.7	0.0	0.0	0.0	0.0	0.0	0.0	0.0
41	18	49	0	0	5	18	55	86	59	14	0	0	0
41	18.0	5.8	3.1	11.8	22.3	27.1	23.5	15.4	7.8	3.1	1.3	0.0	0.0
42	15	36	2	3	17	37	56	43	15	2	0	0	0
42	15.0	5.1	2.9	9.2	14.6	14.7	10.4	5.4	2.1	0.6	0.2	0.0	0.0

Table III.4.2. (Continued)

web number	S	L	number of chains of length 1	2	3	4	5	6	7	8	9	10	>10
43	20	38	0	16	27	16	4	0	0	0	0	0	0
43	20.3	3.9	4.7	10.3	12.1	9.6	5.6	3.7	0.0	0.0	0.0	0.0	0.0
44	12	29	0	3	19	19	7	0	0	0	0	0	0
44	12.0	5.3	2.3	7.5	11.8	11.3	7.3	4.6	0.0	0.0	0.0	0.0	0.0
45	11	20	1	10	3	0	0	0	0	0	0	0	0
45	11.1	3.9	2.6	5.6	5.9	6.1	0.0	0.0	0.0	0.0	0.0	0.0	0.0
46	19	68	3	12	59	85	84	45	13	2	0	0	0
46	19.0	7.6	2.5	14.0	37.5	64.2	77.8	70.9	50.1	28.0	18.8	0.0	0.0
47	27	50	0	1	10	22	25	0	0	0	0	0	0
47	27.6	3.8	6.5	13.5	15.5	12.3	7.4	5.4	0.0	0.0	0.0	0.0	0.0
48	13	20	0	2	7	8	2	0	0	0	0	0	0
48	13.4	3.2	3.5	5.5	4.7	2.6	1.0	0.3	0.0	0.0	0.0	0.0	0.0
49	12	20	0	8	7	1	0	0	0	0	0	0	0
49	12.3	3.6	3.1	5.6	5.3	3.1	1.7	0.0	0.0	0.0	0.0	0.0	0.0
50	14	23	0	10	8	0	0	0	0	0	0	0	0
50	14.4	3.4	3.6	6.4	6.0	5.9	0.0	0.0	0.0	0.0	0.0	0.0	0.0
51	25	46	0	4	17	9	2	0	0	0	0	0	0
51	25.5	3.8	6.0	12.4	14.2	11.1	6.5	4.5	0.0	0.0	0.0	0.0	0.0
52	20	32	2	19	4	0	0	0	0	0	0	0	0
52	20.7	3.2	5.3	8.7	8.1	8.7	0.0	0.0	0.0	0.0	0.0	0.0	0.0
53	22	31	1	19	0	0	0	0	0	0	0	0	0
53	23.4	2.8	6.5	8.6	12.4	0.0	0.0	0.0	0.0	0.0	0.0	0.0	0.0
54	14	20	1	4	6	1	0	0	0	0	0	0	0
54	14.7	2.9	3.9	5.4	4.1	2.1	1.0	0.0	0.0	0.0	0.0	0.0	0.0
55	12	18	0	7	6	0	0	0	0	0	0	0	0
55	12.4	3.1	3.3	5.0	4.0	3.0	0.0	0.0	0.0	0.0	0.0	0.0	0.0
56	10	14	0	7	2	0	0	0	0	0	0	0	0
56	10.4	3.0	2.9	3.9	2.7	1.6	0.0	0.0	0.0	0.0	0.0	0.0	0.0
57	9	19	0	5	14	10	2	0	0	0	0	0	0
57	9.0	4.7	1.9	5.2	6.6	4.8	2.1	0.7	0.0	0.0	0.0	0.0	0.0
58	17	21	1	3	3	2	3	4	2	0	0	0	0
58	18.7	2.4	5.4	5.7	3.5	1.5	0.5	0.1	0.0	0.0	0.0	0.0	0.0
59	29	61	0	34	17	1	0	0	0	0	0	0	0
59	29.3	4.3	6.3	16.0	22.0	20.6	28.2	0.0	0.0	0.0	0.0	0.0	0.0
60	33	69	1	54	33	0	0	0	0	0	0	0	0
60	33.4	4.3	7.3	18.1	24.7	56.1	0.0	0.0	0.0	0.0	0.0	0.0	0.0
61	8	10	2	3	2	0	0	0	0	0	0	0	0
61	8.5	2.7	2.4	2.7	1.5	0.6	0.0	0.0	0.0	0.0	0.0	0.0	0.0
62	11	12	1	0	3	2	0	0	0	0	0	0	0
62	12.6	2.1	3.7	3.1	1.5	0.5	0.1	0.0	0.0	0.0	0.0	0.0	0.0
63	18	75	2	50	131	100	0	0	0	0	0	0	0
63	18.0	8.8	2.0	13.9	45.1	91.9	516.8	0.0	0.0	0.0	0.0	0.0	0.0
64	19	28	7	14	0	0	0	0	0	0	0	0	0
64	20.0	3.0	5.3	7.6	11.7	0.0	0.0	0.0	0.0	0.0	0.0	0.0	0.0
65	13	25	3	17	0	0	0	0	0	0	0	0	0
65	13.1	4.1	3.0	6.9	18.3	0.0	0.0	0.0	0.0	0.0	0.0	0.0	0.0
66	10	18	0	4	8	3	0	0	0	0	0	0	0
66	10.1	4.0	2.4	5.1	5.2	3.2	1.7	0.0	0.0	0.0	0.0	0.0	0.0
67	21	62	1	8	30	48	30	6	0	0	0	0	0
67	21.0	6.2	3.4	14.2	30.1	41.3	40.9	30.9	32.2	0.0	0.0	0.0	0.0
68	22	32	4	8	20	3	0	0	0	0	0	0	0
68	23.2	2.9	6.4	9.0	7.3	4.1	2.6	0.0	0.0	0.0	0.0	0.0	0.0
69	29	73	6	4	37	36	19	2	0	0	0	0	0
69	29.1	5.2	5.5	18.0	31.3	36.6	31.9	21.8	20.8	0.0	0.0	0.0	0.0
70	14	28	0	19	18	0	0	0	0	0	0	0	0
70	14.1	4.3	3.1	7.6	9.4	13.8	0.0	0.0	0.0	0.0	0.0	0.0	0.0

Table III.4.2. (Continued)

web number	S	L	number of chains of length										
			1	2	3	4	5	6	7	8	9	10	>10
71	16	32	0	1	7	17	28	25	11	0	0	0	0
71	16.2	4.2	3.6	8.6	10.8	8.8	5.1	2.2	0.7	0.2	0.0	0.0	0.0
72	17	32	0	3	6	19	10	0	0	0	0	0	0
72	17.3	3.9	4.0	8.7	10.1	7.7	4.2	2.5	0.0	0.0	0.0	0.0	0.0
73	10	15	2	6	8	0	0	0	0	0	0	0	0
73	10.3	3.2	2.7	4.2	3.3	2.2	0.0	0.0	0.0	0.0	0.0	0.0	0.0
74	21	36	2	14	8	2	0	0	0	0	0	0	0
74	21.6	3.5	5.3	9.8	10.1	7.0	5.7	0.0	0.0	0.0	0.0	0.0	0.0
75	9	14	1	3	6	2	0	0	0	0	0	0	0
75	9.2	3.4	2.4	4.0	3.3	1.6	0.6	0.0	0.0	0.0	0.0	0.0	0.0
76	14	17	1	4	5	2	0	0	0	0	0	0	0
76	15.4	2.4	4.5	4.6	2.8	1.1	0.4	0.0	0.0	0.0	0.0	0.0	0.0
77	13	24	1	3	9	13	6	0	0	0	0	0	0
77	13.2	3.9	3.1	6.6	7.3	5.1	2.5	1.1	0.0	0.0	0.0	0.0	0.0
78	16	27	0	5	8	6	1	0	0	0	0	0	0
78	16.4	3.5	4.1	7.4	7.3	4.8	2.2	1.1	0.0	0.0	0.0	0.0	0.0
79	21	29	0	4	8	7	3	0	0	0	0	0	0
79	22.4	2.7	6.3	8.0	6.1	3.2	1.2	0.5	0.0	0.0	0.0	0.0	0.0
80	27	70	3	16	18	33	8	0	0	0	0	0	0
80	27.1	5.4	4.9	17.0	30.7	37.0	33.0	45.3	0.0	0.0	0.0	0.0	0.0
81	12	19	0	6	7	2	0	0	0	0	0	0	0
81	12.3	3.4	3.2	5.3	4.6	2.6	1.3	0.0	0.0	0.0	0.0	0.0	0.0
82	10	14	0	0	3	3	1	0	0	0	0	0	0
82	10.4	3.0	2.9	3.9	2.7	1.2	0.3	0.1	0.0	0.0	0.0	0.0	0.0
83	25	67	2	31	25	2	0	0	0	0	0	0	0
83	25.1	5.6	4.4	16.1	30.2	37.6	82.9	0.0	0.0	0.0	0.0	0.0	0.0
84	12	23	0	4	10	11	6	0	0	0	0	0	0
84	12.1	4.1	2.8	6.4	7.3	5.2	2.5	1.1	0.0	0.0	0.0	0.0	0.0
85	27	49	2	6	27	35	13	0	0	0	0	0	0
85	27.6	3.7	6.6	13.2	14.9	11.5	6.7	4.7	0.0	0.0	0.0	0.0	0.0
86	16	37	0	0	13	43	16	2	0	0	0	0	0
86	16.1	4.9	3.2	9.6	14.5	14.2	9.8	5.0	2.7	0.0	0.0	0.0	0.0
87	11	17	1	5	11	5	0	0	0	0	0	0	0
87	11.3	3.3	3.0	4.8	4.0	2.1	0.9	0.0	0.0	0.0	0.0	0.0	0.0
88	16	42	3	59	0	0	0	0	0	0	0	0	0
88	16.0	5.6	2.8	10.3	73.3	0.0	0.0	0.0	0.0	0.0	0.0	0.0	0.0
89	18	32	0	6	18	3	0	0	0	0	0	0	0
89	18.4	3.7	4.4	8.8	9.4	6.7	5.4	0.0	0.0	0.0	0.0	0.0	0.0
90	22	39	6	32	0	0	0	0	0	0	0	0	0
90	22.5	3.6	5.5	10.6	27.2	0.0	0.0	0.0	0.0	0.0	0.0	0.0	0.0
91	10	13	0	2	4	2	0	0	0	0	0	0	0
91	10.6	2.7	3.0	3.5	2.2	0.9	0.3	0.0	0.0	0.0	0.0	0.0	0.0
92	18	18	3	5	3	0	0	0	0	0	0	0	0
92	22.0	1.7	6.7	4.6	2.0	0.8	0.0	0.0	0.0	0.0	0.0	0.0	0.0
93	26	70	1	51	8	0	0	0	0	0	0	0	0
93	26.1	5.6	4.6	16.8	31.7	130.3	0.0	0.0	0.0	0.0	0.0	0.0	0.0
94	12	19	1	4	8	6	4	0	0	0	0	0	0
94	12.3	3.4	3.2	5.3	4.6	2.6	1.0	0.3	0.0	0.0	0.0	0.0	0.0
95	10	12	1	3	3	1	0	0	0	0	0	0	0
95	10.9	2.4	3.0	3.1	1.7	0.6	0.2	0.0	0.0	0.0	0.0	0.0	0.0
96	9	16	1	11	0	0	0	0	0	0	0	0	0
96	9.1	4.0	2.2	4.6	8.3	0.0	0.0	0.0	0.0	0.0	0.0	0.0	0.0
97	11	17	1	14	1	0	0	0	0	0	0	0	0
97	11.3	3.3	3.0	4.8	4.0	3.0	0.0	0.0	0.0	0.0	0.0	0.0	0.0
98	17	39	1	11	21	35	10	0	0	0	0	0	0
98	17.1	4.9	3.4	10.1	15.2	14.9	10.4	8.5	0.0	0.0	0.0	0.0	0.0

Table III.4.2. (Continued)

web number	S	L	number of chains of length										
			1	2	3	4	5	6	7	8	9	10	>10
99	48	138	14	115	98	21	0	0	0	0	0	0	0
99	48.1	5.9	8.1	31.8	66.1	93.7	315.8	0.0	0.0	0.0	0.0	0.0	0.0
100	22	59	3	27	28	28	16	3	0	0	0	0	0
100	22.0	5.6	3.9	14.2	26.6	32.7	29.2	20.0	18.2	0.0	0.0	0.0	0.0
101	6	5	1	2	0	0	0	0	0	0	0	0	0
101	8.1	1.4	2.4	1.1	0.3	0.0	0.0	0.0	0.0	0.0	0.0	0.0	0.0
102	9	27	0	7	19	24	16	6	1	0	0	0	0
102	9.0	6.7	1.3	6.3	13.0	14.9	10.3	4.3	1.0	0.1	0.0	0.0	0.0
103	23	133	1	46	260	602	769	856	621	285	88	12	0
103	23.0	12.1	1.9	19.2	92.7	284.8	624.1	1036.8	1355.1	1427.1	1230.6	877.9	937.4
104	27	62	2	21	17	22	7	0	0	0	0	0	0
104	27.2	4.7	5.5	15.8	24.4	25.4	19.6	21.2	0.0	0.0	0.0	0.0	0.0
105	10	22	0	3	6	11	4	0	0	0	0	0	0
105	10.0	4.9	2.0	5.9	8.1	6.5	3.3	1.4	0.0	0.0	0.0	0.0	0.0
106	35	73	7	44	22	6	2	0	0	0	0	0	0
106	35.4	4.2	7.7	19.1	26.1	24.7	17.6	17.4	0.0	0.0	0.0	0.0	0.0
107	10	14	1	2	5	0	0	0	0	0	0	0	0
107	10.4	3.0	2.9	3.9	2.7	1.6	0.0	0.0	0.0	0.0	0.0	0.0	0.0
108	14	20	0	11	4	0	0	0	0	0	0	0	0
108	14.7	2.9	3.9	5.4	4.1	3.1	0.0	0.0	0.0	0.0	0.0	0.0	0.0
109	21	57	0	18	40	10	0	0	0	0	0	0	0
109	21.0	5.7	3.6	13.7	25.9	32.1	66.4	0.0	0.0	0.0	0.0	0.0	0.0
110	13	23	3	7	5	0	0	0	0	0	0	0	0
110	13.2	3.8	3.2	6.4	6.6	7.2	0.0	0.0	0.0	0.0	0.0	0.0	0.0
111	19	36	2	15	17	0	0	0	0	0	0	0	0
111	19.3	3.9	4.5	9.8	11.4	17.3	0.0	0.0	0.0	0.0	0.0	0.0	0.0
112	14	17	3	8	1	0	0	0	0	0	0	0	0
112	15.4	2.4	4.5	4.6	2.8	1.5	0.0	0.0	0.0	0.0	0.0	0.0	0.0
113	11	12	1	6	2	0	0	0	0	0	0	0	0
113	12.6	2.1	3.7	3.1	1.5	0.6	0.0	0.0	0.0	0.0	0.0	0.0	0.0

We compute the difference between data and predictions by one of two measures: the sum of squared differences,

$$d_1(\boldsymbol{D}, \boldsymbol{E}) = \sum_{h=1}^{M+1} (D_h - E_h)^2 \, ,$$

or a Pearson χ^2 measure,

$$d_2(\boldsymbol{D}, \boldsymbol{E}) = \sum_{h=1}^{M+1} (D_h - E_h)^2 / E_h \, .$$

Large values of these measures of difference confound two distinct kinds of discrepancies between $\boldsymbol{D}$ and $\boldsymbol{E}$: differences in the expected and observed total numbers of chains, and differences in the expected and observed proportions of all chains that are of given lengths. However, both measures are useful in that low values of either measure signify good agreement between observation and expectation in both total numbers of chains and proportions of each length.

low values of either measure signify good agreement between observation and expectation in both total numbers of chains and proportions of each length.

To measure how likely the difference d_m, $m = 1, 2$, is to arise by chance alone according to the cascade model, we generate random strictly upper triangular adjacency matrices according to the cascade model. For S, the size of the matrix, we use the integer part of the value of S obtained by the iterative procedure in the appendix of Chap. III.3. In most but not all cases, the size of the matrix is identical to the observed number of species in the web. For c, we use exactly the value of c obtained by the iterative procedure in the appendix of Chap. III.3. Rounded values of S and c for each web are given in Table III.4.2. For each randomly generated adjacency matrix, we compute the frequency distribution of chain lengths (see the appendix of this chapter). We then combine the frequencies of all chains longer than M and compute the difference between the resulting $(M + 1)$-vector of simulated frequencies and E. Call this difference $d_m^{(i)}$ for the ith simulated web.

We take our null hypothesis to be that the difference, d_m, between the observed and expected frequency distributions is greater than 95% of randomly chosen values of $d_m^{(i)}$, i.e. that the cascade model provides a description of observed chain lengths that is poor enough to reject at the 5% level of significance. If our simulations show that a sufficiently small proportion of the simulated differences satisfy $d_m^{(i)} < d_m$, then we can reject the null hypothesis and conclude that the cascade model could not be rejected at the 5% level, and hence describes the data on chain lengths.

For each observed web, we test the goodness of fit between E and D as follows. We generate 20 random webs according to the cascade model and find the number, X_{20}, of those simulated webs for which $d_m^{(i)} < d_m$. We then consult a table (previously calculated and stored) of the binomial cumulative distribution function with parameters $N = 20$ and $p' = 0.95$ to find the probability, P, of X_{20} or fewer successes. If this probability P is less than or equal to 0.01, we reject the null hypothesis that the difference d_m between the observed and expected frequency distributions is greater than or equal to 95% of randomly chosen values of $d_m^{(i)}$ and accept E as describing D. In this case, we then go on to the next observed web. However, if $P > 0.01$, we generate another 20 random webs according to the cascade model and find the cumulative number, X_{40}, of the 40 simulated webs for which $d_m^{(i)} < d_m$. We then consult the table of the cumulative binomial distribution with parameters $N = 40$ and $p' = 0.95$ to find the probability P of X_{40} or fewer successes. Once again, if $P \leq 0.01$, we stop and accept the cascade model. If $P > 0.01$, we continue to generate additional batches of 20 random webs, up to a total of 100 random webs, until either we find a $P < 0.01$ and accept the cascade model or we are left with $X_{100}/100$ as the estimated fraction of random webs that satisfy $d_m^{(i)} < d_m$.

For every web, we record the number, N of simulated webs generated, the number, X_N, of 'successes' among the simulated webs, and either the probability P (provided $P \leq 0.01$) of X_N or fewer successes from a binomial distribution

new random webs for each observed web, with d_2, to see whether the choice of difference measure affects our conclusions.

This procedure tests the goodness of fit of E to D for an observed individual web without making any assumption that D for one observed web is independent of D for another observed web.

5. The Original Batch of 62 Webs

By using the sum-of-squares measure, d_1, of difference between observed and predicted frequency distributions, we find that 40 of 62 observed webs (65%) reject at the 0.01 significance level the null hypothesis that the cascade model's expectations fit the data worse than 95% of random webs generated by the cascade model. For brevity, we say that the cascade model describes the observed frequency distributions of chain lengths well in 40 of 62 webs. In 11 of 62 webs (18%), more than 95% of the generated random webs had chain length distributions that were closer to expectation than is the observed chain length distribution. For brevity, we say that the cascade model describes badly the observed frequency distribution of chain lengths in 11 of 62 webs (serial numbers 10, 21, 30, 37, 41, 42, 47, 53, 58, 59, 60). For the remaining 11 ($= 62 - 40 - 11$) webs, we say that the cascade model describes chain lengths moderately well (serial numbers 3, 5, 6, 9, 34, 35, 38, 39, 43, 52, 62). Figure III.4.2 plots the frequency histogram of X_N/N for the 62 webs, where (as above) N is the number of random webs generated for a given web and X_N is the number of these random webs with a chain length distribution closer to the theoretical expectations than is the observed chain length distribution. Evidently a majority of webs have X_N/N greater than or equal to 0.6.

According to the χ^2 measure, d_2, of difference between observed and predicted frequency distributions, 43 of 62 observed webs (69%) have frequency distributions that are described well by the cascade model, and 7 have frequency distributions that are described moderately well (serial numbers 3, 6, 9, 27, 47, 52, 59). The cascade model describes badly the observed frequency distribution of chain lengths in 12 of 62 webs (serial numbers 10, 21, 30, 35, 37, 38, 39, 41, 42, 53, 58, 60). In this batch of webs, the measure of difference chosen makes very little difference to the overall performance of the cascade model.

The frequency distributions of chain lengths that are described badly by the cascade model are of at least three kinds. First, in some webs, the number of chains is so small that it is not clear whether to take seriously any measure of fit (e.g. web 10 has only one chain of length 2). Second, in some webs, most of the observed chains are shorter than most of the predicted chains (e.g. webs 53, 60). Third, in some webs, most of the observed chains are longer than most of the predicted chains (e.g. webs 21, 30, 41, 42, 58).

We conclude that, when webs are considered one at a time, the cascade model predicts the observed frequency distributions of chain length well or moderately well in 50 or 51 of the 62 webs in our original batch, although no information

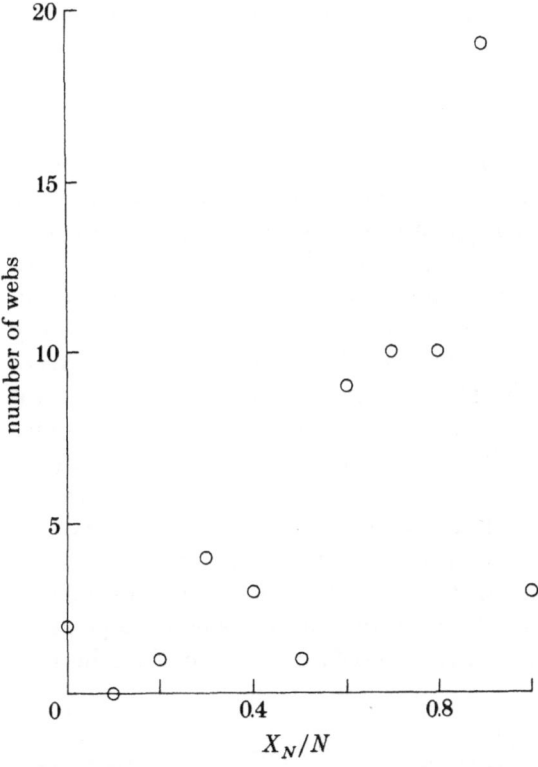

Figure III.4.2. Frequency histogram of X_N/N for 62 webs previously studied: the number of webs with X_N/N in the interval $[0.1i, 0.1(i+1))$, for $i = 0, 1, 2, \ldots, 10$. Here N is the number of random webs generated for each real web and X_N is the number of those random webs with chain length distributions closer (using d_1) to that expected from the cascade model than is that of the real web

about chain length was used in developing the cascade model or in estimating its parameters.

6. A Fresh Batch of 51 Webs

The finding that 11 or 12 of the 62 webs in the first batch have frequency distributions of chain length that the cascade model describes badly shows that there is no logical necessity for the cascade model to describe well, or moderately well, the chain lengths of an individual web. However, such bad fits do not exclude the possibility that the cascade model describes chain lengths, at least in part, because the cascade model also describes, for most webs, the other major features of web structure considered in Chaps. III.2–3. One of us therefore assembled and edited a fresh batch of 51 community webs (described in detail in Chap. IV) and extracted, for each web, the observed number, S', of species, the observed number, L', of links, and the observed frequencies D_n, $n = 1, \ldots, M$, of chain length. These webs provide a strong test of the ability of the cascade model to describe new observations.

6.1 Checking the Assumptions of the Cascade Model

One central structural assumption of the cascade model is that species are arranged in a hierarchy so that (ignoring cannibalism, as in Chap. III.2) cycles should be absent. The 51 new webs contain only one cycle of length 2 (in the web numbered 100 in the serial numbering of Briand) and no longer cycles. Cycles are rare enough that the assumption of a hierarchy is a reasonable assumption.

A second structural assumption of the cascade model is that the probability of a link from one species to another above it in the hierarchy varies inversely as the number of species in the web. This assumption implies that the total number of links in a web should be directly proportional to the total number of species: this is the species-link scaling law. Figure III.4.3 plots the number, L', of observed links as a function of the number, S', of observed species for the 51 webs in the new batch. Apart from two clear outliers with 75 and 133 links (webs numbered 63 and 103), the points appear to fall along a straight line through the origin. Web 63 is an extended version of the River Rheidol subweb depicted by Jones (1950). High connectance aside, nothing special appears to distinguish this web from the others. Web 103, one of three webs in the collection

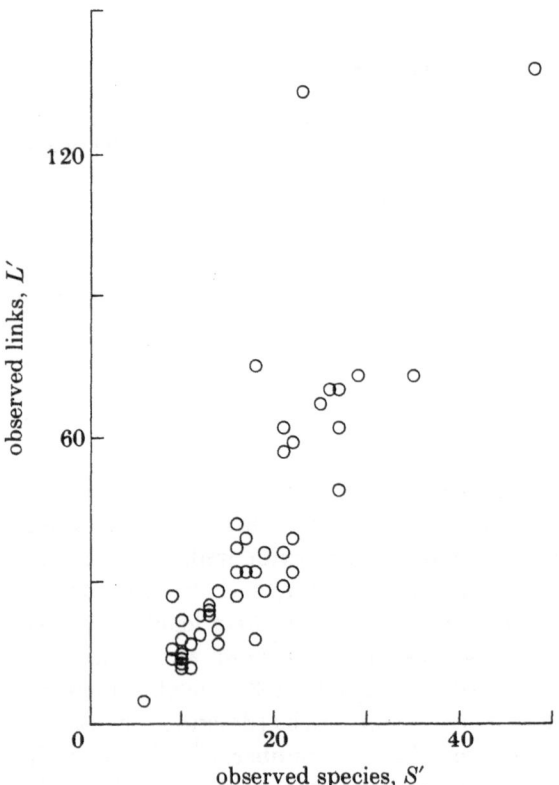

Figure III.4.3. Observed number, L', of links as a function of the observed number, S', of species in 51 webs not previously studied

of 113 provided by Petipa (1979), describes a tropical plankton community in the Pacific Ocean. This web contains the longest chain, with ten links, in the entire collection of 113 webs.

The cascade model implies that the variance of the number, L, of links is asymptotically (for S considerably greater than c) proportional to S, and Fig. III.4.3 makes it plausible that the variance of the number, L', of observed links is proportional to the number, S', of observed species. When this is true (see, for example, Snedecor & Cochran 1967, p. 168), the least squares estimate of the slope of the line through the origin is the ratio of the total number of links to the total number of species. The standard error of the slope may be estimated by a formula, also given by Snedecor & Cochran.

In the 51 webs of this batch, there are 1878 links and 874 species, giving an estimated slope of 2.1487 with an estimated standard error of 0.1220. If webs 63 and 103 are omitted, there remain 1670 links and 833 species, giving an estimated slope of 2.0048 with an estimated standard error of 0.0801. For comparison, Cohen & Briand (Chap. II.3) report, in the first batch of 62 webs, that L' is approximately proportional to S' with slope 1.8559 and estimated standard error 0.0740. Figure III.4.4 plots links, L', versus species, S', for all 113 ($= 62+51$) webs. The lack of marked difference between the slopes 1.86 ± 0.07 for the old batch of 62 webs and 2.00 ± 0.08 for the new batch of 49 webs (51 minus the two outliers), and the lack of clear separation between the old and the new sets of data points in Fig. III.4.4, suggest that underlying both batches of webs is a common direct proportionality between numbers of species and numbers of links, with a constant of proportionality near 2. Combining all 113 webs gives 1908 species, 3797 links and an estimated slope of 1.9900 ± 0.0697. Without webs 63 and 103, the slope is 1.9223 ± 0.0546.

Cohen & Briand (Chap. II.3) remark that the 62 webs available to them do not exclude a slightly nonlinear relation, as noted by Briand (Chap. II.5), between species and links, i.e. a relation of the form $E(L) = aS^b$ with b slightly different from 1. They find that a graph of $L'^{3/4}$ against S' looks very nearly linear through the origin. The same caveat and observation hold here. The use of the ordinary linear least squares method to regress $\log L'$ on $\log S'$ for all 113 webs gives the allometric model $L = 0.6713S^{1.3559+\varepsilon}$, where ε is the error term, or (taking $1/1.3559 \approx \frac{3}{4}$) $L^{3/4}$ proportional to S. The parameters obtained by this procedure are not the least squares estimates for the nonlinear allometric model in the original scales of L and S. Scatter plots (not shown) of the residuals (observed links L' minus predicted) as a function of S' show very little difference between the fitted allometric model and the linear model $L = 1.9900S + \varepsilon$. The sum (rounded to three significant figures) of the absolute residuals of the allometric model, namely 897, is smaller than the corresponding sum for the linear model, namely 999. The sum of the squared residuals of the allometric model (20 000) is also smaller than the sum of the squared residuals of the linear model (21 400). The data thus suggest that a relation between $E(L)$ and S that is mildly nonlinear for the observed range of species may be more precise than a simple proportionality. The exact relation between $E(L)$ and S deserves further empirical and theoretical investigation.

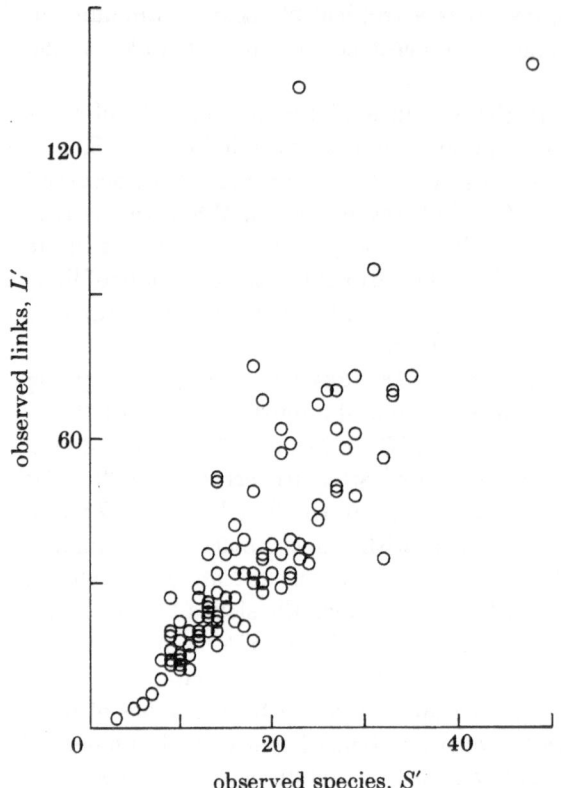

Figure III.4.4. Observed number, L', of links as a function of the observed number S', of species in all 113 webs

However, taking $E(L)$ as proportional to S does not do serious violence to the data. Moreover, in this paper, we estimate c independently for each web rather than assuming c to be constant for all webs. Hence this empirical test of the cascade model is less sensitive to how many links there are than to how the links that do occur are connected into chains.

6.2 Testing the Predictions of Chain Length

On the basis of the rarity of cycles and the near-proportionality shown in Fig. III.4.3, we conclude that the underlying assumptions of the cascade model are approximately satisfied by (nearly all of) the new batch of webs. As with the old batch, for each web in the new batch, we estimate the parameters S and c (given in Table III.4.2 after rounding), compute the expected frequency of chains of each length, and measure the goodness of fit between observed and predicted frequencies by the procedure described in section 4.

From the sum-of-squares measure, d_1, of difference between observed and predicted frequency distributions, we find that the cascade model describes well 36 of 51 observed webs (71%) and moderately well 10 webs (serial numbers 63, 68, 70, 72, 77, 85, 86, 93, 96, 103). In 5 of 51 webs (18%), the cascade model describes the observed frequency distribution of chain lengths badly (serial numbers 65, 71, 88, 90, 97). The outlying webs 63 and 103 are *not* among these badly described

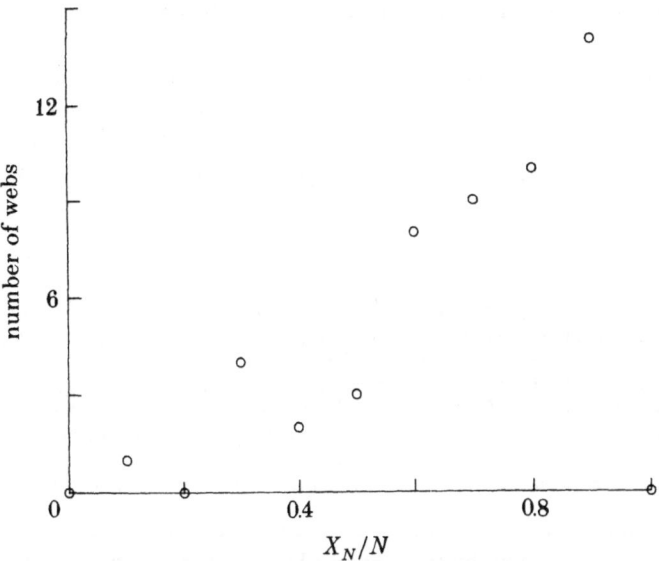

number of webs

X_N/N

Figure III.4.5. Frequency histogram of X_N/N for 51 webs not previously studied: the number of webs with X_N/N in the interval $[0.1i, 0.1(i+1))$, for $i = 0, 1, 2, \ldots, 10$. Here N is the number of random webs generated for each real web and X_N is the number of those random webs with chain length distributions closer (using d_1) to that expected from the cascade model than that of the real web

five webs. Figure III.4.5 plots the frequency histogram of X_N/N for the 51 webs, where (as above) N is the number of random webs generated for a given web and X_N is the number of these random webs with a chain length distribution closer to theoretical expectations than is the observed chain length distribution. As in Fig. III.4.2, a majority of the webs have X_N/N greater than or equal to 0.6.

According to the χ^2 measure, d_2, of difference between observed and predicted frequency distributions, 34 of 51 observed webs (67%) have frequency distributions of chain length that the cascade model describes well. Twelve webs have frequency distributions that the cascade model describes moderately well (serial numbers 63, 65, 68, 70, 72, 77, 86, 87, 96, 97, 99, 106). The cascade model describes the observed frequency distribution of chain lengths badly in 5 of 51 webs (serial numbers 71, 85, 88, 90, 93). In this batch of webs, as in the first, which measure of difference we choose makes very little difference to the overall performance of the cascade model.

As in the original batch of 62 webs, in this new batch sometimes more short chains are observed than expected (e.g. webs 65, 88, 90) and sometimes more long chains are observed than expected (e.g. webs 71, 85).

Table III.4.3 lists, for all 113 webs, the number of random webs generated and the number of those random webs with chain length distributions closer to the expected than that of the real web. For the sum-of-squares measure of difference, d_1, all 74 real webs for which fewer than 100 random webs were generated fitted the cascade model's predictions well. In addition, webs 48 and

87, for each of which 100 random webs were generated, also fitted the cascade model's predictions well.

We conclude that, considering webs one at a time, the cascade model predicts the observed frequency distributions of chain length well, or moderately well, in 46 of the 51 webs in a new batch of webs not previously used to calibrate the model. This success rate is slightly higher than that of the cascade model with the original batch of 62 webs.

Table III.4.3. Characteristics of 113 webs (Web numbers are identified in Chap. IV, and are the same in all previous joint publications of Briand & Cohen. d_1 measures the difference between observed and predicted frequency distributions of chain length by the sum of squared differences; d_2, by a Pearson χ^2 function; see text. N is the number of random webs generated. X is the number of random webs with frequency distributions of chain length closer to that predicted theoretically than is the observed distribution. Variability: 0, unclassified; 1, fluctuating; 2, constant. Dimension: 0, unclassified; 2, two-dimensional; 3, three-dimensional. Productivity: 0, unclassified; 1, low productivity; 2, high productivity. Man: 0, absent from web; 1, present in web.)

web number	d_1 N	X	d_2 N	X	variability	dimension	productivity	man
1	20	6	20	10	0	0	0	1
2	40	33	40	33	1	0	0	0
3	100	94	100	92	1	2	0	0
4	20	5	20	11	1	0	0	0
5	100	92	60	51	0	0	2	0
6	100	94	100	93	1	0	0	1
7	60	50	20	14	0	0	0	1
8	20	6	20	13	1	0	2	1
9	100	92	100	93	0	0	0	0
10	100	100	100	100	1	2	0	0
11	40	33	60	49	1	2	0	0
12	20	12	20	12	1	2	0	0
13	20	8	20	12	1	2	0	0
14	20	13	20	11	0	0	0	0
15	20	14	20	10	1	0	0	0
16	40	30	20	14	1	0	2	0
17	20	14	20	14	0	3	0	0
18	20	7	40	30	0	0	0	1
19	40	31	40	32	1	3	1	0
20	20	13	20	15	0	3	1	0
21	100	99	100	100	0	3	0	0
22	20	11	20	14	1	0	0	0
23	20	14	40	32	1	2	0	0
24	20	14	40	31	1	3	0	0
25	40	29	20	15	1	3	0	0
26	20	0	20	0	1	0	0	0
27	60	51	100	94	1	3	2	0
28	40	30	40	33	1	0	0	0
29	40	31	60	51	0	3	1	0
30	100	96	100	99	0	3	1	1
31	20	12	40	28	0	3	0	0
32	20	13	20	13	2	3	0	0
33	60	52	40	32	2	0	0	0
34	100	92	40	33	2	2	0	0
35	100	92	100	97	0	2	0	0
36	20	13	80	69	0	0	0	0

Table III.4.3. (Continued)

web number	d_1		d_2		variability	dimension	productivity	man
	N	X	N	X				
37	100	98	100	96	2	0	0	0
38	100	94	100	96	2	0	0	0
39	100	94	100	96	2	0	0	0
40	60	51	40	33	2	3	0	0
41	100	100	100	100	2	3	1	0
42	100	100	100	100	2	3	2	0
43	100	94	80	67	2	3	0	0
44	40	27	20	13	2	0	2	0
45	60	49	20	15	2	2	0	0
46	20	6	20	9	0	3	1	0
47	100	96	100	92	2	0	0	0
48	100	88	60	52	1	0	0	1
49	20	8	20	13	1	0	0	1
50	20	13	20	15	1	2	0	0
51	20	14	20	13	0	0	0	0
52	100	92	100	91	1	2	0	0
53	100	99	100	96	1	2	0	0
54	20	8	20	6	0	0	0	0
55	20	13	40	30	1	2	2	0
56	60	50	40	31	1	2	0	0
57	60	52	20	15	0	0	2	0
58	100	96	100	100	1	0	0	0
59	100	98	100	95	1	3	0	0
60	100	97	100	99	1	3	0	0
61	20	0	20	1	1	2	1	0
62	100	93	100	84	1	2	1	0
63	100	91	100	93	0	2	0	0
64	60	52	60	52	0	2	0	0
65	100	97	100	93	0	2	0	0
66	20	7	20	6	0	2	0	0
67	40	25	20	11	0	0	0	0
68	100	91	100	91	1	3	0	1
69	20	14	20	8	1	0	0	0
70	100	92	100	92	1	0	0	0
71	100	99	100	99	1	3	0	1
72	100	90	100	95	1	3	0	0
73	20	13	20	15	1	3	0	0
74	20	9	20	9	1	2	0	0
75	20	7	20	6	1	3	0	0
76	20	11	20	10	1	0	1	0
77	100	90	100	89	2	0	0	1
78	20	3	20	8	2	0	2	1
79	40	33	60	48	1	0	0	0
80	20	15	20	12	1	0	0	0
81	20	11	20	9	0	0	1	0
82	20	15	60	51	1	0	0	0
83	60	50	60	51	1	0	1	0
84	20	14	60	50	1	0	0	0
85	100	95	100	97	1	0	2	0
86	100	93	100	94	1	3	0	1
87	100	87	100	93	0	0	1	0
88	100	96	100	98	0	2	0	0
89	60	51	40	30	0	3	0	0
90	100	97	100	97	1	2	0	0
91	20	13	40	28	1	3	0	0

Table III.4.3. (Continued)

web	d_1		d_2		variability	dimension	productivity	man
number	N	X	N	X				
92	20	12	20	7	0	2	1	0
93	100	93	100	96	1	2	1	0
94	20	15	60	51	1	2	1	0
95	20	7	20	7	1	2	1	0
96	100	94	100	93	1	2	1	0
97	100	96	100	93	1	2	1	0
98	60	51	20	14	0	2	1	0
99	60	52	100	93	0	2	1	0
100	20	8	20	12	0	2	1	0
101	20	7	20	9	1	0	0	0
102	20	12	20	12	2	3	1	0
103	20	12	40	33	2	3	1	0
104	20	15	20	12	0	2	0	0
105	20	12	20	7	1	2	0	0
106	100	89	100	93	1	2	0	0
107	20	15	20	10	1	2	0	0
108	60	49	60	52	1	2	0	0
109	20	15	40	32	1	2	0	0
110	20	11	20	11	1	2	0	0
111	20	14	40	32	1	2	0	0
112	20	13	20	12	1	0	0	0
113	40	32	20	11	1	0	0	0

7. Does the Cascade Model Predict the Moments of Chain Length?

After examining Table III.4.2 in a previous draft of this chapter, S. L. Pimm
(personal communication, 3 September 1985) suggested that the cascade model
does not predict adequately the variance and kurtosis of the distribution of chain
lengths. He allowed that the cascade model may predict roughly the mean chain
length, according to Table III.4.2.

Direct comparisons of the mean and variance of the observed chain lengths
with the corresponding quantities calculated from the expected numbers of
chains of each length shown in Table III.4.2 confirm Pimm's observations regarding the first two moments. However, we claim that to evaluate the cascade
model's ability to predict the moments of chain length the expected numbers in
Table III.4.2 may not be the right numbers to compare with the observed. We
will explain what calculations are required, although they remain to be done.

In computing numerically the mean and variance from the observed and expected numbers of chains of each length, separately for each web in Table III.4.2,
we truncate (i.e. ignore) all predicted frequencies for chains of length 9 or greater.
This truncation lowers the predicted mean and variance of chain length. The effect is small for all webs other than the exceptional web 103 because, for the
remaining 112 webs, the expected number of chains of each length greater than
or equal to 9 is less than 0.05. (We do not cumulate all predicted frequencies

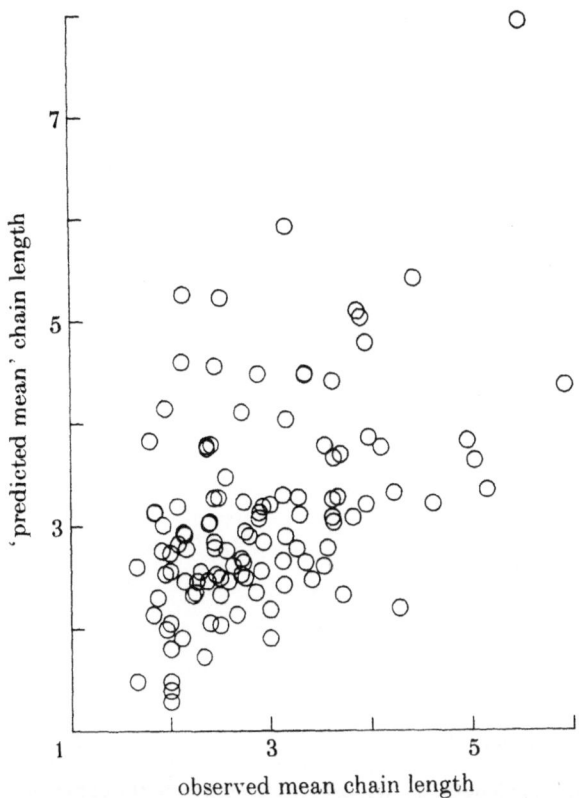

Figure III.4.6. 'Predicted mean' chain length, i.e. mean calculated from the expected numbers of chains of each length according to the cascade model, as a function of the observed mean chain length in 113 webs. The points fall about a line of slope one through the origin. See text for an explanation of why the 'predicted mean' is not the mean chain length predicted by the cascade model

of chains longer than the largest observed, as we did in testing goodness of fit between observed and predicted frequencies.)

Temporarily, we shall call the mean calculated from the theoretically expected numbers of chains of each length the 'predicted mean', and the variance calculated from the theoretically expected numbers of chains of each length the 'predicted variance'. The terminology is misleading, for reasons we shall explain.

The scatter plot (Fig. III.4.6) of 'predicted means' against the observed means clusters around a line of slope one through the origin. The observed mean chain lengths exceed the 'predicted means' in 50 of 113 webs. The 'predicted means' of the cascade model do reasonably well in predicting the observed mean chain length, as Pimm conceded.

In contrast to the acceptable performance of the 'predicted mean', the observed variance of chain length exceeds the 'predicted variance' in only two of 113 webs. Most points in the scatter plot (Fig. III.4.7) of 'predicted variance' against observed variance lie well above a line of slope one through the origin. This finding confirms Pimm's suggestion that chain lengths observed for a single web generally have a smaller variance than the 'predicted variance'.

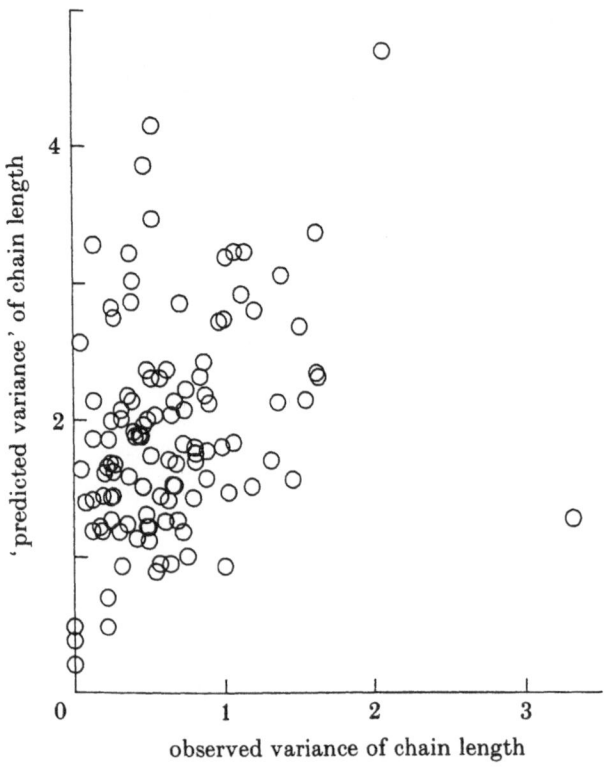

Figure III.4.7. 'Predicted variance' of chain length, i.e. variance calculated from the expected numbers of chains of each length, as a function of the observed variance of chain length in 113 webs. All but two of the points fall above a line of slope one through the origin. See text for an explanation of why the 'predicted variance' is not the variance of chain lengths predicted by the cascade model

However, this finding does not imply that the cascade model predicts the variance of chain lengths badly. Also, unfortunately, the acceptable performance of the 'predicted mean' does not imply that the cascade model predicts the mean of chain lengths well. It is not possible to infer the mean or variance of chain length in one realization of the cascade model with a finite number of species from the expected numbers of chains of each length, averaged over all realizations, which are given in Table III.4.2.

The 'predicted mean' and 'predicted variance' are (except for the truncation of chains of length 9 or greater) the mean and variance of a distribution in which the relative frequency of chains of length n is

$$E(C_n)/E(C) ,$$

where, as before, C_n is the number of chains of length n and C is the total number of chains. As explained in Chap. III.5.3, for finite S this distribution does not describe the chain length distribution of a single web randomly generated by the cascade model, but rather describes the distribution of the pooled chains from many webs generated by the cascade model with a fixed c and S.

The proper theoretical mean to compare with the observed mean chain length is (again ignoring truncation and conditional on $C > 0$)

$$E\left[\sum_k kC_k/C\right] .$$

The proper theoretical variance to compare with the observed variance is (ignoring truncation and assuming $C > 0$)

$$E\left[\sum_k k^2 C_k/(C-1) - \left(\sum_k kC_k\right)^2 /[C(C-1)]\right] .$$

In addition to the difference between $E(C_n/C)$ and $E(C_n)/E(C)$, there are correlations between C_m and C_n, $m \neq n$, illustrated by Table III.4.1, which influence the theoretical varance of chain length but not the 'predicted variance'. This additional discrepancy may explain why the 'predicted variance' (Fig. III.4.7) does worse in describing the variance of observed chain lengths than the 'predicted mean' (Fig. III.4.6) does in describing the mean of observed chain lengths.

It follows from the results of Chap. III.5.4 that the corresponding theoretical and 'predicted' moments have the same limit for large S. However, for any finite S, the corresponding theoretical and the 'predicted' moments need not agree. We are not able analytically to compute the theoretical mean or variance, or higher moments, of chain length according to the cascade model for finite S. It may be impossible to do so. Simulation, observed web by observed web, would make it possible to compare the observed mean and variance of chain lengths with the mean and variance in each of, say, 100 simulations. We have yet to carry out this computation.

Because of the success of the cascade model according to the measures of goodness of fit that we have used so far, we expect that the observed moments should not fall far in the tail of the distributions of the simulated moments. The theoretical moments could not be systematically and grossly different from the observed if the simulated distributions of chain lengths are usually near the observed distributions of chain lengths. However, we have not conclusively demonstrated that the moments of chain lengths according to the cascade model correspond well to the moments of observed chain lengths in real webs.

8. Trying to Explain the Cascade Model's Failures

In this section, we seek characteristics of webs that explain why the cascade model's predictions sometimes fit badly the observed frequency distributions of chain length. We find that bad fits occur far more often than expected among webs in which the mean length of chains is either unusually large (more than four links) or unusually small (less than two links). Twenty-one other characteristics do not appear to be associated with bad fits.

First, we explain why we do not use the conventional statistical tools of hypothesis testing; we then present our descriptive analyses.

Throughout, we have been sceptical of the assumption that our observed webs are a random sample from some statistical ensemble of webs. One reason for scepticism is that webs reported by the same author sometimes share idiosyncrasies that differentiate them from webs reported by others. Sixty-one of our 113 webs were described by distinct observers or teams (two sets of observers are considered distinct here if they have no member in common). The remaining 52 webs were reported by 20 distinct observers or teams, each contributing between two and five webs; there is therefore likely to be dependence among the webs.

A second reason for scepticism is that field ecologists with special training in some taxon (birds or insects or fishes) or in some habitat (lacustrine or marine intertidal or tropical montane) pick communities in which their special training can be used, rather than at random. Until it is shown that the properties of webs are invariant with respect to major taxa, habitats and other characteristics that may bias ecologists' choices of webs to study, it seems implausible a priori to regard any given batch of webs as a random sample of webs from the world.

If the webs *were* a random sample from a cascade model ensemble, then the frequency histograms in Figs. III.4.2 and 5 should approximate histograms sampled from the uniform distribution, which is a horizontal straight line. Under the assumption of random sampling of webs, it would be valid to use the Kolmogorov-Smirnov test (Kendall & Stuart 1973, p. 469) to assign a probability value to the deviation between the sample and uniform cumulative distribution functions. Denoting the test statistic by D (do not be confused with our notation above for the observed total number of chains), with a subscript that gives the sample size, we compute for the first batch of webs $D_{62} = 0.4403$, for the second batch $D_{51} = 0.4127$ and for all webs combined $D_{113} = 0.4142$. These values are all far beyond the 0.01 critical values for the corresponding sample sizes. Because we regard the assumption of random sampling with scepticism, we also regard with scepticism the 'significance' of this rejection of the fit of predicted to observed chain length distributions in the collection of webs as a whole.

Nevertheless, 16 or 17 of 113 webs (11 or 12 in the first batch, 5 in the second) individually have chain lengths that the cascade model describes badly. We now seek a simple explanation for these bad fits in terms of the characteristics of webs.

S. L. Pimm (personal communication, 3 September 1985) suggested that the cascade model describes worse the chain length distributions of webs with large numbers of species. To examine this suggestion, we identified the 45 webs with more than 17 species as 'above average' in size. (The average number of species per web in 113 webs is 16.9.) We also identified the 19 webs with more than 24 species as 'large' in size.

As Fig. III.4.6 shows, most webs have mean chain lengths of two to four links. We defined the 12 webs with mean chain length less than two links (webs numbered 28, 33, 39, 40, 53, 64, 65, 88, 90, 96, 101, and 112) and the 10 webs with mean chain length greater than four links (webs numbered 21, 30, 41, 42, 46, 47, 58, 71, 86, and 103) to be webs with 'extreme mean chain length'.

As Fig. III.4.7 shows, most webs have a variance of chain length that is less than 1. We defined the 22 webs with variance greater than or equal to 1 to be webs with 'high variance of chain length'. We also defined the 17 webs with variance less than 0.25 to be webs with 'low variance of chain length'.

For all 113 webs, we determined four characteristics in addition to trophic structure: dimension, variability and productivity of the environment, and the presence of man in the web (Table III.4.3).

A web is classified as having dimension 2 if it occurs in an environment that is essentially flat, such as grassland, a sea or lake bottom, a stream bed or the rocky intertidal zone. A web is classified as having dimension 3 if it occurs in a solid environment, such as the pelagic water column or forest canopy. Webs that could not clearly be assigned dimension 2 or 3 are shown in Table III.4.3 as having dimension 0.

As in Chap. III.3, the variability of a web's environment is classified as 'fluctuating' or 'constant'. The environment is 'fluctuating' if the original report indicates temporal variations of substantial magnitude in temperature, salinity, water availability or any other major physical parameter. The magnitude, and not the predictability, of the variations is the criterion of classification. In this paper we apply stricter criteria than previously for deciding whether an environment is fluctuating or constant. Whereas previously webs 1 to 28 and 48 to 62 were classified as from fluctuating environments, while webs 29 to 47 were considered to be from constant environments, we now regard a number of webs from each former category as unclassified. These are shown by 0 in Table III.4.3.

In several instances, the original observers measured and reported the net primary productivity of the ecosystems they studied. For such cases, we classify the productivity of a web as *low* if it falls below $100 \, \mathrm{g \, C \, m^{-2} \, a^{-1}}$, and as *high* if it exceeds $1000 \, \mathrm{g \, C \, m^{-2} \, a^{-1}}$. When productivity is unknown or has an intermediate value, we treat it as unclassified (shown by 0).

Man is present in a web if explicitly recorded as one of the species, and is absent otherwise.

We then cross-classified the webs by 22 pairs of dichotomous criteria. One member of each pair was bad fit between predicted and observed frequency distributions of chain length ($X_N/N > 0.95$, with the sum-of-squares measure of difference, d_1) against not a bad fit. Another member of the pair was selected from this list of dichotomies: above-average number of species (more than 17 observed species) versus average or below number of species (17 or fewer observed species); large number of species (more than 24 observed species) against not large number of species (24 or fewer observed species); high value (greater than 3.6) of the parameter c against low value ($c \leq 3.6$); extreme mean chain length against not extreme; high variance of chain length against not high; low variance of chain length against not low; man absent against man present; dimension unclassified against dimension known; dimension 2 against dimension not 2; dimension 3 against dimension not 3; dimension 2 against dimension 3; environment not classified against environment fluctuating or constant; environment fluctuating against environment not fluctuating; environment constant against environment not constant; environment fluctuating against environment constant; productivity unclassified against productivity low or high; productivity low against productivity not low; productivity high against productivity not high; productivity low against productivity high; dimension 2 and fluctuating against dimension 3 and constant; no basal-top links against one or more basal-top links; one or fewer basal-top links against more than one basal-top link. (The last two dichotomies explore the possibility that the webs with anomalously few basal-top links, apparent in Figs. A.3.2–3 on pp. 35–36, might also be those badly described here by the cascade model.) Some of these cross-classifications involve all 113 webs; others involve fewer (for example, only 34 webs are either dimension 2 and fluctuating or dimension 3 and constant).

For each cross-classification, we compute the χ^2 measure of association corrected for continuity (Snedecor & Cochran 1967, p. 217). If we could accept the doubtful assumption that the webs are a random sample, we could assign a level of statistical significance to the computed values of χ^2 with one degree of freedom. Under this assumption, the critical value for significance at the (very weak) 10% level is 2.71. Only three of the 22 values of χ^2 exceed this level: $\chi^2 = 4.55$ for the cross-classification with dimension not classified, $\chi^2 = 5.45$ for the cross-classification with low variance of chain length, and $\chi^2 = 25.33$ for the cross-classification with extreme mean chain length. The first two of these χ^2 values do not exceed the one percent significance level. The third is very large. Table III.4.4 shows the counts of bad and not bad fits cross-classified according to whether or not the mean chain length is extreme.

Table III.4.4. Cross-classification of 113 webs according to fit (based on d_1) between observed and predicted frequency distributions of chain length, and extreme values of mean chain length. (χ^2 with one degree of freedom (corrected for continuity) $= 25.3265$)

goodness of fit	mean chain length	
	≥ 2 and ≤ 4	< 2 or > 4
not bad	86	11
bad ($X_N/N > 0.95$)	5	11

When we carry out the same 22 cross-classifications with bad fit based on d_2, which is the χ^2 measure of difference between observed and predicted chain length distributions, only two of the 22 values of the association χ^2 exceed the 10% critical value: $\chi^2 = 4.60$ for the cross-classification with high c, and $\chi^2 = 16.92$ for the cross-classification with extreme mean length of chains. The former value does not exceed the 2.5% significance level. The latter value far exceeds the 1% significance level.

We conclude that a single dichotomy, extreme mean lengths of chains, explains at least partly why the cascade model's predictions sometimes fit badly the observed frequency distributions of chain length. This finding does not exclude the possibility that a more elaborate stratification of webs by combinations of other characteristics could yield another, and perhaps better, explanation of the bad fits (Mantel 1982). However, we have not explored possible explanations based on more elaborate combinations of characteristics. Table III.4.3 provides raw data for a more sophisticated analysis.

We now speculate briefly on how the deviations between the observed and predicted frequency distributions of chain lengths could arise. To explain the excess numbers of observed long chains relative to the numbers expected, suppose that, instead of describing all species and links in a community, as we assume, an observer initially samples a link at random and then follows a chain containing that link up to a top species and down to a basal species; and then samples another link at random from those not previously recorded and repeats the procedure. The longer a chain is, the more links it contains, and therefore the more likely it is to be sampled by this procedure. This sampling procedure would produce an observed excess of long chains compared to sampling in which each chain is sampled randomly.

To explain the excess numbers of observed short chains relative to the numbers expected, suppose that, as above, an observer picks a link at random and finds some of the other (if any) links in the same chain but, wary of the bias of sampling chains in proportion to their length, interrupts recording the entire chain after a small number of links. This hypothetical procedure would selectively sample long chains at first and would then selectively break the long chains into short chains, producing an observed excess of short chains compared to sampling in which each chain is sampled randomly.

A plausible model of the process of observation that would not explain an observed excess of either long or short chains is to suppose that an observer attempts to record all links, but has a probability ε (for 'error'), $0 < \varepsilon < 1$, of failing to observe or record any given link, independently and identically for all links. The recorded web will then be identical to that of a cascade model in which the true probability $p = c/S$ of an edge is replaced by the recorded probability $p' = p(1 - \varepsilon)$. The mean length of chains will be reduced by these errors of omission, but conditional on the net probability, p', that a link occurs and is recorded, the distribution of the expected number of chains of each length will be as predicted by the cascade model with parameter p'.

The original reports of webs rarely describe the sampling procedures by which the links are determined. Different investigators may use different sampling procedures. It is not possible to prove, from the original reports, either of the above explanations for deviations from the predictions of the cascade model. Still, it is some comfort that simple explanations exist.

9. Discussion and Conclusion

Here we review the accomplishments of this chapter, relate them to previous work, and indicate some useful further efforts.

9.1 Accomplishments of This Chapter

From an exact analysis of the cascade model, we derive the expected number of chains of each length in a web with any finite number, S, of species. Simulations of the cascade model demonstrate substantial dependence among the numbers of chains of different lengths. Because of the dependence, we develop a Monte Carlo method of evaluating the goodness of fit between the numbers of chains observed in an individual web and the numbers expected from the cascade model.

Without fitting any free parameters, and with the use of no direct information about chain lengths other than that implied by the total number of species and the total number of links in a web, the cascade model describes acceptably the observed numbers of chains of each length in all but 16 or 17 of 113 real webs. The cascade model describes well, in the technical sense defined in section 5, the chain lengths of 40 or 43 of the 62 webs previously used to test the cascade model, and well or moderately well, again in the technical sense, the chain lengths of all but 11 or 12 of these webs. In a fresh batch of 51 webs, the numbers of links are very nearly proportional to the numbers of species (apart from two outlying webs). The constant of proportionality is consistent with that in the original 62 webs. This finding independently verifies the species-link scaling law (Cohen & Briand 1984; Chap. II.3). The cascade model describes well the chain lengths of 34 or 36 of the 51 webs, and well or moderately well all but 5 of these webs. When the collection of webs is viewed as a whole, the cascade model describes adequately the mean chain lengths.

The poor fit of the cascade model to 16 or 17 webs is associated with one characteristic of the webs, namely, an unusually large (more than four links) or an unusually small (fewer than two links) mean length of chains.

In Chaps. III.2–3, we evaluated the cascade model's fit to the data on the proportions of each kind of species and link largely by visual inspection of graphical displays. Even measured by that very crude procedure, the fit between predictions and observations was not always good, e.g. for the proportions of basal-top links. Here, in Chap. III.4, we examine a much finer aspect of web structure than in Chaps. III.2–3, namely, the frequency distribution of chain lengths, and we use far more delicate measures of goodness of fit. A priori, the apparent performance of the cascade model should be worse than in Chaps. III.2–3. We consider it significant that the approximation between observed and predicted frequency distributions of chain length, though far from perfect, is as good as it is.

9.2 Relation to Previous Work

This chapter offers three novelties in ecological theory. First, this chapter presents, to our knowledge, the first exactly derived theory of the length of food chains. The only previous quantitative model to predict chain length (Pimm 1982) has been simulated but not analysed mathematically. Secondly, this chapter represents, we believe, the first instance in which an ecological model that was initially developed to explain an aspect of webs different from chain length (namely the proportions of species and links of various kinds) is used to predict chain lengths quantitatively. Thirdly, this chapter gives the first quantitative predictions (obtained either by simulation or by analysis) of the entire frequency distribution of chain length. Pimm (1982, Chap. 6) considers only the modal trophic level of top species.

Although the cascade model is the first to be analysed exactly in the detail given here, it is one of a family of similar models that have been proposed for webs. Cohen's (1978, p. 60) model 5 proposes that webs be generated by constructing a matrix with a number of rows equal to the observed number of prey (basal plus intermediate species), a number of columns equal to the observed number of predators (intermediate plus top species), and a number of 1-elements equal to the observed number of links, all other elements of the matrix being 0. According to this model 5, the positive elements of the 'predation matrix' (a condensed adjacency matrix) are to be distributed randomly.

From comparisons of real food webs with simulations of model 5 and other similar models, Cohen (1978, p. 92) 'concluded that the high observed frequency of arrangements of niche overlap that can be represented in a one-dimensional niche space does not result from the operation, within the framework of several plausible models, of chance alone', i.e. that the species' feeding relations have a one-dimensional ordering.

The null model of Pimm (1982, Appendix 6A) adds to Cohen's model 5 the constraints that each prey have a predator and each predator a prey, and that the intermediate species be in a strict hierarchy or cascade. Such a hierarchy or cascade is a natural interpretation of Cohen's finding that feeding relations have a one-dimensional ordering. Sugihara (1982, 1984, §3.1.2) also discusses the importance of a hierarchical ordering in assembly rules for food webs, but does not analyse the lengths of food chains.

When we proposed the cascade model (Chap. III.2), we had not read Appendix 6A of Pimm (1982) because we were considering questions other than the length of chains. Whereas Pimm's null model takes as given the numbers of links and of basal, intermediate and top species, the cascade model takes as given the total number of species and the number of links. The cascade model *predicts* the fractions of species that are basal, intermediate and top and the numbers of links of each of four kinds. Pimm's null model could be viewed as a conditional version of the cascade model: given numbers of links and of basal, intermediate and top species produced by the chance mechanisms of the cascade model, the distribution of these links among pairs of species in the cascade model is identical to that in Pimm's null model (ignoring the negligible probabilities in the cascade model that top species are not proper top and basal species are not proper basal).

Cohen (1978) and Pimm (1982) propose the models just described as 'null' models, models that would describe how webs should look in the absence of interesting biological structure. Here we consider the cascade model as a 'theory'. We suggest that between 'null models' and 'theories' is a continuum of increasingly sophisticated and successful models. The null models at one extreme are models that do *not* describe much of nature well. 'Theories', at the other extreme, provide a unifying and quantitatively successful view of diverse phenomena. The cascade model provides explanations for some aspects of webs that Cohen's (1978) and Pimm's (1982) models take as given and describes with moderate success the observed frequency distributions of chain lengths. Whether the cascade model should continue to be dignified as theory depends on its success in describing other aspects of real webs.

9.3 Further Work Required

How well does the cascade model describe the variance and higher moments of the distribution of chain length? A key difficulty in answering this question, which was raised by S. L. Pimm, is the dependence among the numbers of chains of different lengths. Attacks via mathematical analysis and via numerical simulation are both desirable.

Why does the cascade model fail to predict 16 or 17 observed frequency distributions of chain length? One possibility is that, like a straight line tangent to a parabola, the predictions of the cascade model are systematically of the wrong shape but are locally good approximations in a certain neighbourhood. According to this possibility, a better model could explain all the observed frequency distributions of chain length, as well as explain better the other features of webs that are described approximately by the cascade model. As noted in Chap. III.3, some assumptions underlying the cascade model are unrealistic. For example, the model assumes that the species at the top of the cascade is equally likely to prey on all other species in the community, and that the prey species a predator eats are chosen statistically, once and for all, independently of the abundance of the prey species and of the existence of other links. A better model might replace these assumptions by more realistic ones. However, we cannot provide and analyse a better model at this point.

A second possibility is that the bad fits of the cascade model are associated with some combination of the characteristics of webs. According to this possibility, the cascade model is acceptable for a large class of webs, e.g. those with mean chain length between two and four links, but for another relatively small class of webs a different model is required.

A third possibility is that the original data are wrong; that links have been overlooked, or that inconsistent criteria have been used for reporting links, or that stomach contents have been misidentified and mistaken links have been reported, or that error has crept into the process of writing, publishing and transcription.

The consequences for action of these three possible explanations are different. If the cascade model is only an approximation to a better global model, then one should try to construct a better global model. If combinations of characteristics could identify exactly webs for which the cascade model fails, one should try to discriminate the webs where the cascade model succeeds from those where it fails. If the reported frequency distributions of chain length are materially wrong, one should go back into the field and do better field work and reporting. There is no shortage of opportunities for diverse skills.

The empirical successes of the cascade model are great enough to encourage the hope that efforts in all three directions may yield further successes.

The present successes of the cascade model also justify attempts to exploit the model further as it stands. Can the cascade model describe or explain yet other aspects of webs, such as the frequency of omnivory, i.e. predation on different trophic levels (S. L. Pimm, personal communication, 3 September 1985), however 'trophic levels' are to be defined? Can the cascade model account for the relative importance of predation against competition (Schoener 1982), the occurrence of compartments (Pimm 1982), and the frequency of intervality (Cohen 1978)?

Appendix: Computing Algorithms

This appendix describes procedures for computing the frequency distributions of chain length and the length of the longest chain of a given acyclic web.

The Frequency Distribution of Chain Length

A digraph (directed graph) with S vertices (species) and L edges (links) may be represented by its $S \times S$ adjacency matrix, A. The elements of A are $a_{ij} = 1$ if (i, j) is an edge, $a_{ij} = 0$ if (i, j) is not an edge, $1 \leq i, j \leq S$.

An easily programmed, but inefficient, way to compute the number of n-chains, C_n, from the adjacency matrix A of an acyclic web uses the powers A^n of A. If S_B and S_T are the subsets of $\{1, 2, \ldots, S\}$ that contain the labels of, respectively, the basal and the top species, then

$$C_n = \sum_{i \in S_B} \sum_{j \in S_T} (A^n)_{ij}, \quad n = 1, 2, \ldots, S - 1 .$$

If each power is computed by $O(S^2)$ multiplications, then the computation of the frequency distribution of chain length $\{C_n\}$ requires $O(S^3)$ multiplications.

A much more efficient algorithm that requires $O(S^2)$ steps (additions or multiplications) was outlined in conversation (1984) by P. H. Sellers. Assume that the adjacency matrix A is strictly upper triangular, so that the vertices are numbered from 1 at the bottom of the web to S at the top of the web, i.e. edges point from vertices with lower numbers to vertices with higher numbers. The following algorithm requires as input the adjacency matrix A and returns as output an $(S-1)$-vector, C, with nth element C_n, the number of n-chains.

Step 1. Set $I = 1$ and set V to be an $S \times S - 1$ matrix with all elements 0. (After completion of the loop on I below, $V(I, J)$ will hold the number of maximal J-walks that terminate at vertex I, i.e. the number of J-walks that originate at some basal species and terminate at species I.)

Step 2. Increment I by 1. If the result exceeds S, go to step 8.

Step 3. Set $H = 0$.

Step 4. Increment H by 1. If the result equals I, go to step 2. (We are going to compute for each J the contribution, to the number of maximal J-walks terminating at vertex I, of maximal $(J-1)$-walks terminating at vertex H, for every $H < I$.)

Step 5. If $A(H, I) = 0$, go to step 4. (If there is no edge from H to I, then walks terminating at H either do not pass through I at all or must pass through some other vertex on their way to I.)

Step 6. If the sum of the Hth row of V is positive, then for $J = 2, \ldots, S - 1$, set $V(I, J) = V(I, J) + V(H, J - 1)$. Then go to step 4. (Each maximal $(J - 1)$-walk that terminates at a vertex H that is connected by an edge to vertex I determines a maximal J-walk that terminates at vertex I.)

Step 7. Otherwise, increment $V(I, 1)$ by 1. Then go to step 4. (If no walks terminate at vertex H but H is joined to I by an edge, then there is a maximal 1-walk terminating at I.)

Step 8. For $J = 1, \ldots, S - 1$, set C_J equal to the sum of $V(I, J)$ over only those I such that the Ith row sum of A is 0. (The chains are the maximal walks that terminate at top vertices. Vertex I is a top vertex if and only if the Ith row sum of A is 0. After all the maximal walks terminating at all the vertices have been counted, the number of J-chains is the total number of maximal J-walks that terminate at top vertices.)

We programmed both the algorithm based on powers and Sellers' algorithm in APL, with the APL68000 interpreter running on the WICAT 150-6, a microprocessor that uses the Motorola 68000 chip. For the 14×14 adjacency matrix of Chap. IV's web number 31, the algorithm based on powers required approximately 10 s to produce the frequency distribution of chain length, whereas Sellers' algorithm required approximately 5 s. For a 50×50 adjacency matrix generated according to the cascade model with $c = 3.71$, the powers algorithm required approximately 25.5 min and Sellers' algorithm required approximately 0.6 min.

The Length of the Longest Chain

For a digraph with a strictly upper triangular adjacency matrix A, finding the *height*, i.e. the length M of the longest chain, is a standard problem in network theory. For example, Gibbons (1985, pp. 121–122) gives a recursive algorithm for finding the longest path from a specified vertex to every other vertex. The following algorithm for finding the longest path from any vertex to any other, which requires in general $O(S^2)$ multiplications, was outlined in conversation (1985) by F. R. K. Chung. The algorithm requires as input the adjacency matrix A and returns as output the height M.

Step 1. Set V equal to an S-vector with all elements 0, and set $I = 0$. (After completion of the loop on I below, $V(I)$ will hold the length of the longest walk terminating at vertex I.)

Step 2. Increment I by 1. If the result exceeds S, go to step 4.

Step 3. Set $V(I) = \max\{A(H, I)(V(H) + 1) \mid 1 \le H \le I - 1\}$. Then go to step 2. (The length of the longest walk terminating at vertex I is 1 greater than the maximum over all $H < I$ with an edge from H to I of the length of the longest walk terminating at H.)

Step 4. Set $M = \max\{V(I) \mid 1 \le I \le S\}$. (The longest chain is as long as the longest of the maximal walks.)

For a 50×50 adjacency matrix, generated according to the cascade model with $c = 3.71$, independently of the matrix used in the previous example, the computation of $M = \max\{n \mid C_n > 0\}$ based on Sellers' algorithm for C required 38 s and Chung's algorithm required only 6 s.

As S gets large, the number of positive elements in adjacency matrices generated by the cascade model increases only as $O(S)$ rather than as $O(S^2)$. The number of multiplications and the amount of memory required by the preceding algorithm may be reduced from $O(S^2)$ to $O(S)$ as S gets large by representing the digraph by an $L \times 2$ matrix that lists, in some order, the initial and final vertex of each of its L edges. Step 3 above is then modified to pay attention only to those vertices $H < I$ for which there is an edge from H to I. By using this modified algorithm, we simulated webs of S species where S^2 far exceeded the words of memory available in our microprocessor.

§5. Theory of Food Chain Lengths in Large Webs

Charles M. Newman and Joel E. Cohen

1. Introduction

The purpose of this chapter is to develop a theory of the length of food chains that is derived from a mathematical model of community food webs called the cascade model. Cohen & Newman (1985, hereafter referred to as Chap. III.2) and Cohen et al. (1985, hereafter referred to as Chap. III.3) showed that the predictions of the cascade model describe, to a first approximation, several major characteristics of a collection of 62 real webs: the proportions of all species that are top, basal and intermediate, and the proportions of all links from basal to intermediate species, from basal to top species, from intermediate to intermediate species, and from intermediate to top species. Cohen et al. (1986, hereafter referred to as Chap. III.4) showed that the cascade model describes the frequency distribution of the length of food chains observed in a large majority of 113 real webs. In the light of this empirical support for the cascade model, it is desirable to analyse the properties of the model further. This chapter determines what the cascade model implies for the frequency distributions of the length of a typical food chain and of the length of the longest chain, primarily in the limit as the number of species in the web becomes arbitrarily large.

Section 2 presents terminology for chains and reviews the cascade model. Section 3 derives a generating function for the expected number of chains of each length and moments of the chain length distribution for webs with a finite number of species. Section 4 describes the frequency distribution of chain lengths in the limit as the number of species in a web gets large. Section 5 describes the length of the longest chain in a web with a finite number of species. Section 6 describes the length of the longest chain as the number of species in a web gets (very) large. Section 7 analyses the sensitivity of the asymptotic behaviour of the longest chain derived in section 6 to the assumptions of the cascade model. The results in sections 3-7 are obtained by mathematical analysis. Numerical simulations of the cascade model in section 8 confirm and amplify the prior analytical results concerning the length of the longest chain. Section 9 reviews what has been achieved in this chapter, and the concluding section 10 identifies some tasks that remain.

We shall accept the mathematical convention of setting off every proof with *Proof* at the beginning and ■ at the end. Readers may defer or skip proofs with no loss of continuity.

2. Terminology; The Cascade Model

This section reviews and introduces terminology, then describes the cascade model.

A *food web* is a set of kinds of organisms and a relation that shows which, if any, kinds of organisms each kind of organism in the set eats. A *community food web* is a food web whose vertices are obtained by picking, within a habitat or set of habitats, a set of kinds of organisms (hereafter called *species*) on the basis of taxonomy, size, location or other criteria, without prior regard to the eating relations (specified by trophic *links*) among the organisms (Cohen 1978, pp. 20–21). Hereafter 'web' means 'community food web'. A *basal* species is a species that eats no other species, and a *top* species is a species that is eaten by no other species.

In the representation of a web by a directed graph or digraph (see Chap. III.2.2), each vertex corresponds to a (lumped trophic) species. An edge (always directed) (a, b) from vertex a to vertex b corresponds to a link from species a to species b, meaning that species b eats species a. An example of a walk in a digraph is the sequence $a, (a, b), b, (b, c), c$ of alternating vertices and edges. The *length* of a walk is the number of edges in it. An n-walk is a walk of length n. The digraph of any web generated by the cascade model is acyclic, so no vertex (or species) can figure more than once in a walk in such a web. A *chain* is a walk from a basal species to a top species. An n-chain is a chain of length n, i.e. a chain with n links or equivalently $n + 1$ species. The *height* of a web is the longest chain in it.

Let S be the number of species in a web, and let C_n be the number of n-chains in an acyclic web, $n = 1, 2, \ldots, S - 1$. The *frequency distribution of chain length* is the vector $(C_1, \ldots, C_{S-1}) \equiv C$. The total number of chains in the web will be denoted

$$C \equiv \sum_{n=1}^{S-1} C_n \,.$$

As usual, $E(.)$ and var$(.)$ denote the expectation (or mean) and variance, respectively, of the random variable enclosed in parentheses. For any function f of any real or integer variable t, we write $f(t) = O(t)$ if $f(t)/t$ stays less than some fixed finite positive constant as $t \to \infty$, and $f(t) = o(t)$ if $\lim_{t\to\infty} f(t)/t = 0$.

The *cascade model* assumes that the $S \geq 2$ species of a web may be labelled from 1 (at the bottom, subject to predation by all other species) to S (at the top, subject to predation by no other species). The probability that species j feeds on species i is 0 if $j \leq i$. If $i < j$, then j feeds on i with probability $p = p(S)$, i.e., with a probability between 0 and 1 that depends on S, and does not feed on i with a probability $q = 1 - p$, independently for all $1 \leq i < j \leq S$. Unless a contrary assumption is explicitly given, it will be assumed that, for some finite positive real number $c < S$, $p = c/S$, where c is a constant independent of S. (Some results below require only the weaker assumption that $Sp(S) \to \gamma$, for some constant γ, as $S \to \infty$.)

According to the cascade model with probability p of a random link, the expected number of n-chains in a web with S species is (Chap. III.4):

$$E(C_n) = p^n q^{S-1} \sum_{k=n}^{S-1} (S - k) \binom{k-1}{n-1} q^{-k}, \quad n = 1, 2, \ldots, S-1 \, .$$

3. Moments of the Frequency Distribution of Chain Length in Finite Webs

To find an average chain length predicted by the cascade model, we need to compute $\sum_n n f_n$, where f_n is the probability density of n-chains according to the cascade model. There are two, not one, natural candidates for f_n. The first corresponds to 'expected relative frequency', and the second corresponds to 'relative expected frequency'. To compute the first, which we denote u_n, find, for each random web, the fraction of all chains that are n-chains, and then average over all webs. The expected relative frequency of chain length n is

$$u_n = E(C_n/C), \quad n = 1, 2, \ldots, S-1 \, .$$

To make this well defined, C_n/C may be set to zero whenever $C = 0$. For typical S and p, the probability that $C = 0$ is very small. To compute the second candidate for f_n, find the expected number of n-chains, averaged over all webs, and then express that average as a proportion of the sum of the averages of all lengths. The relative frequency of chain length n is

$$v_n = E(C_n)/E(C), \quad n = 1, 2, \ldots, S-1 \, .$$

Both u_n and v_n depend on S. A random variable H_S with probability density u_n can be obtained by taking a random web and measuring the length of a single chain chosen at random, with all of the web's chains equally likely. A random variable L_S with probability density v_n can be obtained by taking (in the limit) a very large collection of webs and picking a single chain randomly from the pooled chains of all the webs, each chain again being equally likely.

We shall compute $E(L_S) = \sum_n n v_n$ and higher moments of L_S by means of a generating function, defined as

$$f_S(t) = \sum_{n=1}^{S-1} E(C_n/S) t^n, \quad 0 < t < \infty .$$

According to the cascade model,

$$f_S(t) = t S p q^{S-2} \Big\{ [1 + (p/q)(1+t)]^S - 1$$
$$- (Sp/q)(1+t) \Big\} / \Big\{ (Sp/q)(1+t) \Big\}^2 .$$

Proof. Using first the formula for $E(C_n)$ and then the identity

$$\sum_{n=1}^{S-1} \sum_{k=n}^{S-1} = \sum_{k=1}^{S-1} \sum_{n=1}^{k} ,$$

we compute

$$f_S(t) = S^{-1} \sum_{n=1}^{S-1} (pt)^n q^{S-1} \sum_{k=n}^{S-1} (S-k) \binom{k-1}{n-1} q^{-k}$$

$$= S^{-1} \sum_{k=1}^{S-1} \sum_{n=1}^{k} \binom{k-1}{n-1} (pt)^n q^{S-1} q^{-k} (S-k)$$

$$= pt q^{S-2} \sum_{k=1}^{S-1} q^{-(k-1)} [(S-k)/S] \sum_{h=0}^{k-1} \binom{k-1}{h} (pt)^h$$

$$= pt q^{S-2} \sum_{k=1}^{S-1} (1 - k/S) q^{-(k-1)} (1+pt)^{k-1}$$

and letting

$$r \equiv (1+pt)/q ,$$

$$f_S(t) = pt q^{S-2} \left(\sum_{k=1}^{S-1} r^{k-1} - S^{-1} \sum_{k=1}^{S-1} k r^{k-1} \right)$$

$$= pt q^{S-2} [(1 - r^{S-1})/(1 - r) - S^{-1}(d/dr)\{(1 - r^S)/(1 - r)\}]$$

which, upon further elementary calculation, becomes

$$= pt q^{S-2} S^{-1} [r^S - 1 - S(r-1)]/(r-1)^2 ,$$

which eventually simplifies, with $r = (1+pt)/q$, to the formula given. ∎

It follows from the generating function $f_S(t)$ that, setting $z = 2Sp/q$,

$$E(L_S) = (z/2)[(1 + z/S)^{S-1} - 1]/[(1 + z/S)^S - 1 - z],$$

or

$$E(L_S) = Sp[(1+p)^{S-1} - (1-p)^{S-1}]/[(1+p)^S - (1-p)^S - 2Sp(1-p)^{S-1}]$$

and

$$\text{var}(L_S) = \frac{\frac{1}{4}\{[(S-1)/S]z^2(1+z/S)^{S-2} - 4z(1+z/S)^{S-1} + 6(1+z/S)^S - 6 - 2z\}}{(1 + z/S)^S - 1 - z}$$
$$- [E(L_S)]^2 + 3E(L_S) - 2.$$

Proof. $E(L_S) = f'_S(1)/f_S(1)$. The two versions of $E(L_S)$ are equivalent because $1 + z/S = (1 + p)/(1 - p)$. The formula for $\text{var}(L_S)$ follows from a very long, but elementary, simplification of the result of substituting

$$E(t^{L_S}) = f_S(t)/f_S(1)$$

into

$$\text{var}(L_S) = \{(d/dt)^2 E(t^{L_S}) + (d/dt)E(t^{L_S}) - [(d/dt)E(t^{L_S})]^2\}|_{t=1}. \quad \blacksquare$$

Figure III.5.1 plots the mean of L_S and the mean plus or minus one standard deviation (corresponding roughly to a two-thirds confidence interval) for values of $p = c/S$ and S typical of the observed webs analysed in Chaps. III.2–3. With increasing S and fixed c, the mean and confidence interval stabilize for webs with more than 30 species, but change noticeably for smaller webs.

4. Limiting Frequency Distribution of Chain Length in Large Webs

We now describe the behaviour of C_n/S, predicted according to the cascade model, as S gets large, assuming that, for large S, $p(S)$ declines like γ/S or more precisely that $\lim_{S \to \infty} Sp(S) = \gamma$. When $p(S) = c/S$, then $\gamma = c$.

Define the generating function (which does not depend on S) for $0 < \gamma < \infty$:

$$g(t) = t\gamma e^{-\gamma}\{e^{\gamma(1+t)} - 1 - \gamma(1+t)\}/\{\gamma(1+t)\}^2.$$

The coefficients $K_{n-1}, n = 1, 2, \ldots$ of the (convergent) power series expansion

$$g(t) = K_0 t + K_1 t^2 + \ldots$$

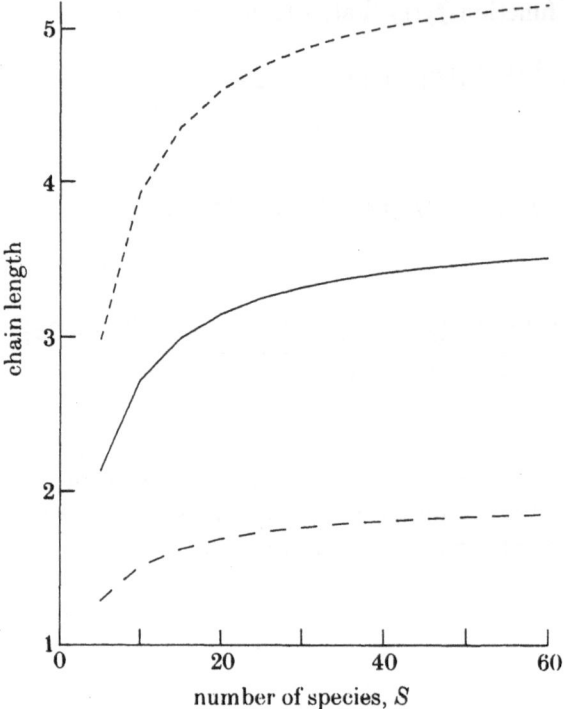

Figure III.5.1. Mean (—), and mean plus (- - -) or minus (— — —) one standard deviation, of chain length, L_S, as a function of the number of species, S, according to the cascade model with $c = 3.71$

have the meaning

$$\lim_{S \to \infty} E(C_n)/S = K_{n-1}, \quad n = 1, 2, \ldots ,$$

provided $0 < \gamma < \infty$. K_{n-1} may be computed explicitly from

$$K_{n-1} = (d/dt)^n g(t)|_{t=0}/n!$$

or from

$$K_{n-1} = [\gamma^n e^{-\gamma}/(n-1)!](d/d\gamma)^{n-1}[(e^\gamma - 1 - \gamma)\gamma^{-2}] .$$

The limit of the mean total number, $E(C)$, of chains satisfies

$$\lim_{S \to \infty} E(C)/S = g(1) = \sum_{n=1}^{\infty} K_{n-1} .$$

Hence

$$\lim_{S\to\infty} v_n \equiv \lim_{S\to\infty} E(C_n)/E(C)$$

$$= K_{n-1}/\sum_{h=1}^{\infty} K_{h-1}$$

$$= \{\gamma^{n-1}(d/d\gamma)^{n-1}[(e^{\gamma}-1-\gamma)\gamma^{-2}]\}$$
$$/[(n-1)!\,(e^{2\gamma}-1-2\gamma)(2\gamma)^{-2}]\,.$$

The moments and factorial moments of the length of chains, according to the random variable L_S with probabilily density $\{v_n\}_1^{S-1}$, approach the limits

$$\lim_{S\to\infty} E(L_S(L_S-1)\ldots(L_S-(k-1))) = g^{(k)}(1)/g(1), \quad k=1,2,\ldots$$

$$\lim_{S\to\infty} E(L_S^k) = (d/du)^k[g(e^u)]|_{u=0}/g(1), \quad k=1,2,\ldots$$

In particular,

$$\lim_{S\to\infty} E(L_S) = \gamma(e^{2\gamma}-1)/(e^{2\gamma}-1-2\gamma)$$

and, letting

$$h(t) = (e^t - 1 - t)/t^2\,,$$
$$\lim_{S\to\infty} \text{var}(L_S) = \gamma - \tfrac{1}{2} + [4h(2\gamma)]^{-1}[3 - 2\gamma - 1/h(2\gamma)]\,.$$

It follows that

$$\lim_{\gamma\to\infty} [\lim_{S\to\infty} E(L_S) - \gamma] = 0$$

$$\lim_{\gamma\to\infty} [\lim_{S\to\infty} \text{var}(L_S) - (\gamma - \tfrac{1}{2})] = 0\,.$$

Proof. To establish all the above limits, it suffices to prove that $f_S(t)$ and its derivatives converge to $g(t)$ and its corresponding derivatives for $t \in [0,1]$. Because $Sp \to \gamma$, $q^S \to e^{-\gamma}$ and so on, f_S converges to g on $(-\infty, +\infty)$. To show that the derivatives of f_S converge to those of g, it suffices, by standard arguments in the theory of analytic functions, to show that $f_S(w)$ is uniformly bounded in some neighbourhood of $[0,1]$ in the complex plane, as $S \to \infty$. But that follows because, for complex w,

$$|[(1 + w/S)^S - 1 - w]/w^2| \le (e^{|w|} - 1 - |w|)/|w|^2\,.$$

This last inequality may be established by comparing the power series expansions of both sides term by term. Given convergence of the generating function and its derivatives, $\lim E(L_S) = g'(1)/g(1)$ and $\lim \text{var}(L_S) = g''(1)/g(1) +$

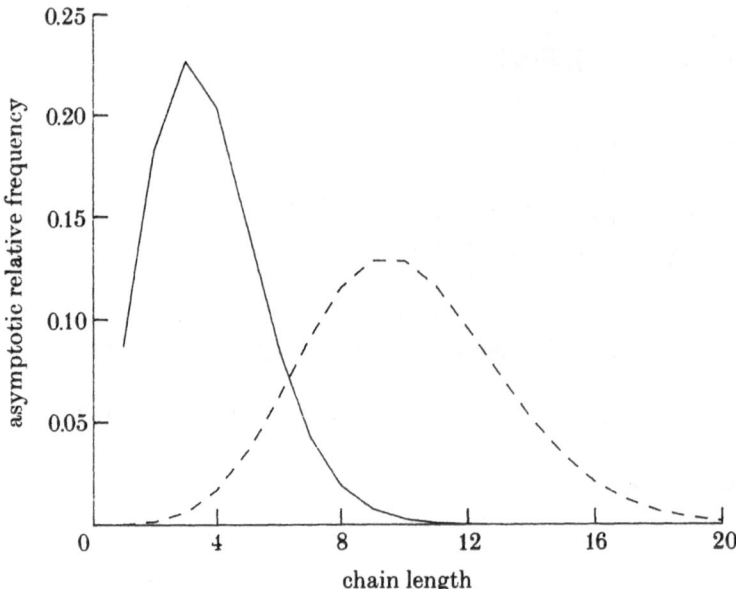

Figure III.5.2. Asymptotic relative expected frequency of chains of each length, in webs with an arbitrarily large number of species, according to the cascade model with $\gamma = 3.71$ (—) and with $\gamma = 10$ (– – –). When $p(S) = c/S$, then $\gamma = c$

$g'(1)/g(1) - [g'(1)/g(1)]^2$. The formulae given then follow by long but elementary calculations. ∎

Figure III.5.2 plots $\lim_{S\to\infty} v_n$, the limiting relative expected frequency of chains of length n, as a function of n, for $\gamma = 3.71$, a value suggested by the data of Chaps. III.2–3, and for $\gamma = 10$. The graph for $\gamma = 3.71$ is very similar to the theoretical and simulated graphs for $\gamma = c = 3.75$ and finite $S = 17$ given in Fig. III.4.1. In effect, for this value of γ, $S = 17$ is 'large'. The graph in Fig. III.5.2 for $\gamma = 10$ illustrates the general numerical observation that, for large values of γ, $\lim_{S\to\infty} v_n$ increases monotonically up to a value of n very near γ and then decreases very nearly symmetrically, in a shape closely resembling a normal distribution. We conjecture that $\lim_{S\to\infty} v_n$ is maximal for n equal to the largest integer less than γ or one less than the largest integer less than γ. We have numerical examples in which either of these two values of n makes $\lim_{S\to\infty} v_n$ maximal. The approximate normality, for large γ, of this limiting distribution (with mean approximately γ and variance approximately $\gamma - \frac{1}{2}$) can be proved mathematically. We do not present the proof, since typical values of γ (e.g. 3.71 or 4) are too small for the approximate normality to hold, and we have no significant application of the result for larger values of γ.

Figure III.5.3 plots $\lim_{S\to\infty} E(L_S)$ and $\lim_{S\to\infty} E(L_S)$ plus or minus $[\lim_{S\to\infty} \text{var}(L_S)]^{1/2}$ (corresponding roughly to a two-thirds confidence interval) as function of γ, for a range of γ likely to include that suggested by the largest observed webs in Chaps. III.2–3. Figure III.5.3 shows that $\lim_{S\to\infty} E(L_S)$ approaches the asymptotic (for large γ) limit γ quite rapidly, even within the

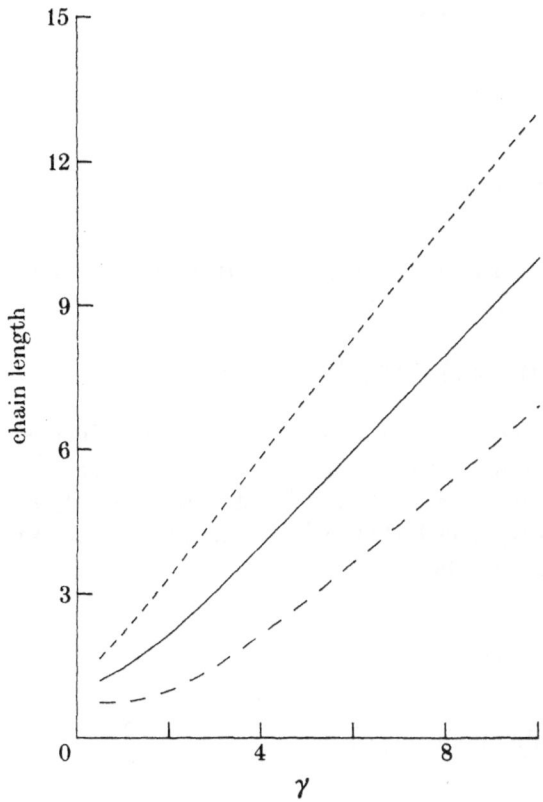

Figure III.5.3. Asymptotic mean (—), and asymptotic mean plus (- - -) or minus (— — —) one asymptotic standard deviation, of chain length, L_S, as a function of γ, according to the cascade model. When $p(S) = c/S$, then $\gamma = c$

range estimated from the data in Chaps III.2–3. The cascade model thus implies a simple rule of thumb: in webs with a large number of species, the mean length of a chain roughly equals the mean of the numbers of predators and prey of any species in the web (i.e. the mean total number of links that enter and leave any vertex or, in graph theoretic jargon, the mean in-degree plus the mean out-degree).

For any fixed length n, the standard deviation of the number C_n of n-chains vanishes relative to S as S gets large. Equivalently, for fixed length n in large webs the variance of C_n/S vanishes as S gets large. That is, for fixed $n \geq 1$,

$$\lim_{S \to \infty} \text{var}(C_n/S) = 0 \quad \text{if} \quad \lim_{S \to \infty} Sp(S) = \gamma < \infty .$$

Proof. For $1 \leq i_0 < i_1 < \ldots < i_n \leq S$, let W_i denote the indicator random variable of the event that $i \equiv i_0, (i_0, i_1), i_1, \ldots, (i_{n-1}, i_n), i_n$ is a chain. Thus $W_i = 1$ if i is a chain, and $W_i = 0$ if not. Then $C_n = \sum_i W_i$, where the summation covers all possible n-chains. Then

$$
\begin{aligned}
\mathrm{var}(C_n) &= \mathrm{var}\left(\sum_i W_i\right) \\
&= \sum_i \sum_j \mathrm{cov}(W_i, W_j) ,
\end{aligned}
$$

where the covariances are summed over all n-chains i and j. If the chains i and j share exactly m links, $0 \leq m \leq n$, then

$$
\mathrm{cov}(W_i, W_j) \leq P(W_i = 1, W_j = 1) \leq p^m p^{2(n-m)} = p^{2n-m} .
$$

If $m = 0$, i.e. i and j share no links, and if in addition none of $i_0, \ldots, i_n$ coincides with any of $j_0, \ldots, j_n$, then W_i and W_j are independent so $\mathrm{cov}(W_i, W_j) = 0$. Let Q_m be the number of ordered pairs (i, j) such that i and j share exactly m links. Let Q be the number of pairs (i, j) such that at least one species (vertex) of the chain i is a vertex of the chain j. Then

$$
\mathrm{var}(C_n) \leq \sum_{m=1}^{n} Q_m p^{2n-m} + Q p^{2n} .
$$

Now Q_n is just the number of possible n-chains, so

$$
Q_n = \binom{S}{n+1} \leq S^{n+1} .
$$

For $m < n$, if i and j share m links, they will share at least $m+1$ species; but i and j could share $m+1$ species without necessarily sharing $m+1$ links; so Q_m does not exceed the number of pairs (i, j) in which i and j have $m+1$ species in common. Therefore $Q_m \leq S^{m+1} S^{2(n+1-(m+1))} = S^{2n+1-m}$. Similarly $Q \leq S^{2n+1}$. Therefore

$$
\begin{aligned}
\mathrm{var}(C_n) &\leq \sum_{m=0}^{n} S^{2n+1-m} p^{2n-m} \\
&= \sum_{m=0}^{n} O(S^{2n+1-m} S^{-(2n-m)}) = O(S)
\end{aligned}
$$

and thus $\mathrm{var}(C_n/S) = O(1/S)$, which tends to zero as S tends to ∞. ∎
  It follows that, for any fixed $n = 1, 2, \ldots$,

$$
C_n/C \to K_{n-1} \left/ \sum_{h=1}^{\infty} K_{h-1} = g^{(n)}(0)/[n!\, g(1)] \right.
$$

$$
\text{in probability as } S \to \infty ,
$$

$$\lim_{S \to \infty} E(C_n/C) \to K_{n-1} \Big/ \sum_{h=1}^{\infty} K_{h-1} \ .$$

As in section 3, C_n/C is set to zero when $C = 0$.

Proof. We have already proved that $E(C_n/S) \to K_{n-1}$ and $\mathrm{var}(C_n/S) \to 0$. It follows that, for each fixed n, $C_n/S \to K_{n-1}$ in mean square (i.e. in L^2) and hence in probability, and therefore that, for any fixed positive integer M,

$$\sum_{n=1}^{M} C_n/S \to \sum_{n=1}^{M} K_{n-1} \ \text{in probability} \ .$$

Our next goal is to prove that this implies

$$C/S \to \sum_{n=1}^{\infty} K_{n-1} \ \text{in probability} \ .$$

Since

$$P\left(\left| \sum_{n=1}^{M} C_n/S - C/S \right| \geq \varepsilon \right) = P\left(\sum_{n=M+1}^{S-1} C_n/S \geq \varepsilon \right)$$

$$\leq \varepsilon^{-1} \sum_{n=M+1}^{S-1} E(C_n)/S$$

$$= \varepsilon^{-1} \left(E(C)/S - \sum_{n=1}^{M} E(C_n)/S \right)$$

$$\to \varepsilon^{-1} \sum_{n=M+1}^{\infty} K_{n-1} \ \text{as} \ S \to \infty \ ,$$

we have

$$P\left(\left| C/S - \sum_{n=1}^{\infty} K_{n-1} \right| \geq 3\varepsilon \right) \leq P\left(\left| C/S - \sum_{n=1}^{M} C_n/S \right| \geq \varepsilon \right)$$

$$+ P\left(\left| \sum_{n=1}^{M} C_n/S - \sum_{n=1}^{M} K_{n-1} \right| \geq \varepsilon \right)$$

$$+ P\left(\left| \sum_{n=1}^{M} K_{n-1} - \sum_{n=1}^{\infty} K_{n-1} \right| \geq \varepsilon \right) \ .$$

Taking $\limsup_{S \to \infty}$ of this last inequality and choosing M large enough, we have

$$\limsup_{S \to \infty} P\left(\left|C/S - \sum_{n=1}^{\infty} K_{n-1}\right| \geq 3\varepsilon\right) \leq \varepsilon^{-1} \sum_{n=M+1}^{\infty} K_{n-1} + 0 + 0 \,.$$

Letting $M \to \infty$ establishes that

$$C/S \to \sum_{n=1}^{\infty} K_{n-1} \text{ in probability} \,.$$

Hence

$$C_n/C = (C_n/S)/(C/S) \to K_{n-1} \Big/ \sum_{h=1}^{\infty} K_{h-1} \text{ in probability} \,.$$

Since $|C_n/C| \leq 1$, all moments of C_n/C converge. In particular,

$$E(C_n/C) \to K_{n-1} \Big/ \sum_{h=1}^{\infty} K_{h-1} \,. \qquad \blacksquare$$

Since u_n and v_n converge to the same limit for large S, it makes no difference, for large enough S, which probability density is used to describe typical chain lengths. Of course, for finite S (and all observed webs have finite S), the two probability densities $\{u_n\}$ and $\{v_n\}$ are different; we have obtained exact formulae only for the latter.

5. The Longest Chain in Finite Webs

We now show that the cascade model explains remarkably well the qualitative observation, frequently made (see, for example, Hutchinson 1959), that the length of the longest chain, and hence the height, of a web is small compared to the number of species in the web.

In a web with S species, define M_S to be the height. For random webs generated by the cascade model, M_S is a random variable. For brevity, we henceforth drop the subscript S, bearing in mind that the distribution of M does depend on S.

To investigate the distribution of M, given S and $p = c/S$, we find, for a positive integer m, upper bounds for $P(M \geq m)$ and $P(M < m)$.

First, for any positive integer $m \leq S - 1$ with $p = c/S$,

$$P(M \geq m) \leq L_1(S, m) \equiv 1 - (1 - p^m)^{\binom{S}{m+1}}$$

$$\leq L_2(S, m) \equiv \binom{S}{m+1} p^m .$$

$$\leq L_3(S, m) \equiv c^m S/(m+1)! .$$

Proof. Let B_n be the number of n-walks, $n = 1, 2, \ldots, S-1$. Such walks may or may not be chains, which are walks from basal to top species. For $1 \leq i_0 < i_1 < \ldots < i_n \leq S$, let V_i denote the indicator random variable of the event that there is a walk $i \equiv i_0, (i_0, i_1), i_1, \ldots, (i_{n-1}, i_n), i_n$. Then $B_n = \sum_i V_i$, where the summation covers all possible n-walks. The V_i's are non-decreasing functions of the independent random variables that determine whether individual links are present and hence are associated random variables (Harris 1960; Esary et al. 1967). This justifies the inequality in the computation (where we set $n = m$)

$$P(M \geq m) = P(B_m > 0) = 1 - P(B_m = 0)$$

$$= 1 - P \text{ (for every } i \text{ of length } m, \ V_i = 0)$$

$$\leq 1 - \prod_i P(V_i = 0)$$

[the product taken over all m-walks i]

$$= 1 - \prod_i (1 - p^m)$$

$$= 1 - (1 - p^m)^{\binom{S}{m+1}} .$$

The last step holds because there are exactly $\binom{S}{m+1}$ possible m-walks in the cascade model.

For any positive integer n and any $x \in (0, 1)$, $1 - (1-x)^n \leq nx$, as it is easy to show by comparing derivatives with respect to x. Setting $x = p^m$ and $n = \binom{S}{m+1}$ gives $L_1 \leq L_2$. Finally,

$$\binom{S}{m+1} (c/S)^m = [c^m S/(m+1)!][(S-1)!/\{(S-1-m)! S^m\}]$$

$$\leq c^m S/(m+1)! \qquad \blacksquare$$

On the other side of M,

$$P(M < m) \leq U_1(S, m) \equiv \binom{S}{m+1}^{-1} \left[\sum_{k=1}^{m} (m-k+1) \binom{S-k-1}{m-k} (p^{-k}-1) \right.$$

$$\left. + \sum_{k=1}^{m-2} \binom{m}{k} \binom{S-k-2}{m-k-1} (p^{-k}-1) \right]$$

$$\leq U_2(S,m) \equiv \left[\binom{S}{m+1}p^m\right]^{-1}\left[\sum_{k=o}^{m}(m-k+1)(Sp)^{m-k}/(m-k)!\right.$$

$$\left. + \sum_{k=0}^{m}\binom{m}{k}S^{m-k-1}p^{m-k}\right]$$

$$\leq U_3(S,m) \equiv [c^m S(1-1/S)\dots(1-m/S)/(m+1)!]^{-1}$$

$$\times\left[\sum_{j=0}^{\infty}(j+1)(Sp)^j/j! + S^{-1}\sum_{j=0}^{m}\binom{m}{j}(Sp)^j\right],$$

and if $\frac{1}{2}m(m+1) < S$,

$$\leq U_4(S,m) \equiv [c^m S(1-m(m+1)/(2S))/(m+1)!]^{-1}$$

$$\times [e^{Sp} + Spe^{Sp} + S^{-1}(1+Sp)^m]$$

$$= 2(m+1)![(1+c)e^c + (1+c)^m/S]/[c^m(2S-m(m+1))]$$

Proof. In the notation used in the previous proof,

$$P(M < m) = P(B_m = 0) \leq P[|B_m - E(B_m)| \geq E(B_m)],$$

and now, from Chebychev's inequality,

$$\leq \operatorname{var}(B_m)/[E(B_m)]^2 .$$

We now seek an upper bound for $\operatorname{var}(B_m) = \sum_i \sum_j \operatorname{cov}(V_i, V_j)$, where the summations cover all m-walks. If i and j have exactly k links in common, then

$$\operatorname{cov}(V_i, V_j) = P(V_i = 1 \text{ and } V_j = 1) - P(V_i = 1)P(V_j = 1)$$

$$= p^{2m-k} - p^{2m} .$$

Let Q_k be the number of ordered pairs (i,j) of m-walks (not chains now) such that i and j have exactly k links in common. Then

$$\operatorname{var}(B_m) = \sum_{k=0}^{m} Q_k(p^{2m-k} - p^{2m})$$

$$= \sum_{k=1}^{m} Q_k(p^{2m-k} - p^{2m}) .$$

Define Q_k^α to be the number of ordered pairs (i,j) of m-walks with exactly k links in common in which the k common links form a k-walk. Define Q_k^β to be the number of ordered pairs (i,j) of m-walks with exactly k links in common in which the k common links do not form a single k-walk. Clearly, $Q_m^\beta = Q_{m-1}^\beta = 0$

and

$$Q_k = Q_k^\alpha + Q_k^\beta \ .$$

Moreover,

$$Q_k^\alpha \le \binom{S}{m+1}(m-k+1)\binom{S-k-1}{m-k}, \quad k=1,\ldots,m$$

because there are $\binom{S}{m+1}$ ways to choose the m-walk i, there are $m-k+1$ ways to choose a subwalk of i of length k, the links of which will be the links in common with j, and since each common subwalk of length k determines $k+1$ vertices of j, there are not more than $\binom{S-(k+1)}{(m+1)-(k+1)} = \binom{S-k-1}{m-k}$ ways to choose the remaining $(m+1)-(k+1)$ vertices of j. Also,

$$Q_k^\beta \le \binom{S}{m+1}\binom{m}{k}\binom{S-k-2}{m-k-1}, \quad k=1,\ldots,m-2$$

because (again) there are $\binom{S}{m+1}$ ways to choose i, there are at most $\binom{m}{k}$ to choose the k links of i that will be the links in common with j, and since these links form at least two subwalks which determine not less than $k+2$ vertices of j, leaving at most $(m+1)-(k-2) = m-k-1$ vertices to be determined, there are at most $\binom{S-k-2}{m-k-1}$ ways to choose the remaining vertices of j. This last step depends on the observation that $\binom{S-h}{m+1-h}$ is a non-increasing function of $h = 0,\ldots,m+1$.

Since $E(B_m) = \binom{S}{m+1}p^m$, collecting all the inequalities gives

$$P(M < m) \le \sum_{k=1}^{m}(Q_k^\alpha + Q_k^\beta)(p^{2m-k} - p^{2m}) \Big/ \left[\binom{S}{m+1}p^m\right]^2$$

$$\le \binom{S}{m+1}^{-1}\left[\sum_{k=1}^{m}(m-k+1)\binom{S-k-1}{m-k}(p^{-k}-1)\right.$$

$$\left.+ \sum_{k=1}^{m-2}\binom{m}{k}\binom{S-k-2}{m-k-1}(p^{-k}-1)\right].$$

This establishes $P(M < m) \le U_1(S,m)$. The remaining approximations follow by elementary calculations. ∎

These inequalities imply bounds on any quantile of the distribution of the height M. For example, to bound the median of M, we determine numerically m_1, the smallest integer m such that $L_2(S,m) < \frac{1}{2}$, and m_2, the largest integer m such that $U_1(S,m) < \frac{1}{2}$. Then

$$m_2 \le \text{median of } M \le m_1 - 1 \ .$$

(Why do we use $L_2(S, m)$ to determine m_1 rather than $L_1(S, m)$, which is a sharper bound? For moderately large values of S and m, when p^m becomes very small, e.g. less than 10^{-15}, a computer approximates $1 - p^m$ by 1 and the results become nonsense. $L_2(S, m)$ and $L_3(S, m)$ avoid the problem of subtracting numbers of very different orders of magnitude.)

For a fixed value of c typical of observed webs and a broad range of values of S, Table III.5.1 gives the lower and upper bounds on the median height. In other examples, in a web of 20 species with $c = 3.71$, the median height is between 2 and 7 links. In a web with 250 times as many, or 5000, species, the median height is between 8 and 13 links. The upper bound on the median has increased by less than a factor or two.

Table III.5.1. The length of the longest chain in large webs according to the cascade model ($c = 3.71$)

number of species	bounds on the median		limiting value, m^*	asymptotic value $\ln S / \ln (\ln S)$
	lower	upper		
10^2	4	10	11	3.0
10^4	8	14	14	4.1
10^6	12	17	17	5.3
10^8	15	19	20	6.3
10^{10}	18	22	22	7.3
10^{12}	21	24	25	8.3

6. Asymptotic Behaviour of the Length of the Longest Chain in Large Webs

In the cascade model with a fixed $c > 0$, as the number of species, S, gets very large (as we shall see, far larger than the number of species on Earth), the limiting behaviour of the height, the length of the longest chain, is simple. For each S, there is a positive integer m^* (which depends on S, but we drop the subscript S for brevity), such that the probability that the height is m^* or $m^* - 1$ approaches one as S gets large. Thus in a web generated by the cascade model the height is either m^* or $m^* - 1$, with probability approaching one for large S.

A qualitatively similar phenomenon has been observed elsewhere in the theory of random graphs. Bollobás & Erdös (1976) and, according to them, D. W. Matula independently proved that the size of the maximal complete subgraph (clique) in a random graph takes one of at most two values (that depend on the size of the random graph) with a probability that approaches 1 as the random graph gets large, when the edge probability is held fixed, independent of the number of vertices.

For very, very large numbers, S, of species, m^* grows at a rate that is essentially independent of c or $p = c/S$ (provided $c > 0$) and depends only on S. For extremely large S, m^* is approximately $\ln S / \ln (\ln S)$ in the sense that their ratio approaches 1. By contrast, according to Bollobás & Erdös (1976), the asymptotic behaviour of the (at most two) possible values for the size of the largest clique does depend on the fixed probability that there is an edge between any two given vertices.

We now describe more precisely the height M in very large webs according to the cascade model. Define m^* to be the smallest positive integer m such that

$$c^{m+1}S/(m+2)! \leq (m+2)^{-1/2} .$$

Then, for large enough S, m^* is a non-decreasing sequence such that

$$\lim_{S\to\infty} m^*/[\ln S/\ln(\ln S)] = 1$$

and

$$\lim_{S\to\infty} P(M = m^* \text{ or } M = m^* - 1) = 1 .$$

However, the estimated rate of convergence of $P(M = m^* \text{ or } M = m^* - 1)$ to 1 is very slow, namely,

$$1 - P(M = m^* \text{ or } M = m^* - 1) = O(m^{*-1/2}) .$$

Proof. We begin by establishing that, for every positive integer S and every $c > 0$, there *is* a positive integer m such that $c^{m+1}S/(m+2)! < (m+2)^{-1/2}$ or equivalently $(m+2)^{1/2}c^{m+1}S/(m+2)! \leq 1$. (If m exists, then m^* exists.) Stirling's approximation may be written

$$n! = (2\pi)^{1/2}n^{n+1/2}e^{-n}(1 + O(n^{-1})) .$$

Substituting into $(m+2)^{1/2}c^{m+1}S/(m+2)!$ shows that this quantity approaches 0 as $m \to \infty$, so the desired m exists. The least such m, namely m^*, satisfies $S \leq (m^* + 2)!/[c^{m^*+1}(m^* + 2)^{1/2}]$ and must not decrease as S increases. The next question is: how fast does m^* increase?

Pick any $\varepsilon > 0$. If

$$m(S) \sim (1 + \varepsilon) \ln S/\ln(\ln S) ,$$

then

$$\begin{aligned}
\ln m(S) &= \ln(\ln S) - \ln(\ln(\ln S)) + O(1) \\
&\sim \ln(\ln S)
\end{aligned}$$

so

$$-m(S)\ln m(S) \sim -(1 + \varepsilon)\ln S .$$

Now, by using Stirling formula and dropping ineffectual constants,

$$\ln[m(S)^{1/2}c^{m(S)}S/m(S)!] \sim m(S)\ln c + \ln S - \ln[m(S)!] + \tfrac{1}{2}\ln m(S)$$
$$\sim \ln S - m(S)\ln m(S)$$
$$\sim \ln S - (1+\varepsilon)\ln S$$
$$\sim -\varepsilon \ln S \to -\infty \text{ as } S \to \infty \,.$$

Consequently

$$m(S)^{1/2}c^{m(S)}S/m(S)! \to 0 \quad \text{as} \quad S \to \infty \,,$$

so $m^* < (1+\varepsilon)\ln S/\ln(\ln S)$ for large enough S. On the other hand, if

$$m(S) \sim (1-\varepsilon)\ln S/\ln(\ln S) \,,$$

then the same argument shows that

$$m(S)^{1/2}c^{m(S)}S/m(S)! \to \infty \quad \text{as} \quad S \to \infty \,,$$

so $m^* > (1-\varepsilon)\ln S/\ln(\ln S)$ for large S. This establishes that

$$m^* \sim \ln S/\ln(\ln S) \,.$$

So m^* increases without bound (but very slowly) as $S \to \infty$.

We can now prove $\lim_{S\to\infty} P(M = m^* \text{ or } m^* - 1) = 1$. By the inequalities established for $P(M \geq m)$ and $P(M < m)$ for finite S, with S large enough that $m^*(m^* + 1)/2 < S$,

$$P(M \geq m^* + 1) \leq c^{m^*+1}S/(m^* + 2)!$$
$$\leq (m^* + 2)^{-1/2} \to 0 \quad \text{as} \quad S \to \infty \,,$$
$$P(M < m^* - 1) \leq 2m^*![(1+c)e^c + (1+c)^{m^*-1}/S]$$
$$/[c^{m^*-1}(2S - m^*(m^* - 1))]$$
$$\sim (1+c)e^c m^*!/[Sc^{m^*-1}]$$
$$= (1+c)e^c[c^{m^*}S/(m^* + 1)!]^{-1}c/(m^* + 1) \,.$$

But m^* is the smallest m such that $c^{m+1}S/(m + 2)! < (m + 2)^{-1/2}$. Therefore $c^{m^*}S/(m^* + 1)! > (m^* + 1)^{-1/2}$ and hence $[c^{m^*}S/(m^* + 1)!]^{-1} < (m^* + 1)^{1/2}$. As $S \to \infty$, $P(M < m^* - 1)$ is therefore of order of magnitude not greater than $(1 + c)ce^c(m^* + 1)^{-1/2}$, which approaches 0 as $O(m^{*-1/2})$. ∎

For each value, m, of m^*, there is a range of values of S such that m^* for that S is m. When S is large and at the upper end of this range of values, then the height equals m^* with a probability that approaches 1. When S is large and at the lower end of this range of values, the height equals $m^* - 1$ with a probability that approaches 1. When S is in the middle of this range, it can happen that both the event that the height equals m^* and the event that the height equals $m^* - 1$ occur with non-negligible probabilities.

Proof. For any positive integer m, let S_m^+ be the greatest integer less than or equal to $(m+2)!/[c^{m+1}(m+2)^{1/2}]$ and let $S_m^- = S_{m-1}^+ + 1$ (with $S_0^+ = 0$). Then $S_m^+ \sim m^{1/2}(m+1)!/c^{m+1}$, $S_m^-/S_m^+ \sim c/m$ and the range of values of S such that $m^* = m$ is precisely $\{S_m^-, S_m^- + 1, \ldots, S_m^+\}$. Suppose S_m is a sequence satisfying $S_m \leq S_m^+$ and $m^{1/2} S_m / S_m^+ \to \infty$; then with $S = S_m$, we find by using $U_4(S, m)$ that

$$P(M = m - 1) \leq P(M < m) = O[(m+1)!/(c^m S_m)]$$
$$= O[S_m^+/(m^{1/2} S_m)] \to 0 \text{ as } m \to \infty,$$

so that $P(M = m^*) \to 1$ for such a sequence S_m. Similarly if S_m satisfies $S_m \geq S_m^-$ and $S_m/(m^{1/2} S_m^-) \to 0$, or equivalently $m^{1/2} S_m / S_m^+ \to 0$, then with $S = S_m$, we find, by using $L_3(S, m)$, that

$$P(M = m) \leq P(M \geq m) = O[c^m S_m/(m+1)!]$$
$$= O[S_m/(m^{1/2} S_m^-)] \to 0 \text{ as } m \to \infty,$$

so that $P(M = m^* - 1) \to 1$ for such a sequence S_m.

We now consider S_m^0 in the middle of the range from S_m^- to S_m^+. Define S_m^0 to be the greatest integer less than or equal to $2(1+c)e^c(m+1)!/c^m$. Then for large m, $S_m^- < S_m^0 < S_m^+$ and for $S = S_m^0$ we have $m^* = m$. In this case,

$$P(M = m - 1) \leq P(M < m)$$
$$\leq [c^m S(1 + o(1))/(m+1)!]^{-1}[(1+c)e^c + o(1)]$$
$$\to [2(1+c)e^c]^{-1}(1+c)e^c < \tfrac{1}{2} \text{ as } m \to \infty,$$

and $L_1(S, m)$ gives

$$P(M = m) \leq P(M \geq m) \leq 1 - (1 - (c/S)^m)^{\binom{S}{m+1}}$$
$$\to 1 - \exp[-2(1+c)e^c] < 1 \text{ as } m \to \infty.$$

Since $P(M = m \text{ or } m - 1) \to 1$ as m and $S = S_m^0$ increase without bound, we conclude that

$$\liminf_{m \to \infty} P(M = m - 1) > 0,$$
$$\liminf_{m \to \infty} P(M = m) > 0$$

for $S = S_m^0$. ∎

To find m^* numerically for various numbers, S, of species, we find the smallest integer m such that

$$S \leq (m+2)!/[c^{m+1}(m+2)^{1/2}].$$

For a range of values of S, Table III.5.1 gives the calculated values of m^* as well as the values of the asymptotic expression for m^*, $\ln S/\ln(\ln S)$. For three values of S (10^2, 10^8, 10^{12}), the calculated value of m^* exceeds the upper bound given for the median height. This is consistent with the understanding that the height will be concentrated on m^* or $m^* - 1$ only in the limit as S becomes extremely large. In Table III.5.1, for finite S as large as 10^{12}, evidently m^* is larger than the range of possible values for the median height. Simulations described below for (for example) $S = 1000$ give an estimated median height of 9 links; this height falls between the lower and upper bounds of 6 and 12, respectively, although the calculated value of m^* is 13 (see Table III.5.2).

The values of $\ln S/\ln(\ln S)$, which fall far below m^*, emphasize further that m^* is dependent on c and converges (in ratio) to the c-independent quantity $\ln S/\ln(\ln S)$ only for *very* large S. For values of S in the range considered in Table III.5.1, second and higher order terms in the asymptotic expansion for m^* are evidently influential in addition to the leading term $\ln S/\ln(\ln S)$. Calculations similar to those used above to prove that

$$m^* = [\ln S/\ln(\ln S)](1 + o(1))$$

establish that, to second order,

$$m^* = [\ln S/\ln(\ln S)]\{1 + [\ln(\ln(\ln S))/\ln(\ln S)](1 + o(1))\} \ .$$

For $S = 10^{12}$, $\ln(\ln(\ln S))/\ln(\ln S) = 0.36$, a non-negligible correction. It is interesting that even the second-order term in the expansion of m^*, like the first, depends only on S and is independent of c.

7. Sensitivity Analysis: Anisotropic Cascade Models

If the assumptions of the cascade model are relaxed, what happens? This question arises first from the scepticism expressed in Chap. III.3 about the exact truth of these assumptions. For example, would the ability of the cascade model to explain, qualitatively, the slow growth of the height be destroyed by a small change in the parameter c? No, because for very large webs the height grows very slowly regardless of the value of the parameter c.

If one retained the assumption that the probability p_{ij} of a random link from species i to species j were 0 for $j \leq i$ (this is the 'cascade' assumption) but permitted values for p_{ij} to depend on i and j when $i < j$ (we propose to call all such models anisotropic cascade models), the webs can be qualitatively different from those generated by the (isotropic) cascade model (with $p_{ij} = p > 0$, for all $i < j$). Consider three examples.

First, suppose that the webs were partitioned into what some ecologists call 'compartments', meaning that the adjacency matrix of the web is block diagonal (see Pimm [1982, Chap. 8] for a review). Suppose that each compartment or block contained at most S^* species, where S^* is some fixed finite positive integer. As the total number, S, of species in the web increased, suppose that more and more blocks of size at most S^* were added. Obviously the height will not exceed $S^* - 1$, regardless of S.

Second, consider an anisotropic cascade model with block diagonal (strictly upper) triangular matrix $\{p_{ij}\}$ of edge probabilities and blocks (or compartments) of size $S_1, \ldots, S_n$, where $S_1 + \ldots + S_n = S$. Suppose in the block h, of size S_h, that $p_{ij} = c_h/S_h > 0$ for $i < j$. Then the height, M, satisfies

$$P(M \geq m) \leq 1 - \prod_{h=1}^{n} (1 - (c_h/S_h)^m)^{\binom{S_h}{m+1}}$$

$$\leq \left(\sup_{1 \leq h \leq n} c_h \right)^m S/(m+1)! \, .$$

Proof. Let $M(h)$ be the maximum chain length in the hth block. Then, since different blocks are independent and $M = \sup_h M(h)$, it follows, by using $L_1(S, m)$ for each block, that

$$P(M \geq m) = 1 - P(M < m)$$

$$= 1 - \prod_{h=1}^{n} P(M(h) < m)$$

$$\leq 1 - \prod_{h=1}^{n} (1 - (c_h/S_h)^m)^{\binom{S_h}{m+1}}$$

$$\leq \sum_{h=1}^{n} \binom{S_h}{m+1} (c_h/S_h)^m \leq \sum_{h=1}^{n} (c_h)^m S_h/(m+1)!$$

where the next to last inequality follows from $\prod_i (1 - x_i) \geq 1 - \sum_i x_i$. ∎

Now if $S_1 = \ldots = S_n \sim \ln S$ so that the number, n, of blocks grows as $S/\ln S$ while $c_h = c$ independent of S for $h = 1, \ldots, n$, then, within each block, $p_{ij} \sim c/\ln S$. The expected number of links to and from each species, i.e. the expected number of predators plus the expected number of prey, is asymptotically $c = S_h p_{ij} \sim \ln S(c/\ln S)$. Since here $P(M \geq m) \leq c^m S/(m+1)!$, exactly as in the isotropic cascade model, M cannot grow asymptotically faster than $\ln S/\ln(\ln S)$.

Third and finally, consider an anisotropic cascade model chosen, not for its realism, but to illustrate that without some special structure in the matrix p_{ij} of edge probabilities the height could be asymptotically proportional to S (even when the expected number of links per species is kept fixed), contrary to observation. This example is taken from a study of one-dimensional percolation by Newman & Schulman (1985) and incidentally illustrates that there are interesting connections between percolation models and cascade models.

Suppose, for $j = 2, 3, \ldots$, that $\{y_j\}$ is a fixed sequence of probabilities, independent of S, such that, for some $s < 2$,

$$\liminf_{j \to \infty} j^s y_j > 0 \,,$$

e.g. suppose $y_j \sim K j^{-s}$ as $j \to \infty$, for some $K > 0$ and $s < 2$. For any ρ such that $0 < \rho < 1$, there is an ε, $0 < \varepsilon < 1$, such that if

$$p_{i,i+1} \geq 1 - \varepsilon \text{ for all } S \text{ and } i = 1, \ldots, S-1 \,,$$
$$p_{ij} \geq y_{j-i}, \text{ for } i+2 \leq j \leq S \,,$$
$$p_{ij} = 0 \text{ for } j \leq i \,,$$

then

$$\lim_{S \to \infty} P(M \geq \rho S) = 1 \,.$$

To give a concrete instance of this example, pick some $c > 2$ and define $\{p_{ij}\}$ by

$$p_{ij} = (c-2) \,\Big/\, \left[2 \left(\sum_{k=2}^{\infty} k^{-3/2} \right) (j-i)^{3/2} \right], \quad j \geq i+2$$
$$= 1 - \varepsilon, \quad j = i+1$$
$$= 0, \quad j \leq i \,.$$

Taking, say, $\rho = 0.999$, there is a small enough ε that

$$\lim_{S \to \infty} P(M \geq 0.999 S) = 1$$

even though

$$\sup_{1 \leq i \leq S} E(\text{number of predators and prey of species } i)$$
$$= \sup_{1 \leq i \leq S} \left(\sum_{k=1}^{i-1} p_{ki} + \sum_{j=i+1}^{S} p_{ij} \right)$$
$$\leq 2(1 - \varepsilon) + 2 \sum_{h=2}^{\infty} (c-2) \,\Big/\, \left[2 \left(\sum_{k=2}^{\infty} k^{-3/2} \right) h^{3/2} \right]$$
$$= 2(1 - \varepsilon) + c - 2 < c, \text{ for all } S \,.$$

The example demonstrates that even when the expected number of links per species is kept below a fixed c, not every anisotropic cascade model will explain the observed slow increase in the height of real webs.

8. Simulations of the Cascade Model

The preceding analysis leaves open the question: how good are our theoretical bounds for the median height? We set $c = 3.71$ based on the sample of 62 webs, then generated webs according to the cascade model for each of $S = 50, 100, 150$ and 1000 and found the height of each simulated web (by using an algorithm described in the appendix of Chap. III.4). Table III.5.2 presents the simulated frequency distributions of height, and beneath each simulated distribution the numerical values of our theoretical bounds on the median height. Evidently the bounds on the median do contain the sample median height. The concentration of height on at most two values established above in the limit of unrealistically large S does not occur for the values of S used in these simulations. There is however a suggestion of more concentration for $S = 1000$ than for $S = 50$.

Table III.5.2. Frequency distributions of the length of the longest chains in webs of various sizes, and theoretical estimates of the median (Web sizes were simulated according to the cascade model with $c = 3.71$)

| longest chain | number of species | | | |
| | 50 | 100 | 150 | 1000 |
	relative frequency			
4	0.03	0.00	0.01	0.00
5	0.17	0.24	0.07	0.00
6	0.32	0.31	0.27	0.00
7	0.24	0.24	0.33	0.05
8	0.15	0.14	0.18	0.25
9	0.08	0.05	0.08	0.40
10	0.02	0.02	0.05	0.20
11	0.01	0.00	0.01	0.10
	number of simulations			
	200	100	100	20
	theoretical estimates of median longest chain			
lower bound	3	4	4	6
upper bound	9	10	10	12
m^*	10	11	11	13
$\ln S / \ln (\ln S)$	2.87	3.02	3.11	3.57

9. Achievements of This Theory

This chapter presents the first, to our knowledge, exactly derived theory of the length of food chains in webs with a large number of species. This theory suggests for the first time a (simple) quantitative relation between the mean length of chains and the mean number of predators plus prey per species. The analysis also provides the first quantitative explanation, derived from an explicit model

that is not invented *ad hoc* for the purpose, of why the longest chains are very
short relative to the number of species in a web even when the number of species
is large.

From a generating function for the expected numbers of chains of each length,
we derive the mean and variance of the length of chains by using the relative
expected frequency as the probability density function of chain length. For webs
in which S becomes arbitrarily large, we show that the limiting relative expected
frequency and the limiting expected relative frequency of chain lengths are the
same, so that either may be used to describe the distribution of chain lengths.
We compute the asymptotic distribution and all moments of chain length, giving
explicit closed-form formulas for the asymptotic mean and variance. We show
that the relative frequency of chains of any given length converges in probability
to its expectation as S gets large. The cascade model implies a simple 'rule of
thumb' for large webs: the mean length of chains equals the mean number of
predators plus prey of any species in the web.

We also derive, from the cascade model, upper bounds on the upper and lower
tails of the probability distribution of the height, or length of the longest chain,
of a web. From these, we compute bounds on the median height in webs with
a finite number of species. These bounds show that the median height is a very
slowly increasing function of the number of species in a web, remaining below
20 up to 10^8 species. For webs in which S becomes unrealistically large, the
height equals one of two adjacent integers (that depend on S) with a probability
that approaches 1. For very large S, these integers approximate $\ln S / \ln(\ln S)$,
a function that grows very slowly with S.

By considering variations on the assumptions of the cascade model, we show
that the ability of the cascade model to explain the slow growth of the height is
robust with respect to changes in the probability that one species eats another.
However, if the probability that one species eats another is permitted to depend
on the pair of species concerned, then the height may increase either not at all
or linearly with the total number of species. Hence not every variation on the
cascade model will explain the observed short height, relative to the number of
species, of real webs.

Simulations of the cascade model show that the concentration of the height
on just two integer values, predicted by the asymptotic theory, occurs only in
webs with an unrealistically large number of species.

10. Some Remaining Tasks

Although the cascade model yields to mathematical analysis, the acyclic model
(model 2 in Chap. III.2) resists analysis. We do not know, for example, whether
the median height in the acyclic model grows slowly with S, as demonstrated
here for the cascade model. A solution to this problem might reveal whether
the asymptotic behaviour of the height could be used to discriminate between
different models of webs.

The cascade model and its kin are static models. They describe data that are snapshots, sketches of webs at a single moment. Static models and static data ignore the reality that the species and links of webs may change with the seasons and over longer intervals. It would be highly desirable to develop and test dynamic models of communities that are consistent with the static empirical regularities on which the cascade model is based.

The cascade model and the data it is intended to interpret ignore the numbers of individuals or biomass of each species and the quantities of flows in each link. Far fewer observed webs give quantitative measurements than give, like the webs studied here, all-or-none information about species and links. Thus the whole line of work from Cohen (1978) to this book is only a first step towards a real understanding of webs, because it deals entirely with combinatorial structure rather than with quantities of stocks and flows in webs. However, gross anatomy precedes physiology. This line of work at least offers a coherent theoretical and empirical approach to some aspects of the gross anatomy of webs.

What might be offered by better data and models that will, we hope, replace those we analyse here? Quantitative, predictive models of webs could assist in foreseeing the paths and concentrations of natural and artificial toxins in the environment, and the consequences of the removal and introduction of species. Such models could assist in the design of nature reserves on Earth and closed regenerative ecosystems for supporting humans during prolonged stays in space; the cascade model suggests already that certain proportions of top, intermediate and basal species (or physico-chemical equivalents) need to be provided or else will evolve. Finally, since the webs containing the species man are not notably different in structure from those without man, such models may provide some understanding of man's place in nature. These grand opportunities are an incentive to pursue the hard scientific work that may bring them within reach.

§6. Intervality and Triangulation in the Trophic Niche Overlap Graph

Joel E. Cohen and Zbigniew J. Palka

1. Introduction

When the diets of different organisms overlap in natural communities, the possibility arises that the different consumers may compete for food (Grant 1986) or may interact mutualistically (Kawanabe 1986, 1987). Competitive or mutualistic interactions over food may influence the evolution of the competing or cooperating consumers. Hence overlaps in the diets of different organisms are of both ecological and evolutionary interest.

The diets of organisms, and the relations among the diets of different kinds of organisms, vary greatly from one ecological community to another. The minimum number of variables required to describe or represent the overlaps among consumers' diets has been called the dimension of trophic niche space (Cohen 1978). A priori the dimension of trophic niche space would be expected to vary among communities. If, in a particular community, this dimension were one, then intervals of some single variable, perhaps food size, would be necessary and sufficient to describe when the diet of one species overlaps with another. If the dimension were greater than one, then intervals of no single variable, such as food size alone, would suffice to describe when the diets of consumers overlap. That is, if the dimension exceeded one, then it would be necessary to consider at least two variables, perhaps food size and time of day, to account for or describe the presence or absence of overlaps in the diets of different organisms. The dimension of trophic niche space is one measure of how complex the dietary relations among consumers are in a given community.

This chapter reports new theoretical and empirical information about the overlaps among the diets of organisms in natural communities. On the basis of mathematical calculations, computer simulations and new analyses of 113 community food webs, we shall show that, in nature and in theory, the larger the number of trophic species in a web, the larger the probability that the dimension of a community's trophic niche space exceeds one. Equivalently, the larger the number of species in a web, the smaller the probability that the web is interval.

In addition to intervality, the overlaps among the diets or among the consumers of organisms in natural communities may possess another property (defined below) called triangulation (Sugihara 1982). Again on the basis of mathematical calculations, computer simulations and new analyses of 113 community food webs, we shall show that the larger the number of trophic species in a web, the smaller the probability that either the overlap graph or the resource graph of the web is triangulated.

The mathematical calculations and computer simulations use a stochastic model of community food webs called the cascade model (Cohen and Newman 1985; Chap. III.2).

The remainder of this introduction gives further details on the background to this work. The terms used in this chapter are defined, including consumer or niche overlap graph, resource or common enemy graph, interval graph, interval web, and triangulated web. None of these terms is new; readers familiar with theoretical developments in food webs over the last decade could jump directly to the section on intervality. There the data and theory on the frequency of intervality are described and compared. The description of theoretical results is meant to be intelligible to those who are willing to deal with quantitative concepts but are not interested in the details of proofs, which are provided in an Appendix. The following section compares the observed and predicted frequency of webs with triangulated overlap graphs and triangulated resource graphs. The final section summarizes the results and relates them to previous work. An Appendix analyzes mathematically the cascade model's implications for overlap and resource graphs.

Background

The ecological niche of a species has been defined (Hutchinson 1944) as "a region of n-dimensional hyper-space, comparable to the phase-space of statistical mechanics," that represents all the environmental, including biotic, factors that influence individuals of that species. "Dimension" as used here refers to the minimum number of variables needed to describe the niche, and should not be confused with the physical dimension (e.g., flat or 2-dimensional vs. solid or 3-dimensional) of a habitat (Silvert 1984, pp. 158–161; Chaps. II.6, III.4). Hutchinson's definition raises several questions. What is the (Hutchinsonian) dimension of the niches in a particular community? Equivalently, what is the minimum number of variables required to describe the factors that influence species in a community? Is the dimension the same or different in different communities?

Food webs offer information about the number of *trophic* dimensions in the niches of species in a community (Cohen 1978; Chap. II.4). If two species eat a common food species, then their niches must overlap along the trophic dimensions: otherwise, the two consumers would not have access to the same food. If the dietary overlaps among consumers in a community can be described by the overlaps among intervals of a single variable, the web of the community is said to be an "interval" web. If intervals of more than one variable are required to describe the dietary overlaps among consumers in a community, the web is said to be "non-interval".

In the first collection of webs assembled to investigate the trophic dimension of ecological niches, 22 or 23 of 30 webs were found to be interval (Chap. II.4). The exact number (22 or 23) depended on how the data in the web were edited. The observed numbers of interval webs exceeded markedly the numbers of interval webs predicted by seven simple models of food webs (Cohen 1978; Cohen, Komlós and Mueller 1979). These findings provoked further analyses of the available data (e.g., Critchlow and Stearns 1982; MacDonald 1979; Pimm 1982; Sugihara 1982; Yodzis 1982, 1984). We shall discuss these analyses later.

Two recent changes now make it opportune to re-examine the question of intervality. First, more data are available. Second, a better food web model is available, and can be analyzed.

As for data, the number of webs studied in Chap. II.4 is small. Sugihara (1982) analyzed Briand's 40 webs (Chap. II.5, including 13 of those assembled by Cohen [1978]) and reached conclusions consistent with those of Chap. II.4. Briand has now informed us which of the 113 community food webs listed in Chap. IV of this book are interval.

As for modeling, the models Cohen (1977, 1978) considered were constructed *ad hoc* to match the mean number of dietary overlaps. Some of those models also matched the variance of the number of dietary overlaps. Recently, a better food web model, called the cascade model, has been discovered. The cascade model describes qualitatively and quantitatively the numbers of top, intermediate and basal trophic species and numbers of basal-intermediate, basal-top, intermediate-intermediate and intermediate-top trophic links, when all food webs are considered together (Chap. III.2) or individually (Chap. III.3). The cascade model also describes the numbers of food chains of each length (Chap. III.4) and explains Hutchinson's (1959) observation that food chains are typically very much shorter than the number of species in a web (Chap. III.5). It is natural to ask (as in Chap. III.2 and Chap. III.4) whether the cascade model can account for the observed frequency of intervality, and likewise for triangulation.

For further background on food webs, see Pimm (1982), DeAngelis et al. (1983) and MacDonald (1983).

Terminology

The dietary overlaps of the consumers in a web are described by an overlap graph, short for "trophic niche overlap graph," which is constructed as follows (Chap. II.4). Given the web W (whether W is represented as a digraph or a predation matrix), the vertices of the overlap graph $G(W)$ are the same as those of W, i.e., one vertex for each species in the community. In $G(W)$, there is an undirected edge between distinct vertices i and j (representing an overlap between the diets of species i and species j) if and only if there exists some third vertex k such that, in W, i eats k and j eats k. Thus, two vertices are joined by an edge in $G(W)$ if there are arrows in W from k to i and from k to j, for at least one k; or if at least one row of A has elements equal to 1 in both column i and column j. The overlap graph of a web was originally called the competition graph (Cohen 1968), a name still used by graph theorists, and has also been called the consumer graph (MacDonald 1983, p. 32).

The resource graph, in the terminology of Sugihara (1982), describes which prey share a common predator. The vertices of the resource graph $H(W)$ are the same as those of the web W. In $H(W)$, there is an undirected edge between distinct vertices i and j if and only if there exists some third vertex k such that, in W, k eats i and k eats j. Thus, two vertices are joined by an edge in $H(W)$ if there are arrows in W from i to k and from j to k, for at least one k; or if at least one column of A has elements equal to 1 in both row i and row j.

The resource graph of a web W is the dual of the overlap graph of W, in the sense that the resource graph equals the overlap graph of the web W^* obtained from W by reversing the direction of every link in W, i.e., $H(W) = G(W^*)$. The resource graph was simultaneously and independently invented by Sugihara (1982) and by Lundgren and Maybee (1985, in a paper prepared for a 1982 conference), who called it the common enemy graph. Independently, and prior to either of these graph theoretic constructions, Holt (1977) introduced the notion that two species are in "apparent competition" if there is a consumer that preys on both of them and if a change in the abundance of one species induces a numerical response in the other. The resource graph presents necessary but not sufficient conditions for the relation of "apparent competition" in a community.

Many other graphs can be constructed from a web (Sugihara 1982; Roberts 1988; Cable et al. 1988). We shall discuss primarily the overlap graph and, to a lesser extent, the resource graph.

A graph (with undirected edges) is said to be an interval graph if, for each vertex of the graph, there exists an open interval of the real line such that there is an edge between any two vertices if and only if the two corresponding intervals intersect, that is, overlap. In an interval graph, it is possible to find an interval of the real line corresponding to each vertex of the graph, and the connections among the vertices are exactly represented by the overlaps among the intervals of the line.

A web W is said to be interval if its overlap graph $G(W)$ is an interval graph (Cohen 1978; Chap. II.4). In a web that is interval, the dimension of trophic niche space could be 1, because the range of variation of the diet of each consumer could be identified with an interval of the real line (for example, the range of sizes of food eaten by a consumer), and overlaps among diets of consumers in the web would correspond to overlaps of the intervals on the real line. Lumping trophically equivalent kinds of organisms into trophic species has no effect on whether a web is interval: an unlumped web is interval if and only if the corresponding lumped web is interval.

2. Intervality

Data

The sources and principal characteristics of the 113 webs analyzed here have been presented already (Chap. II.6 and Chap. III.4). A few minor corrections of previously published numbers of species and links are required. The numbers of species and links for web number 37 given in Chap. III.4 are corrected in Chap. II.6; the latter values will be used here. In webs numbered 6, 7, 24, 45, 51, 65, and 93, Briand and Cohen (1987; Chap. II.6) overlooked the possibility of lumping two consumers into a single trophic species. Hence the correct number of trophic species for these webs is one less than the number given by Briand and Cohen (1987; Chap. II.6) and the correct number of trophic links is, respectively, 2, 3, 2, 3, 3, 5 and 7 less than published by Briand and Cohen (1987; Chap. II.6). In calculating these values of species and links in the webs taken from Cohen (1978), matrix elements Cohen (1978) reported as −1 are replaced by 1 and matrix elements reported as −2 are replaced by 0.

According to F. Briand (personal communication, 18 August 1985), all but 16 of the 113 webs are interval, that is, have interval overlap graphs. The non-interval webs have serial numbers 3, 6, 18, 20, 22, 26, 27, 33, 39, 41, 60, 67, 98, 99, 100, and 106. We have not repeated his calculation.

The proportion of all webs that are interval webs is $97/113 = 0.86$. This proportion is higher than the proportion of interval webs among the community webs in Chap. II.4, namely, $9/14 = 0.64$ or $8/14 = 0.57$, depending on the version of the webs used. Using the same 40 webs as Chap. II.5, Sugihara (1982, Chap. 4) identified 73 connected components with more than one species, and found that only 10 of these 73 had overlap graphs that were not interval. The proportion of interval webs in Sugihara's collection of components is $63/73 = 0.86$. Those 40 webs are among the 113 webs analyzed here, and a web is interval if and only if its components are interval, so Sugihara's proportion of intervality and the proportion just found here are not independent. However, excluding the first 40 webs of Briand's collection (those in Chap. II.5 and Sugihara (1982)), only seven of the remaining $73 = 113 - 40$ complete webs (not components, as in Sugihara (1982)) fail to be interval. The proportion of intervality among these 73 webs, namely, $66/73 = 0.90$ is independent of the proportion of intervality among the 73 components studied by Sugihara (1982). Thus, in this collection of 113 webs, the proportion of webs that are interval is as high as, or higher than, the proportion of interval webs observed previously.

Thanks to the large number of webs now available, it is possible to examine how the proportion of intervality covaries with other characteristics of webs. The most fundamental characteristic, which is examined here, is the number S of species. All webs with $S \leq 16$ are interval. Of the five webs with the largest numbers of species (ranging from 32 to 48 species), none is interval. When the observed range of variation of S, from 3 to 48, is divided into four nearly equal intervals, the fraction of interval webs declines steadily from one among webs with 3 to 14 species to zero among webs with 35 to 48 species (Table III.6.1). However, there are only two webs with 35 to 48 species. When the frequency distribution of S is divided into quartiles, so that each group contains, as nearly as possible, one quarter of all the webs, the fraction of interval webs again declines steadily from one among webs with 3 to 11 species to 0.59 among webs with 22 to 48 species (Table III.6.1). In summary, the fraction of webs that are interval is strongly associated with the number of species in the webs, declining from one for small webs toward zero for large webs.

Though quantitative documentation of this finding seems to be new, hints of it appeared earlier. Cohen (1978, p. 40) observed that webs which incorporated multiple habitats were much less likely to be interval than webs from single habitats; multiple-habitat webs also tend to have more species. More explicitly, MacDonald (1979, p. 586) remarked that "[t]he non-interval community webs ... are the webs with the largest" numbers of species. Neither Cohen (1978) nor MacDonald (1979) analyzed the relation between species number and intervality any further, empirically or theoretically.

Our empirical finding that intervality is less frequent among larger webs is consistent with data presented by Sugihara (1982, his Table 4.1, pp. 73–74). The numbers of consumers in his 73 components of overlap graphs range from 2 to 34 species. According to our tabulation of his data, of the 52 component webs with 2 to 10 species, 50 are interval (96%); of the 14 component webs with 11 to 14 species, 12 are interval (86%); and of the seven component webs with 15 to 34 species, one is interval (14%).

Table III.6.1. Observed relative frequency of interval overlap graphs and of triangulated overlap and resource graphs in 113 community food webs as a function of the number of species. Presence or absence of intervality was computed by F. Briand. We computed the presence or absence of triangulation in the overlap and resource graphs from the predation matrices in Chap. IV

Number of species	Number of webs	Fraction of webs that are interval	Fraction of overlap graphs that are triangulated	Fraction of resource graphs that are triangulated
(a) Species divided into four intervals of nearly equal length				
3–14	56	1	1	1
15–24	40	0.775	0.975	0.875
25–34	15	0.667	0.800	0.800
35–48	2	0	0	0.500
(b) Species divided into four intervals of nearly equal frequency				
3–11	28	1	1	1
11–14	28	1	1	1
15–21	28	0.857	1	0.929
22–48	29	0.586	0.793	0.759

Theory

This section describes the cascade model and its predictions regarding the probability that a web is interval.

The cascade model assumes that species in a community are ordered in a cascade such that any species can consume only those species below it in the ordering. Operationally, if there are S species in the web, the cascade model assumes a labeling of the species from 1 to S in such a way that whenever a species labeled i is eaten by a species labeled j, then $i < j$. This assumption excludes the possibility of trophic cycles, e.g., cases where i eats j and j eats i. Moreover, the cascade model assumes that for any two species i and j with $i < j$, the probability that j actually eats i is p, and whether j eats i is statistically independent of all other eating relations in the web. The positive probability p is independent of the particular pair of species i and j. When webs with different total numbers S of species are compared, the cascade model assumes that p depends inversely on S according to $p = c/S$, where c is a positive constant that is independent of S.

Under these assumptions, the probability p of a link is just the expected or average value of the connectance defined by Rejmánek and Starý (1979): $p = E\{L/[S(S-1)/2]\}$.

We calculated explicit formulas for the probability P that a web W is interval, i.e., the probability that the overlap graph $G(W)$ of W is an interval graph, for extremely small S and extremely large S (Appendix, Theorem 5). The probability P that a web is interval depends on, and should not be confused with, the probability p of a link between any two species i, j with $i < j$ in the web.

If $S = 3$, 4, or 5, then $P = 1$. If $S = 6$ and $S = 7$, a lower bound on P is the difference between 1 and a sum of high powers of p (the link probability) times high powers of $1 - p$ (see Appendix). As the product of high powers of p times high powers of $1 - p$ must be small, one expects (and numerical results below will confirm) that this lower bound on P will be very close to one. Thus for low values of S, the probability P that a web is interval is one or close to one.

At the other extreme, the larger S gets, the closer P gets to $\exp(-\lambda)$ where $\lambda = 0.002527(2L/[S-1])^9 S$ (see Appendix, Theorem 5). According to the cascade model, the expected number of links in a web is $pS(S-1)/2 = c(S-1)/2$, so the average of $2L/[S-1]$ is just c. The best current estimate of c, based on aggregate data for all webs, is approximately 4. If we replace $2L/[S-1]$ by 4 in the expression for λ, we obtain approximately $\lambda = 660S$. Thus, for average webs according to the cascade model, P is expected to decline exponentially fast with increasing S, and the coefficient of S in the exponent is large, in excess of 660. Thus for large S, the cascade model predicts a frequency of intervality near zero.

These are the principal results of the Appendix about the probability that a web is interval. In addition, the Appendix establishes other important structural properties predicted by the cascade model for the overlap graphs of large webs. The cascade model predicts that the overlap graph should contain a complete subgraph on n vertices, for any finite n, with probability one as S becomes large. The cascade model predicts that the overlap graph should contain an induced tree on n vertices, for any finite n, with probability one as S becomes large. The probability that the overlap graph is a unit interval graph approaches zero as S becomes large.

Table III.6.2 below reports simulations that establish an upper bound on the probability that a web is interval when S is 10, 20, 30, 40 and 50 species. These simulations establish that the probability of intervality predicted by the cascade model is essentially 0 by the time S is as large as 40.

Because of the duality between the overlap graph and the resource graph, with a corresponding duality in the probability distribution of edges according to the cascade model (see Appendix), all the preceding analytical and numerical results in this section remain valid if "overlap graph" is replaced by "resource graph."

Confronting Data and Theory

This section compares the data on intervality with the cascade model's quantitative predictions about the probability that a web is interval.

To do so, it is necessary first to estimate either of the parameters $p = c/S$ or $c = pS$ of the cascade model. The parameters may be estimated in two different ways: using data on all webs simultaneously (Chap. III.2, Chap. III.4), and using data from each web separately (Chap. III.3, Chap. III.4).

Using data on all webs simultaneously, c is twice the estimated slope of a straight line through the origin fitted to the data points (S, L), where L is the number of links in a web with S species. In the 113 webs analyzed here, that slope is 1.99 ± 0.07 (standard error), so c is very nearly 4 (Chap. III.4). With this value of c, the cascade model makes sense only for webs with $S \geq 4$, since by definition $p \leq 1$.

Using data from a single web with S species and L links, a reasonable estimate of p is $L/[S(S-1)/2]$, which is the connectance; the numerator is the observed number of feeding relations, and the denominator is the number of possible feeding relations given the assumption of ordering. (Estimating p by the connectance $L/[S(S-1)/2]$ overlooks the omission of isolated species from the data. A more complex estimate (Chap. III.3) allows for the omission of isolated species. Since the number of isolated species is small, the error introduced by estimating p from connectance is also small.)

As a preliminary, we now check the cascade model's assumption that the link probability p depends on the number S of species according to c/S. This assumption implies that if the connectance or p is estimated separately for each web, then the points $(S, 1/p)$ should fall around the straight line $S/c = S/4$ derived from the aggregated data. The agreement in Fig. III.6.1 between the individual points and the predicted straight line justifies further testing of the cascade model. Each web in Fig. III.6.1 is represented by the symbol "1" (for one-dimensional) if the web is interval, or by the symbol "2" (for $\geq$ 2-dimensional) if the web is not interval.

When the link probability p is estimated separately for each web, the cascade model predicts a probability of intervality $P = 1$ for $S = 3, 4, 5$, as already mentioned, $P \geq 0.9999$ for $S = 6$, and $P \geq 0.9986$ for $S = 7$. All webs with $S \leq 16$ (the webs plotted in the left third of Fig. III.6.1) are interval. Thus for very small numbers of species S, the data are consistent with the predicted probability P that a web is interval.

For intermediate numbers of species, the observed fraction of webs that are interval declines as shown in Table III.6.1. An upper bound on the predicted probability that a web is interval is given by the predicted probability that a web is triangulated (see next section). Table III.6.2 reports estimates of the probability of triangulation for $S = 10, 20, 30, 40$ and 50, based on 100 simulations for each value of S. The predicted probability of triangulation (Table III.6.2), and therefore the predicted probability of intervality, appears to decline with increasing S more rapidly than the observed frequency of intervality. That is, there is still an excess frequency of intervality not explained by the cascade model. But the cascade model does predict correctly the existence and the location of a range of S over which the probability of intervality declines smoothly from near 1 to near 0.

For very large S, the cascade model predicts asymptotically a probability of intervality P lying between $\exp(-39)$ and $\exp(-3.2 \times 10^8)$, according to the theory developed in the Appendix. In these calculations, the link probability p is estimated separately for each web. The simulations in Table III.6.2 suggest that the asymptotic theory becomes relevant when the number of species is between

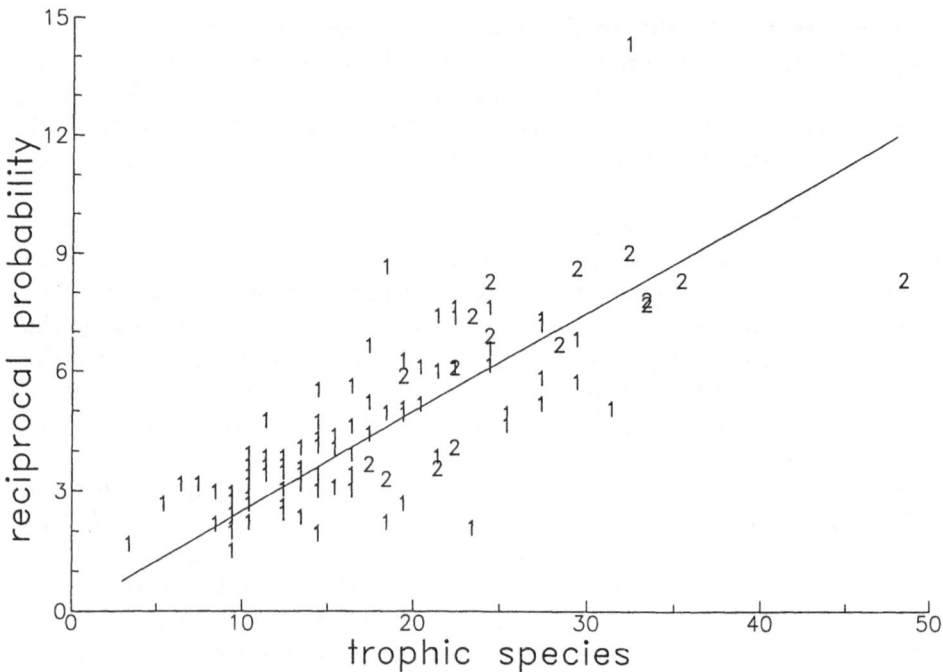

Figure III.6.1. Reciprocal of the link probability or reciprocal of connectance as a function of the number of species in 113 community food webs. If S is the number of trophic species and L is the number of trophic links in a web, the ordinate is $(S[S-1]/[2L])$ and the abscissa is S. "1" signifies that the web is interval, "2" that the web is not interval. The solid line plots $S/4$ as a function of S

30 and 40. Consistent with these analytical and computational predictions, the five largest webs, with S ranging from 32 to 48, are all non-interval.

Overall, there is good qualitative agreement, and reasonable quantitative agreement, between the observed frequency of interval webs and the frequency of interval webs predicted by the cascade model. For intermediate numbers of species, more interval webs are observed than are predicted by the simulations of the cascade model. It remains to be determined whether this excess identifies a deficiency of the cascade model or a deficiency of the data on trophic links or both.

An upper bound on the fraction of webs with interval *resource* graphs is given by the fraction of webs with triangulated resource graphs. The relative frequencies of triangulated resource graphs are given in Table III.6.1 and are discussed in the next section.

3. Triangulation

A web is said to be triangulated if its overlap graph is triangulated. A graph is triangulated if it has no induced cycles of four or more edges; that is, whenever four or more vertices in the overlap graph make a cycle, there is an edge that cuts across the cycle, reducing the cycle to a composition of triangles. Lekkerkerker and Boland (1962) showed that a graph is interval if and only if it is triangulated and it contains no asteroidal triples. Thus the probability that a graph is triangulated is an upper bound on the probability that it is interval.

Sugihara (1982) showed that the frequency of intervality in simulated webs could largely be accounted for by requiring the overlap graphs to be triangulated. As part of a more extensive theory that will not be reviewed here, he proposed that triangulation was a more fundamental property of webs than intervality.

Sugihara (1982, p. 118) simulated a dynamical Lotka-Volterra model with random interaction coefficients, and allowed species to go extinct until the hypothetical community was "feasible." His model communities started with 15 species and the final number of species ranged from six to nine. In 18 of 20 simulations, the final communities had triangulated niche overlap graphs. Sugihara noted that "the high frequency of rigidity [equivalent to triangulation] may simply be an artifact of generating relatively small final communities," i.e., communities with a small number of species. Though it was not Sugihara's preferred interpretation of the high frequency of triangulation, this possibility is consistent with the following analyses of data and the cascade model.

Data

We determined the triangulation of the overlap graph and the resource graph of each of the 113 webs in Chap. IV by constructing these graphs from the predation matrices. The most efficient algorithms to determine whether a graph is triangulated are LEX P and FILL of Rose, Tarjan and Lueker (1976), based on lexicographic breadth-first search. We programmed their algorithms using a description by Booth (1975, p. 126) and verified the performance of our program in numerous examples.

All of the webs with non-triangulated overlap graphs (numbers 6, 18, 33, 39, 60, 99, 100 and 106) are also non-interval, as is logically required by the theorem of Lekkerkerker and Boland (1962). This consistency provides a check, albeit weak, on our independent computations. Nine webs have non-triangulated resource graphs (numbers 6, 18, 33, 60, 63, 67, 69, 99 and 100).

George Sugihara (personal communication, 25 October 1988) provided proposed corrections to several of Briand's predation matrices. When these corrections are made, webs 6 and 18 have triangulated overlap graphs. This change does not alter the general trends in the data. For consistency, we shall use the predation matrices as furnished by Briand (Chap. IV).

Table III.6.1 shows the relative frequency of triangulated overlap graphs and triangulated resource graphs in 113 community food webs as a function of the number of species. For both overlap and resource graphs, the frequency of triangulation declines from 1 for the smallest observed webs to much smaller values for the largest observed webs. All four webs of more than 32 species have non-triangulated overlap graphs and two of those four webs have non-triangulated resource graphs.

Theory

The predictions of the cascade model regarding triangulation are obtained by
mathematical analysis (see Appendix) and simulation. Analytically, the probability that an overlap or resource graph is triangulated is one whenever the
number of species in the web is five or smaller, and is very close to one for
six and for seven species. For large numbers of species and a link probability
$p = 4/S$, the cascade model predicts asymptotically that the probability that a
resource or overlap graph is triangulated is very near zero (Appendix Theorem
7). For intermediate numbers of species (Table III.6.2), the simulated probability
of triangulation according to the cascade model declines rapidly with increasing
S.

Table III.6.2. Simulated relative frequency of triangulated overlaps graphs or resource graphs
predicted by the cascade model, according to 100 simulations for each number of species.
The 95% confidence interval incorporates the correction for continuity, and negative lower
confidence limits for 30, 40 and 50 species were set to 0

Number of species	Fraction of triangulated overlap or resource graphs	Lower 95% confidence limit	Upper 95% confidence limit
10	0.91	0.85	0.97
20	0.26	0.17	0.35
30	0.03	0	0.07
40	0	0	0.005
50	0	0	0.005

Confronting Data and Theory

The cascade model's predictions are consistent with observation for very small
numbers of species and for large numbers of species. For intermediate numbers
of species, the simulated probability of triangulation appears to decline with
increasing numbers of species more rapidly than the observed relative frequency
of triangulation. But the cascade model does predict correctly the existence and
location of a range of S over which the probability of triangulation declines
smoothly from near one to near zero. The difference between the observed and
simulated relative frequency of triangulation for intermediate numbers of species
may be due to imperfections of the data or of the cascade model or both.

4. Discussion and Conclusions

Major Findings

The main accomplishments of this chapter are three. First, while confirming empirically the overall high relative frequencies of interval and triangulated overlap
graphs found previously, we observe that the relative frequencies of interval and
triangulated webs are strongly associated with web size, as measured by the
number of species. All overlap graphs of webs with small numbers of species

(16 or fewer in our data) are observed to be interval and triangulated, while no overlap graphs of webs with large numbers of species (33 or greater in our data) are observed to be interval or triangulated. Between these extremes, a steady downward trend is observed in the fraction of interval and triangulated overlap graphs. The pattern of triangulated resource graphs is similar. Broadly, the larger the number of species in a community, the less likely it is that a single dimension suffices to describe the community's trophic niche space, and the less likely it is that there are no "holes" in the overlap graph or resource graph.

There are two ways to look at this finding. One possibility is that webs with small numbers of species come from specially simple communities; the simplicity gives the communities a small number of species as well as a very small number of dimensions of trophic niche space, namely, just one. Another possibility, which we favor, is that most webs with small numbers of species are very incomplete descriptions of real communities. When communities are described in detail, reported webs contain larger numbers of species and are less likely to be interval and triangulated.

This interpretation is consistent with the empirical finding of Schoener (1974) that the "separation [of species in niche space] appears generally to be multidimensional" (p. 29), although he recognized that "the dimensions that ecologists recognize are rarely independent" (p. 32). In 81 studies of niche relations in groups of three or more species, when the dimensions originally reported are classified into the broad categories of food, space and time, most niches are separated by two dimensions (Schoener 1974). Other studies of the dimension of ecological niches are reviewed by Cohen (1978, pp. 97–100).

This interpretation leads to a concrete prediction. If webs reported in the future are consistent with the trends in the existing data and if they are reported in greater detail than most present webs, they will display much lower relative frequencies of intervality and triangulation than do the existing webs with small numbers of species, even in the communities with webs presently reported as interval or triangulated. As the fidelity and detail of the description of communities improve and the numbers of species in reported webs increase, we expect the relative frequencies of intervality and triangulation to decline.

Second, we calculate the predictions of the cascade model about the probabilities that the overlap graph and resource graph are interval and triangulated, for both very small and very large numbers of species. For very small webs, the predicted probability that either graph is interval or triangulated approximates one. For a web with a very large number S of species, and with approximately twice as many links as species (in accordance with the empirical link-species scaling law), the predicted probability that either graph is interval falls as approximately $\exp(-660S)$, i.e., extremely rapidly with increasing S. The predicted probability that either graph is triangulated also falls exponentially fast.

We do not know of any previous analytical (as opposed to numerical) calculations of the probability of interval or triangulated overlap or resource graphs starting from a model of webs. The calculations constitute nontrivial new mathematics.

Third, comparing data and theory, we show that the predictions of the cascade model account quantitatively for the observed relative frequencies of interval and triangulated overlap graphs and triangulated resource graphs for webs with seven or fewer and 33 or more species. The cascade model also predicts correctly the existence and location of a range of numbers of species over which the relative frequencies of interval and triangulated overlap and resource graphs decline smoothly from near one to near zero. Our simulations of the cascade model reveal, however, that there are more interval and triangulated overlap graphs and more triangulated resource graphs observed than expected in webs with intermediate numbers of species. This difference may be due to imperfections of the data or of the model or both.

The cascade model's successful prediction of the existence and location (though not the exact rate) of declines in the relative frequencies of intervality and triangulation with increasing numbers of species suggests that the relative commonness or rarity of interval and triangulated webs may be a statistical consequence of the general ecological processes posited in the hypotheses of the cascade model, rather than a consequence of special constraints (of whatever origin) acting directly on the dimension of trophic niche space or the homological structure of the overlap graph or the resource graph.

Related Prior Work

There have been several previous attempts to explain the relative frequency of intervality. Cohen (1978) simulated six simple web models and found that they predicted fewer interval overlap graphs than were observed. Cohen, Komlós and Mueller (1979) calculated the probability that a random graph is interval when the random graph is constructed with an edge probability that is the same for every pair of vertices, i.e., according to the classical model of Erdös and Rényi (1960). That model also failed to account for the observed frequency of intervality. (By contrast with the model of Erdös and Rényi, when the overlap graph is derived from the cascade model, the probability of a dietary overlap between two species, or of an edge between the corresponding vertices in the overlap graph, is much higher for two species high in the ordering than for two species low in the ordering.)

Critchlow and Stearns (1982) showed that the predation matrices of the real webs analyzed by Cohen (1978) were divided into block submatrices much more than were the simulated predation matrices generated by Cohen's model 5, and that in general the real webs had fewer dietary overlaps (or edges in the overlap graph) than webs simulated according to model 5 with the same number of predators, prey and links. Critchlow and Stearns showed that both the deficit of block submatrices and excess of dietary overlaps in the simulated webs helped to explain why model 5 underpredicted the observed frequency of intervality.

Yodzis (1984) formulated assembly rules, based on energetic constraints, for the hypothetical construction of an ecosystem from species that arrive sequentially. These assembly rules generate model webs that describe well many structural features of 25 of the 28 webs from fluctuating environments in Briand's

(1983; Chap. II.5) collection of 40 webs, and 3 of the 12 webs from constant environments in Briand's collection. In particular, when Yodzis's model describes well most other structural features of a real web, it also describes well the presence or absence of an interval overlap graph.

Yodzis reported his model's expected intervality for the 28 webs well described by his assembly rules (Yodzis 1984, p. 122, his Table 1). For these webs, we graphed his expected intervality as a function of the observed number of trophic species for all the webs (graph not shown). We found Yodzis's expected intervality near one for the webs with the smallest number of species; a hint, amid much scatter, of a declining trend in Yodzis's expected intervality with an increasing number of species; and the smallest values of Yodzis's expected intervality for the webs with the largest number of species. Yodzis did not remark this association between his expected intervality and the number of species in a web.

Yodzis's assembly rules provide an alternative explanation for the trend we have reported here in the frequency of intervality as a function of number of species. But this explanation may be limited to webs from fluctuating environments. By contrast, the cascade model deals equally well with webs from fluctuating and constant environments. Whereas Yodzis's assembly rules so far have been analyzed only by computer simulation, the cascade model is tractable to explicit analysis. In spite of (what we view as) the advantages of the cascade model, the parallels between its predictions and those of Yodzis's assembly rules suggest that it would be worthwhile in the future to determine whether there are deeper connections between the two models.

Sugihara (1982, Chap. 4, p. 65) explained the high frequency of interval graphs in terms of different assembly rules that prevent the appearance of "homological holes" in communities. He considered the highly frequent, but not universal, appearance of intervality in real webs to be a consequence of a more fundamental requirement that real webs be triangulated. The data (Table III.6.1) indicate that larger webs are less likely to be interval and triangulated. If these trends are not an artifact of faulty data, then the absence of homological holes in the overlap graph is not a universal feature of food webs. An independent theory, such as the cascade model, is required to explain the frequencies of both intervality and triangulation.

The history of data and theory on the intervality and triangulation of the niche overlap graph may be caricatured simply. Initially, the high average proportion of interval webs came as a surprise, and could not be explained by the available models (Cohen 1977 [Chap. II.4], 1978). Subsequently, various explanations were offered for the high average proportion of intervality, including compartmentalization (Critchlow and Stearns), energetic constraints on community assembly (Yodzis), and triangularity (Sugihara). Though, in retrospect, the data then available and some of these explanations hinted at a decline in the frequency of intervality with an increasing number of species, it seems fair to say that any such decline remained unremarked. The data presented here provide unambiguous evidence of a decline in the relative frequency of intervality and triangulation with increasing numbers of species. These data seem to us to weaken or obliterate the claim that trophic niche overlap graphs and resource graphs are interval or triangulated (always or at a constant high frequency) regardless of the number of species in a web. The cascade model predicts accurately the existence of this decline in intervality and triangulation. The cascade model also predicts the range in numbers of species where this decline occurs. However, the cascade model predicts that the relative frequencies of intervality and triangulation will decline more rapidly, with increasing numbers of species, than they actually do. Excess proportions of interval and triangulated overlap and resource graphs remain to be explained.

5. Summary

We report new empirical and theoretical information about the intervality and triangulation of overlap graphs and resource graphs in community food webs. In 113 community food webs, the overall proportion of webs that are interval is as high as, or higher than, the proportion of interval webs observed previously. However, the fraction of webs that are interval is strongly associated with the number of species in the webs. The fraction of interval webs declines from one for small webs (16 or fewer species) toward zero for large webs (33 or more species). According to new mathematical and numerical calculations presented here, the cascade model predicts, as observed, that the probability that a web is interval is near one for webs with under 10 species, declines as the number of species increases from 10 to 30 or 40, and is very near zero for larger numbers of species. However, in the range from 10 to 40 species, the cascade model predicts a more rapid decline in the relative frequency of intervality than is observed.

Using the predation matrices of the same 113 webs, we determined which webs have triangulated overlap graphs and triangulated resource graphs. The empirical, mathematical and computational results on the relative frequency of triangulation parallel those on intervality.

The broad ecological interpretation of our findings is that the larger the number of species in a community, the less likely it is that a single dimension suffices to describe the community's trophic niche space, and the less likely it is that there are no "homological holes" (in the sense of Sugihara 1982) in the overlap graph and resource graph. Most reported webs with small numbers of species are very incomplete descriptions of real communities. If future webs have larger numbers of species and are described in greater detail, we predict that those webs will have smaller relative frequencies of being interval and triangulated.

6. Appendix: Mathematical Analysis

Basic Concepts

The cascade model W_p assumes that $S \geq 2$ species (vertices) of a web may be labeled from 1 to S. If $i < j$, then j feeds on i (there is a link from i to j) with probability p and j does not feed on i with probability $q = 1 - p$, independently for all $1 \leq i < j \leq S$. The probability that species j feeds on species i is 0 if $j \leq i$. The probability p is assumed to depend on S so that $p = p(S) \to 0$ as $S \to \infty$.

By replacing each link of W_p by an undirected edge, one obtains the usual random graph model G_p, i.e., an undirected simple graph on the vertex set $\{1, 2, \ldots, S\}$ in which each edge appears with probability p, independently of all other edges. A simple graph is one that has neither loops nor multiple edges. The structure of G_p when p changes from 0 to 1 has been studied extensively since the fundamental paper of Erdös and Rényi appeared in 1960 (see e.g., Bollobás 1985). The greatest discovery of Erdös and Rényi (1960) was that many important properties of graphs appear quite suddenly. We shall use such facts about G_p here.

We shall say that *almost every* G_p has property π if the probability that G_p has π tends to 1 as $S \to \infty$. If we pick a function $p = p(S)$ then, in many cases, either almost every graph G_p has property π or else almost every graph fails to have property π. More precisely, for many properties there is a *threshold function* $p^* = p^*(S)$ such that

$$\lim_{S \to \infty} P(G_p \text{ has property } \pi) = \begin{cases} 0 & \text{if } p/p^* \to 0 \,, \\ 1 & \text{if } p/p^* \to \infty \,. \end{cases}$$

As examples, here are two facts from Erdös and Rényi (1960) which we shall use later.

Fact 1. The threshold function that G_p contains a complete subgraph K_n on n vertices is $p^* = S^{-2/(n-1)}$.

Fact 2. The threshold function that G_p contains a cycle on n vertices is $p^* = S^{-1}$ for any fixed $n \geq 3$.

If G is a simple graph on the vertices $V = V(G)$ and F is another simple graph on the vertices $V(F)$, we say that F is an induced subgraph of G if $V(F) \subset V(G)$, and the edges of F contain all the possible edges from the edges of G, i.e., if $v_i, v_j \in V(F)$ and $\{v_i, v_j\}$ is an edge of G, then $\{v_i, v_j\}$ is an edge of F.

There are some properties of a random graph G_p that suddenly appear, then hold when p increases and at some point suddenly disappear. For example, consider the property that G_p contains an induced cycle on a fixed number of vertices. By Fact 2, such a cycle appears with probability 1 when $p = w(S)S^{-1}$, where $w(S) \to \infty$ (arbitrarily slowly) as $S \to \infty$. However, when p is very close to 1, then the cycle is no longer induced. Thus in our investigations we focus on the *appearance function* of a given subgraph of G_p, which describes when such a subgraph first appears as p increases. Of course, when one considers subgraphs (but not induced subgraphs) of G_p, then the appearance function and the threshold function coincide.

The concepts of threshold and appearance functions also apply to the cascade model W_p and to the overlap graph $G(W_p)$, defined as follows (Cohen 1978; Chap. II.4).

The trophic niche overlap graph $G(W_p)$ is defined as an undirected simple graph on the vertices of W_p. Two consumers are joined by an undirected edge when there is at least one prey that both consumers eat. That is, $\{v_j, v_k\}$ is an edge in $G(W_p)$ if and only if there exists some v_i in W_p such that both (v_i, v_j) and (v_i, v_k) are links in W_p.

Let G be a simple graph on the set of vertices $V = \{v_1, v_2, \ldots, v_n\}$. G is an *interval graph* when there is a collection $I_1, I_2, \ldots, I_n$ of open, closed, or mixed intervals of the real line such that there is an edge between v_i and v_j, $i \neq j$, if and only if I_i and I_j overlap, that is $I_i \cap I_j \neq \emptyset$. Thus G is an interval graph if and only if G is the intersection graph of some family of intervals of the line. If each interval $I_1, \ldots, I_n$ has length equal to 1, then G is called a *unit interval graph*.

Existence of Some Induced Subgraphs in $G(W_p)$

We now establish the appearance functions of induced subgraphs of various types in a random overlap graph $G(W_p)$. We will find the appearance functions of the properties that $G(W_p)$ contains an induced tree, an induced cycle and an induced asteroidal 1-triangle (see Fig. III.6.A1). These subgraphs determine the intervality of $G(W_p)$ when S is large, which we examine in the following section.

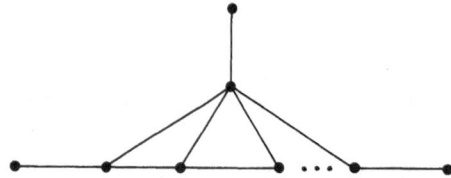

Figure III.6.A1. An asteroidal k-triangle with $k + 5$ vertices. (An asteroidal 1-triangle contains a single central triangle, each vertex of which is joined by an edge to one outlying vertex)

We begin with the existence of a complete subgraph in a random overlap graph $G(W_p)$. There are two reasons for this. First, the threshold function for having a complete subgraph in $G(W_p)$ (which in this case is also the appearance function) is quite different from that in the usual random graph model G_p (see Fact 1). Second, all the proofs in this section rely on the so-called "second-moment method". It is easiest to present this method in the case of complete subgraphs. Thus we present first a rather detailed proof of the threshold function for the existence of a complete subgraph of $G(W_p)$ and then state the remaining results, indicating only the crucial points in their proofs.

Theorem 1 (complete subgraphs). *Let $n \geq 3$ be fixed. The threshold function of the property that $G(W_p)$ contains a complete subgraph K_n on any n vertices is* $S^{-1-1/n}$, *i.e.,*

$$\lim_{S \to \infty} P(G(W_p) \supset K_n) = \begin{cases} 0 & \text{if } pS^{1+1/n} \to 0, \\ 1 & \text{if } pS^{1+1/n} \to \infty. \end{cases}$$

Proof. Denote by X_n the number of all configurations in the cascade model W_p that produce complete subgraphs on n vertices in $G(W_p)$. As an example,

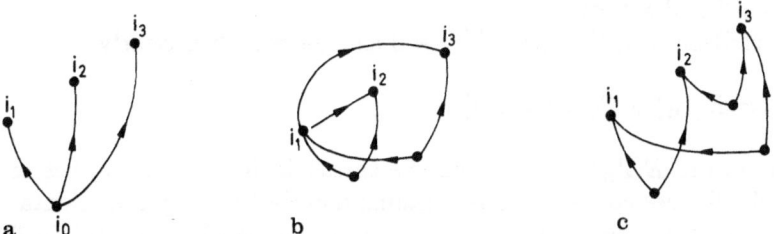

Figure III.6.A2. Three web configurations that produce the complete graph K_3 in the overlap graph

Fig. III.6.A2 presents three types of configurations of W_p that correspond to K_3 in $G(W_p)$.

The graph in Fig. III.6.A2(a) is called a 3-star with root i_0. Generally, a subgraph of W_p on $n+1$ vertices $i_0, i_1, i_2, \ldots, i_n$ where $1 \le i_0 < i_1 < i_2 < \ldots < i_n \le S$ such that (i_0, i_k) is a link for every $k = 1, 2, \ldots, n$ will be called an n-star with root i_0. Let Y_n stand for the number of all n-stars in W_p. Then $X_n = Y_n + Z_n$ where Z_n is the number of configurations other than n-stars that produce a K_n in the overlap graph. (If we forget about the orientation of links, then all those configurations contain at least one cycle.) Elementary calculation shows that

$$E(Y_3) = \binom{S}{4} p^3 = O(S^4 p^3)$$

and

$$E(Z_3) = \sum_{2 \le i_1 < i_2 < i_3 \le S} (i_1 - 1)(i_1 - 2)p^6$$

$$+ \sum_{2 \le i_1 < i_2 < i_3 \le S} (i_1 - 1)(i_1 - 2)(i_2 - 4)p^6$$

$$= O(S^5 p^6) + O(S^6 p^6)$$

$$= O(S^6 p^6) \, .$$

The first sum in $E(Z_3)$ enumerates the expected number of graphs of the form shown in Fig. III.6.A2(b); the second sum refers to Fig. III.6.A2(c). Similarly, for $n \ge 4$

$$E(Y_n) = \binom{S}{n+1} p^n = O(S^{n+1} p^n)$$

and it is not hard to see that in a formula for the expectation of Z_n, the exponent of p is always greater than the exponent of S (only if $n \ge 4$). Consequently,

$$E(Z_n) = O(S^m p^k)$$

for some $m \geq n + 2$ and $k > m$.

Now let $p = p(S)$ be such that $pS^{1+1/n} \to 0$ as $S \to \infty$. Then clearly

$$E(X_n) = E(Y_n) + E(Z_n) = o(1) .$$

(We could have proved $E(Z_n) = o(1)$ from the threshold function for cycles in G_p, because each of these configurations contains a cycle (if we ignore orientation) and from Fact 2 we know that there are no cycles in G_p when $pS \to 0$ as $S \to \infty$, which is satisfied under our assumption on p.) Since $P(X_n \geq 1) \leq E(X_n)$, it follows that, with probability approaching 1 as $S \to \infty$, the cascade model W_p contains no configurations producing a complete subgraph K_n in $G(W_p)$, i.e.,

$$P(G(W_p) \supset K_n) = P(X_n \geq 1) \to 0 \quad \text{as} \quad S \to \infty .$$

Now assume that $pS^{1+1/n} \to \infty$ as $S \to \infty$. We shall show that under this assumption

$$P(Y_n \geq 1) \to 1 \quad \text{as} \quad S \to \infty . \tag{1}$$

Since

$$P(Y_n \geq 1) \leq P(G(W_p) \supset K_n) ,$$

it will then follow that, with probability tending to 1, a random overlap graph $G(W_p)$ contains at least one complete subgraph K_n. For $1 \leq i_0 < i_1 < \ldots < i_n \leq S$, let S_i denote the indicator random variable of the event that there is in W_p an n-star i on the vertices $\{i_0, i_1, \ldots, i_n\}$ with i_0 as the root. Then

$$\text{var}(Y_n) = \sum_i \sum_j \text{cov}(S_i, S_j)$$

where the summations are over all n-stars specified by i and j, respectively. If the stars i and j share exactly m links, $0 \leq m \leq n$, then

$$\text{cov}(S_i, S_j) \leq P(S_i = 1, S_j = 1) = p^{2n-m} .$$

If $m = 0$ and none of $i_0, i_1, \ldots, i_n$ coincides with any of $j_0, j_1, \ldots, j_n$, then S_i and S_j are independent so $\text{cov}(S_i, S_j) = 0$. Let Q_m be the number of ordered pairs (i, j) such that i and j share m links and at least one vertex. Then for $m \geq 1$, clearly the roots i_0 and j_0 coincide and $Q_m \leq S^{2n+1-m}$ while for $m = 0$, $i_0 \neq j_0$ and $Q_0 \leq S^{2n+1}$. Consequently,

$$\text{var}(Y_n) \leq \sum_{m=0}^{n} (Sp)^{2n-m} S .$$

Thus, from Chebyshev's inequality,

$$P(Y_n = 0) \le \text{var}\,(Y_n)/E(Y_n)^2$$

$$= O\left(\sum_{m=0}^{n} (Sp)^{-m} S^{-1}\right)$$

$$= o(1)$$

since under the assumption on p, Sp can be expressed as $Sp = w(S)S^{-1/n}$ where $w(S)$ is a sequence tending to infinity as $S \to \infty$. Thus we proved (1). ∎

Theorem 2 (induced trees). *Let $k \ge 2$ be fixed. The appearance function of an induced tree on k vertices in $G(W_p)$ is $S^{-(2k-1)/(2k-2)}$.*

Proof. If $G(W_p)$ contains an induced tree on vertices (consumers) $i_1, i_2, \ldots, i_k$, where $2 \le i_1 < i_2 < \ldots < i_k \le S$, then there must exist $k-1$ vertices (prey species) $j_1, j_2, \ldots, j_{k-1}$ where $j_1 < i_1$ and $j_{k-1} < i_k$ such that for every $j_m (m = 1, 2, \ldots, k-1)$ there are exactly two links from j_m to two appropriately chosen vertices from $\{i_1, i_2, \ldots, i_k\}$. See Fig. III.6.A3(a). It may happen that some of the consumers are at the same time prey species; see Fig. III.6.A3(b). Figure III.6.A3 presents two examples of configurations in W_p that produce a tree in $G(W_p)$ as shown in Fig. III.6.A4.

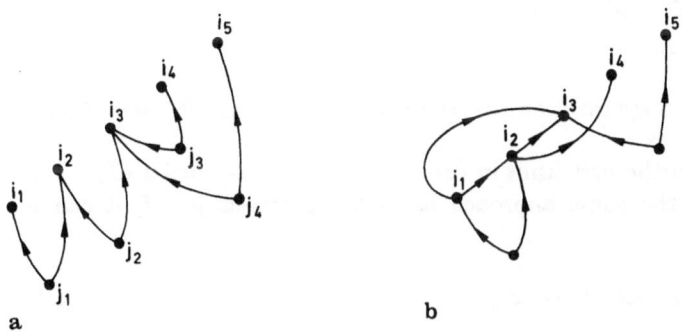

Figure III.6.A3. Two web configurations that produce the tree shown in Fig. III.6.A4 in the overlap graph

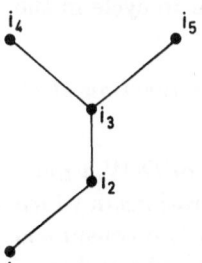

Figure III.6.A4. Tree in the overlap graph produced by the web configurations shown in Fig. III.6.A3

Each configuration of the web W_p that produces an induced tree on k vertices in the overlap graph $G(W_p)$ must have exactly $2(k-1)$ links. The configurations are of two types. In configurations of the first type, $\{i_1, i_2, \ldots, i_{k-2}\} \cap \{j_1, j_2, \ldots, j_{k-1}\} = \emptyset$; this means that none of the vertices $\{i_1, i_2, \ldots, i_{k-2}\}$ is a prey for two consumers from $\{i_2, \ldots, i_k\}$. In configurations of the second type, some of the vertices $i_1, i_2, \ldots, i_{k-2}$ are at the same time consumers and prey. In the latter case, if we ignore the orientation of links, there is always a cycle in the configuration.

Assume that $p = p(S)$ is such that

$$pS^{\frac{2k-1}{2k-2}} \to 0 \quad \text{as} \quad S \to \infty. \tag{2}$$

Since our p is of smaller order than S^{-1}, by Fact 2 almost every G_p has no cycles and consequently almost every W_p has no configurations of the second type. Moreover, each configuration of the first type forms an induced tree of W_p and there is no vertex lying below i_k and different from $\{i_1, i_2, \ldots, i_{k-1}, j_1, j_2, \ldots, j_{k-1}\}$ that is connected with exactly two vertices from $\{i_1, i_2, \ldots, i_k\}$, for such a vertex would destroy the property that the tree in $G(W_p)$ is induced. Thus if T_k denotes the number of configurations of the first type then

$$
\begin{aligned}
E(T_k) &= O\left(\sum_{2 \le i_1 < i_2 < \ldots < i_k \le S} (i_k - 1)^{k-1} p^{2(k-1)}\right) \\
&= O(S^{2k-1} p^{2(k-1)}) \\
&= o(1).
\end{aligned}
$$

Consequently, under the assumption (2), the overlap graph $G(W_p)$ contains no induced tree on k vertices.

On the other hand, if the first limit in (2) is ∞ instead of 0, then $E(T_k) \to \infty$ as $S \to \infty$. Applying the same approach as in the previous proof, it can be shown that

$$P(T_k \ge 1) \to 1 \quad \text{as} \quad S \to \infty,$$

i.e., with probability tending to 1, $G(W_p)$ contains an induced tree on k vertices. ∎

The next result shows that the appearance function of an induced cycle on m vertices in $G(W_p)$ is the same as the threshold function for an m-cycle in the usual random graph model G_p if $m \ge 4$.

Theorem 3 (induced cycles). *Let $m \ge 4$ be fixed. The appearance function of an induced m-cycle in $G(W_p)$ is S^{-1}.*

Proof. Each configuration of W_p producing an induced m-cycle of $G(W_p)$ must contain exactly $2m$ links. As in the case of induced trees, configurations in which none of vertices $i_1, i_2, \ldots, i_{m-2}$ is used in W_p as a prey for any two consumers from $\{i_2, \ldots, i_m\}$ are most likely to occur. Therefore, the expected number of configurations of W_p giving induced m-cycles in $G(W_p)$ is of the order of magnitude $O(S^{2m}p^{2m})$. Now the same ideas as in the proof of Theorem 1 imply our result. ■

The asteroidal 1-triangle plays a special role in the asymptotic probability as $S \to \infty$ that a random overlap graph $G(W_p)$ is an interval graph.

Theorem 4 (asteroidal 1-triangle). *The appearance function of an induced asteroidal 1-triangle in $G(W_p)$ is $S^{-10/9}$.*

Proof. Consider a configuration (see Fig. III.6.A5) in W_p that gives an asteroidal 1-triangle in $G(W_p)$. The expected number of such configurations is $O(S^{10}p^9)$. It is easy to check that the expected number of all other configurations of W_p that produce an asteroidal 1-triangle subgraph of $G(W_p)$ is of an order of magnitude less than $O(S^{10}p^9)$. Thus the same argument as before applies. ■

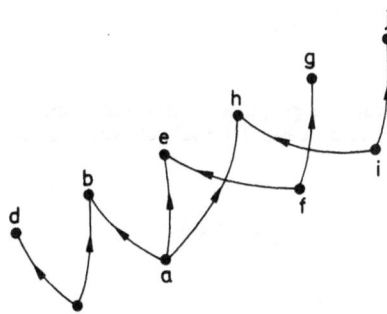

Figure III.6.A5. A web configuration that produces an asteroidal 1-triangle in the overlap graph

Intervality of $G(W_p)$

Lekkerkerker and Boland (1962) showed that a graph G is an interval graph if and only if it contains no induced subgraph of the forms pictured in Figs. III.6.A1 and III.6.A6. This characterization of interval graphs differs from, but is consistent with, the characterization in terms of triangulation and asteroidal triples, which is mentioned in the text.

We now describe the probability that a random overlap graph $G(W_p)$ is an interval graph for $S = 3, 4, 5, 6, 7$, and $S \to \infty$. For $S = 3, 4$ and 5, $P(G(W_p)$ is interval$) = 1$, since the web W_p contains no configurations that could destroy the intervality of $G(W_p)$. If W_p has the vertex set $\{1, 2, 3, 4, 5, 6\}$, then the only possible forbidden subgraph of $G(W_p)$ is an induced 4-cycle, which may appear on vertices $\{3, 4, 5, 6\}$ in four different configurations of W_p as shown in Fig. III.6.A7.

Let X_4 be the number of configurations in W_p on $S = 6$ vertices that produce an induced 4-cycle in $G(W_p)$. Then

$$P(G(W_p) \text{ is not interval}) = P(X_4 \geq 1) \leq E(X_4) .$$

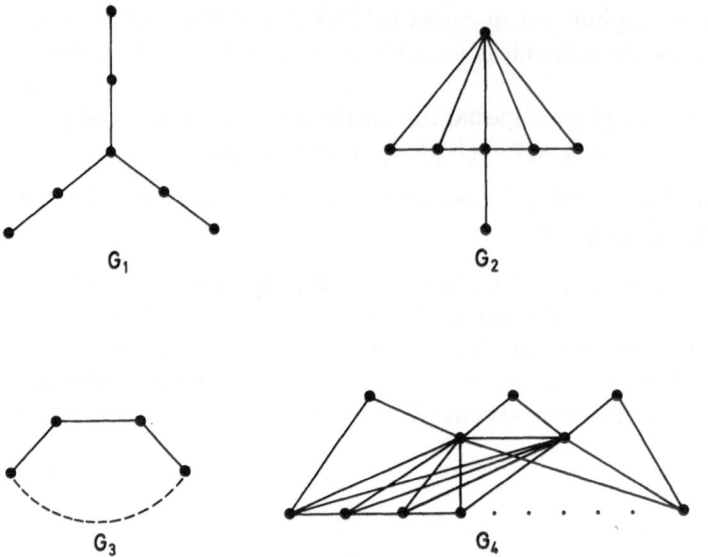

Figure III.6.A6. Forbidden subgraphs of an interval graph: a graph is an interval graph if and only if it contains none of the subgraphs shown here and in Fig. III.6.A1. G_3 contains k vertices, $k \geq 4$. G_4 contains $k + 5$ vertices, $k \geq 1$

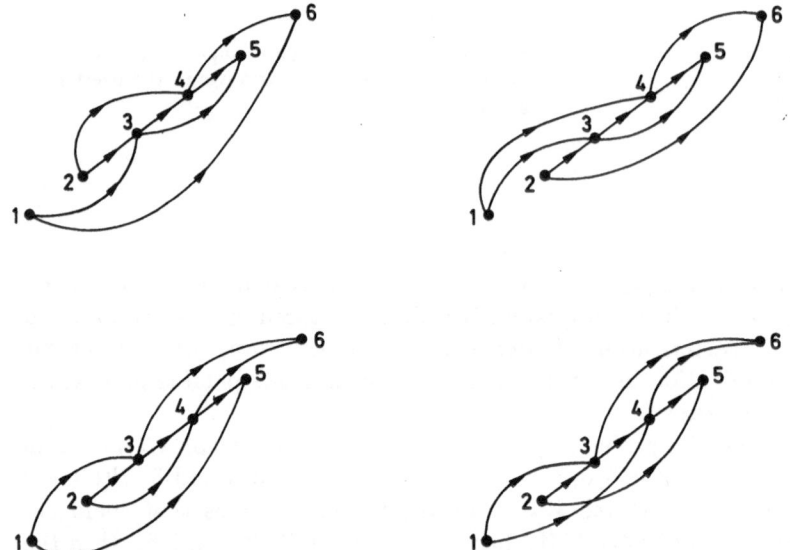

Figure III.6.A7. Four configurations of a web on six vertices that produce an induced 4-cycle in the overlap graph

Since each configuration in Fig. III.6.A7 contains 8 arcs and must exclude 5 arcs, and since there are exactly 4 such configurations,

$$E(X_4) = 4p^8(1 - p)^5$$

so

$$P(G(W_p) \text{ is interval}) \geq 1 - 4p^8(1-p)^5, \quad \text{for} \quad S = 6 .$$

When $S = 7$, the subgraphs of $G(W_p)$ that destroy intervality are induced 4-cycles and induced 5-cycles. There are many different configurations of W_p that produce induced 4-cycles or induced 5-cycles of $G(W_p)$. If Y_4 and Y_5 stand for the number of configurations (of W_p on $S = 7$ vertices) that produce induced 4-cycles and induced 5-cycles, respectively, then a lengthy enumeration of the possibilities yields, with $q = 1 - p$,

$$
\begin{aligned}
E(Y_4) = {} & 36p^8q^7(1-p^2) + 12p^8q^6(2 - 2qp^2 - p^2 - p^3) \\
& + 4p^8q^5(6 + 6q^4 + 24q^3p + 24q^2p^2 - 3qp^2 - p^2 - 2p^3)
\end{aligned}
\tag{3}
$$

and

$$E(Y_5) = 8p^{10}q^9 . \tag{4}$$

We leave the proofs of (3) and (4) to an eager reader as additional entertainment. Consequently, when $S = 7$

$$
\begin{aligned}
P(G(W_p) \text{ is not interval}) &= P(Y_4 \geq 1 \text{ or } Y_5 \geq 1) \\
&= P(Y_4 \geq 1) + P(Y_5 \geq 1) \\
&\quad - P(Y_4 \geq 1 \text{ and } Y_5 \geq 1) \\
&\leq E(Y_4) + E(Y_5) .
\end{aligned}
$$

Note that $P(Y_4 \geq 1 \text{ and } Y_5 \geq 1) = 0$. Thus

$$P(G(W_p) \text{ is interval}) \geq 1 - E(Y_4) - E(Y_5)$$

where $E(Y_4)$ and $E(Y_5)$ are given by (3) and (4), respectively.

We shall not even try to estimate $P(G(W_p)$ is interval) when $S = 8$, since the calculation looks hopeless. Perhaps surprisingly, the calculation becomes much easier when S is large.

Theorem 5 (interval graphs). *Let $p = p(S) \to 0$ so that $pS^{10/9} = d$. Then*

$$
\lim_{S \to \infty} P(G(W_p) \text{ is interval}) =
\begin{cases}
1 & \text{if } d = d(S) \to 0 \\
e^{-\lambda} & \text{if } 0 < d < \infty \\
0 & \text{if } d = d(S) \to \infty
\end{cases}
$$

where

$$\lambda = \frac{9170}{10!} d^9 .$$

Proof. Let

$$pS^{10/9} \to 0 \quad \text{as} \quad S \to \infty . \tag{5}$$

By Theorems 2, 3 and 4, it follows immediately that a random graph $G(W_p)$ contains no induced subgraphs of the forms of G_1, G_3 and asteroidal 1-triangles. For example, in the case of G_1, if

$$pS^{13/12} \to 0 \quad \text{as} \quad S \to \infty , \tag{6}$$

then by Theorem 2, $P(G(W_p) \supset G_1) \to 0$. Clearly (5) implies (6). Next, it is not hard to see that asteroidal k-triangles for $k \geq 2$ as well as G_2 and G_4 are unlikely to occur when p satisfies (5). One need simply estimate the expected numbers of configurations in W_p that produce those subgraphs in $G(W_p)$ and check that under the assumption on p given by (5) these expected values tend to 0 as $S \to \infty$. Consequently, by the Lekkerkerker-Boland characterization of interval graphs, if p satisfies (5) then

$$\lim_{S\to\infty} P(G(W_p) \text{ is interval}) = 1 .$$

Now assume that $pS^{10/9} \to \infty$ as $S \to \infty$. Then by Theorem 4, with probability tending to 1 as $S \to \infty$, a random overlap graph $G(W_p)$ contains at least one induced asteroidal 1-triangle that destroys the intervality of $G(W_p)$.

Finally, let $pS^{10/9} \to d$, $0 < d < \infty$. The same argument as in the first part of our proof shows that in this case the only induced subgraphs that destroy the intervality of $G(W_p)$ are induced asteroidal 1-triangles. Let X denote the number of such subgraphs in $G(W_p)$. We shall show that

$$\lim_{S\to\infty} P(X = k) = \frac{\lambda^k e^{-\lambda}}{k!}, \quad k = 0, 1, 2, \ldots \tag{7}$$

where $\lambda = \frac{9170}{10!}d^9$, i.e., X has asymptotically the Poisson distribution with parameter λ. Define a configuration of type C to be a configuration of the type presented in Fig. III.6.A5. Let Y be the number of configurations of type C that may appear in W_p as an induced subgraph such that none of the vertices lying below $\{a, b, c, d, e, f, g, h, i, j\}$ is connected with exactly two vertices from $\{b, d, e, g, h, j\}$. Then (compare the proof of Theorem 4) the probability distribution of X is asymptotically the same as the distribution of Y. Replace for a moment each link of W_p by an undirected edge. Clearly, the configuration in Fig. III.6.A5 becomes an ordinary tree on 10 vertices. It is well-known (see e.g., Bollobás 1985) that if $pS^{10/9} \to d$ then the number of such trees in G_p has asymptotically the Poisson distribution with parameter $\mu = d^9/A$, where A is the order of the automorphism group of a tree on 10 vertices, i.e., $A = 10!/10^8$. Now if we return to the model W_p then (applying the same approach as in e.g. Bollobás 1985) it can be shown that the number of configurations of type C also has a Poisson distribution but with a parameter $\gamma = d^9/B$, where $B = 10!/\xi$ and

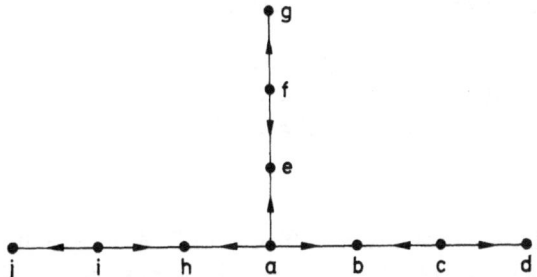

Figure III.6.A8. A redrawing of the web configuration in Fig. III.6.A5

ξ is the number of different ways of labeling 10 given vertices of a configuration of type C. Let us redraw the graph from Fig. III.6.A5 in a different but more useful form (see Fig. III.6.A8).

Since there are at least three vertices lying above vertex a in W_p, we must have $1 \le a \le 7$. Furthermore, b is above a (i.e., $a+1 \le b \le 10$) and c is below b (i.e., $1 \le c \le b-1$) but different from a. Also, d is above c (i.e., $c+1 \le d \le 10$) but different from a and b. Continuing this process up to vertex j, we obtain,

$$3!\xi = \sum_{a=1}^{7} \sum_{b=a+1}^{10} \sum_{c=1}^{b-1} \sum_{d=c+1}^{10} \sum_{e=a+1}^{10} \sum_{f=1}^{e-1} \sum_{g=f+1}^{10} \sum_{h=a+1}^{10} \sum_{i=1}^{h-1} \sum_{j=i+1}^{10} 1$$

where $c \neq a$, $d \notin \{a,b\}$, $e \notin \{b,c,d\}$, $f \notin \{a,b,c,d\}$, $g \notin \{a,b,c,d,e\}$, $h \notin \{b,c,d,e,f,g\}$, $i \notin \{a,b,c,d,e,f,g\}$ and $j \notin \{a,b,c,d,e,f,g,h\}$. Each author independently wrote a computer program in BASIC to compute ξ and each obtained independently $\xi = 55020/3! = 9170$.

The probability that a configuration of type C is an induced subgraph of W_p and that none of the vertices lying below C is connected with two vertices from $\{b,d,e,g,h,j\}$ tends to 1 as $S \to \infty$ (since if $pS^{10/9} \to d$ then almost every G_p has no cycle – see Fact 2). Consequently the random variable Y has asymptotically the Poisson distribution with parameter $\lambda = \frac{9170}{10!}d^9$ and (7) is proved. Under the assumption on p,

$$\lim_{S \to \infty} P(G(W_p) \text{ is interval}) = \lim_{S \to \infty} P(X=0) = e^{-\lambda} . \qquad \blacksquare$$

The numerical value of λ in Theorem 5 may be estimated for an observed web with L links (or arcs) and S species. The maximum likelihood estimate of p is $p = L/[S(S-1)/2] = 2L/[S(S-1)]$. Hence $d = pS^{10/9} = 2LS^{1/9}/(S-1)$; hence $d^9 = (2L/[S-1])^9 S$. As $9170/10! = 0.002527$, we get $\lambda = 0.002527(2L/[S-1])^9 S$ and, for sufficiently large S, the probability that the overlap graph $G(W_p)$ is interval is arbitrarily close to $e^{-\lambda}$.

Finally, we describe the behavior of the probability that an overlap graph $G(W_p)$ is a unit interval graph. Roberts (1969) proved that a graph is unit interval if and only if it is an interval graph and does not contain the bipartite complete graph $K_{1,3}$ as an induced subgraph.

Theorem 6 (unit interval graphs). *Let $p = p(S) \to 0$ so that $pS^{7/6} = d$. Then*

$$\lim_{S\to\infty} P(G(W_p) \text{ is unit interval}) = \begin{cases} 1 & \text{if } d = d(S) \to 0, \\ e^{-\mu} & \text{if } 0 < d < \infty, \\ 0 & \text{if } d = d(S) \to \infty, \end{cases}$$

where $\mu = \frac{48}{7!}d^6$.

Proof. The proof follows the same lines as the proof of Theorem 5. A subgraph of W_p that produces an induced $K_{1,3}$ in $G(W_p)$ is of the form presented in Fig. III.6.A9. If $pS^{7/6} \to d$ for some d such that $0 < d < \infty$, then the number of trees on 7 vertices in G_p has asymptotically the Poisson distribution with parameter $7^5 d^6/7!$. Similarly, the number of configurations of the form in Fig. III.6.A9 in the cascade model W_p has also the Poisson distribution but with parameter $\xi d^6/7!$, where

$$3!\xi = \sum_{a=4}^{7} \sum_{b=1}^{a-1} \sum_{c=b+1}^{7} \sum_{d=1}^{a-1} \sum_{e=d+1}^{7} \sum_{f=1}^{a-1} \sum_{g=f+1}^{7} 1 \,,$$

and $c \neq a$, $d \notin \{b,c\}$, $e \notin \{a,b,c\}$, $f \notin \{b,c,d,e\}$ and $g \notin \{a,b,c,d,e\}$. Again, using computer programs, we obtained $\xi = 48$. ∎

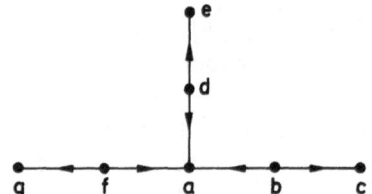

Figure III.6.A9. A subweb that produces an induced $K_{1,3}$ in the overlap graph. This configuration prevents a web from having a unit interval overlap graph

As before, the numerical value of μ in Theorem 6 may be estimated for an observed web with L links and S species. Here $d = pS^{7/6} = 2LS^{1/6}/(S-1)$; hence $d^6 = (2L/[S-1])^6 S$. As $48/7! \doteq 0.0095238$, we get $\mu \doteq 0.0095238(2L/[S-1])^6 S$. For sufficiently large S, the probability that the overlap graph $G(W_p)$ is a unit interval graph is arbitrarily close to $e^{-\mu}$.

Triangulation of W_p

We say that the cascade digraph W_p is triangulated if its overlap graph $G(W_p)$ contains no induced k-cycles for all $k \geq 4$. As in the case of the intervality of $G(W_p)$, for $S = 3$, 4 and 5 the probability that W_p is triangulated equals 1, whereas

$$P(W_p \text{ is triangulated}) \geq \begin{cases} 1 - 4p^8 q^5 & \text{for } S = 6 \\ 1 - E(Y_4) - E(Y_5) & \text{for } S = 7 \end{cases}$$

where $E(Y_4)$ and $E(Y_5)$ are given by (3) and (4), respectively. Now suppose that S is large.

Theorem 7 (triangulated graphs). *Let $p = p(S) \to 0$ so that $pS = d$. Then*

$$\lim_{S \to \infty} P(W_p \text{ is triangulated}) = \begin{cases} 1 & \text{if } d = d(S) \to 0 \\ e^{-\gamma} & \text{if } 0 < d < 1, \\ 0 & \text{if } d \geq 1 \end{cases}$$

where

$$\gamma = \sum_{\substack{k=8 \\ k \text{ even}}}^{\infty} \frac{d^4(-4)^{k/2}}{k!} \sum_{m=1}^{k-1} (-1)^m m! \, S(k-1, m) 2^{-2-m}$$

and $S(k, m)$ are Stirling's numbers of the second kind.

Proof. If $d = d(S) \to 0$ as $S \to \infty$, then by Theorem 3 there is no induced k-cycle for all $k \geq 4$ in $G(W_p)$, so W_p is triangulated with probability approaching to 1. Keeping in mind the remarks made in the proof of Theorem 3, we can focus our attention only on the very special subgraphs of W_p that form induced k-cycles in $G(W_p)$. Those subgraphs (denote their number by Z_k) have $2k$ vertices, $2k$ links (appropriately joining those vertices) and are such that after removing the orientation of links they form induced $(2k)$-cycles in the usual random graph model G_p. Assume that d is a constant, $0 < d < 1$. It is well-known (Bollobás 1985) that in this case almost every random graph G_p is a union of tree components and unicyclic components. Thus each cycle that may appear in G_p is an induced cycle. Let X_k be the number of k-cycles of G_p. Then (Bollobás 1985, p. 79) $X_3, X_4, \ldots, X_k$ are asymptotically independent Poisson random variables with means $\lambda_i = d^i/(2i)$, $i = 3, 4, \ldots, k$. No cycle of odd length contributes to forming an induced cycle of $G(W_p)$. The only cycles of even length in W_p that contribute to forming an induced k-cycle of $G(W_p)$ are $(2k)$-cycles with the property that for each vertex i its neighbors are either both smaller or both larger than i. Ruciński (1988) observed that the number of such cycles that may be formed on a given set of vertices is

$$a_{2k} = (-4)^k \sum_{m=1}^{2k-1} (-1)^m m! \, S(2k-1, m) 2^{-2-m} \,, \quad k = 2, 3, \ldots \,.$$

Furthermore, the same approach as in e.g., Bollobás (1985) shows that the random variable

$$Z = \sum_{\substack{k=8 \\ k \text{ even}}} Z_k$$

has asymptotically a Poisson distribution with parameter

$$\gamma = \sum_{\substack{k=8 \\ k \text{ even}}} \binom{S}{k} a_k p^k$$

$$\sim \sum_{\substack{k=8 \\ k \text{ even}}} \frac{d^4(-4)^{k/2}}{k!} \sum_{m=1}^{k-1} (-1)^m m! \, S(k-1,m) 2^{-2-m} \, .$$

Thus

$$\lim_{S \to \infty} P(W_p \text{ is triangulated}) = \lim_{S \to \infty} P(Z = 0) = e^{-\gamma} \, .$$

Finally, when $d \geq 1$, almost every graph G_p with $p = d/S \to 0$ contains a long induced cycle (see e.g., Bollobás 1985) which can form an induced cycle of $G(W_p)$. ∎

The Resource Graph

The resource graph $H(W_p)$ is defined as an undirected simple graph, with the same vertex set as W_p, such that $\{v_j, v_k\}$ is an edge in $H(W_p)$ if and only if there exists some v_i in W_p such that both (v_j, v_i) and (v_k, v_i) are links in W_p. For $1 \leq i < j \leq S$, define P_{ij} to be the probability of an edge between i and j in the overlap graph $G(W_p)$; then $P_{ij} = 1 - (1 - p^2)^{i-1}$. Similarly, for $k < l$, define Q_{kl} to be the probability of an edge between k and l in the resource graph $H(W_p)$; then $Q_{kl} = 1 - (1 - p^2)^{S-l}$. Now define π to be the permutation $\pi(i) = S + 1 - i$, for $i = 1, \ldots, S$. Then $P_{ij} = Q_{\pi(j), \pi(i)}$ for all $1 \leq i < j \leq S$. This means that the probability of any configuration of edges is the same in $G(W_p)$ as in $H(W_p)$, after relabeling the vertices by π. Hence all the results in this Appendix apply equally to overlap graphs and to resource graphs.

Chapter IV. Data on 113 Community Food Webs

This appendix presents the references to the original sources, the predation matrices, and the lists of organisms in the 113 webs used in the empirical studies in this book.

The first 40 webs have appeared previously, in slightly different versions (J. E. Cohen, 1978, *Food Webs and Niche Space*, Princeton: Princeton University Press; and F. Briand, 1983, Environmental control of food web structure, *Ecology* 64, 253-263). A compilation of the remaining 73 webs has not been published before.

The references are arranged according to the serial numbering of the webs adopted by F. Briand. Webs that appeared in Cohen (1978) with a different number are also identified by that number. Except for the webs with a number assigned by Cohen (1978), these webs were compiled by F. Briand.

The element in the first row and first column of each predation matrix gives the serial number of the web. The other numbers in the first row identify the kinds of organisms that correspond to each column. The numbers, other than the first, in the first column identify the kinds of organisms that correspond to each row. The key to these identifying numbers is the list of organisms that accompanies each predation matrix. The groupings of organisms in the list and in the predation matrix are not necessarily trophic species (defined in Chap. II.2); that is, these are unlumped food webs.

Except for the first row and the first column of each predation matrix, an entry of 1 in the predation matrix means that the kind of organism corresponding to that column eats the kind of organism corresponding to that row. An entry of 0 means that the kind of organism corresponding to that column does not eat the kind of organism corresponding to that row. An entry of −1 or −2 results from a coding scheme of Cohen (1978). All −1 entries should be interpreted as 1 and all −2 entries should be interpreted as 0.

Statistical summaries (such as numbers of trophic species or trophic links, defined in the text) that may be calculated from the following predation matrices will differ slightly for a few webs from those that appear in the chapters of this book (except the last chapter, which is based on this form of the data). The reason is that after preparing these webs in machine-readable form, it became possible to check calculations that had previously been done by hand, and a few errors in the hand calculations were identified. The form of the predation matrices given here represents our best joint effort at the time of publication. Readers are invited to inform us of errors.

Because the original sources of these webs are inconsistent in reporting cannibalism, we have usually suppressed reports of cannibalism from the predation matrices. These webs should not be used to investigate cannibalism or properties of webs that are sensitive to the presence or absence of cannibalism.

A machine-readable form of these and additional food webs is available to qualified academic researchers for purposes of scientific research. Contact Joel E. Cohen, Rockefeller University, 1230 York Avenue, Box 20, New York, NY 10021-6399, U.S.A..

Sources of the Community Food Webs

Web 1 Cochin backwater, India: S. Z. Qazim, Some problems related to the food chain in a tropical estuary. In: Marine Food Chains, J. H. Steele, Ed. (Oliver and Boyd, Edinburgh, 1970), pp. 46-51.

Web 2 Knysna estuary, South Africa: J. H. Day, The biology of Knysna estuary, South Africa. In: Estuaries, G. H. Lauff, Ed. (AAAS Publication 83, Washington, DC, 1967), pp. 397-407.

Web 3 Salt marsh, Long Island, USA: G. M. Woodwell, Toxic substances and ecological cycles, Sci. Am. 216:24-31 (March 1967).

Web 4 Salt marsh, California: R. F. Johnston, Predation by short-eared owls on a Salicornia salt-marsh, Wilson Bull. 68:91-102 (1956).

Web 5 Salt marsh, Georgia: J. M. Teal, Energy flow in the saltmarsh ecosystem of Georgia, Ecology 43:614-624 (1962). Prior number: Cohen (1978) 24

Web 6 Tidal flat, California: G. E. MacGinitie, Ecological aspects of a California marine estuary, Am. Midl. Nat. 16:629-765 (1935).

Web 7 Narragansett Bay, Rhode Island: J. N. Kremer and S. W. Nixon, A Coastal Marine Ecosystem: Simulation and Analysis, Vol. 24 of Ecol. Studies (Springer-Verlag, Berlin, 1978).

Web 8 Salt marsh, Rhode Island: S. W. Nixon and C. A. Oviatt, Ecology of a New England salt marsh, Ecol. Monogr. 43:463-498 (1973).

Web 9 Lough Ine rapids, Ireland: J. A. Kitching and F. J. Ebling, Ecological studies at Lough Ine, Adv. Ecol. Res. 4:197-291 (1967).

Web 10 Exposed rocky shore, New England, USA: B. A. Menge and J. P. Sutherland, Species diversity gradients: synthesis of the roles of predation, competition and temporal heterogeneity, Am. Nat. 110:351-369 (1976).

Web 11 Protected rocky shore, New England, USA: B. A. Menge and J. P. Sutherland, Species diversity gradients: synthesis of the roles of predation, competition and temporal heterogeneity, Am. Nat. 110:351-369 (1976).

Web 12 Exposed rocky shore, Washington: B. A. Menge and J. P. Sutherland, Species diversity gradients: synthesis of the roles of predation, competition and temporal heterogeneity, Am. Nat. 110:351-369 (1976).

Web 13 Protected rocky shore, Washington: B. A. Menge and J. P. Sutherland, Species diversity gradients: synthesis of the roles of predation, competition and temporal heterogeneity, Am. Nat. 110:351-369 (1976).

Web 14 Mangrove swamp 1, Hawaii: G. E. Walsh, An ecological study of a Hawaiian mangrove swamp. In: Estuaries, G. H. Lauff, Ed. (AAAS Publication 83, Washington, DC, 1967), pp. 420-431.

Web 15 Mangrove swamp 3, Hawaii: G. E. Walsh, An ecological study of a Hawaiian mangrove swamp. In: Estuaries, G. H. Lauff, Ed. (AAAS Publication 83, Washington, DC, 1967), pp. 420-431.

Web 16 Pamlico estuary, North Carolina: B. J. Copeland, K. R. Tenore, D. B. Horton, Oligohaline regime. In: Coastal Ecological Systems of the United States, H. T. Odum, B. J. Copeland, E. A. McMahan, Eds. (Conservation Foundation, Washington, DC, 1974) 2:315-357.

Web 17 Coral reefs, Marshall Islands: R. Hiatt and D. W. Strasburg, Ecological relationships of the fish fauna on coral reefs of the Marshall Islands, Ecol. Monogr. 30:65-127 (1960).

Web 18 Kapingamarangi Atoll, Polynesia: W. A. Niering, Terrestrial ecology of Kapingamarangi Atoll, Caroline Islands, Ecol. Monogr. 33:131-160 (1963). Prior number: Cohen (1978) 11

Web 19 Moosehead Lake, Maine: J. L. Brooks and E. S. Deevey, New England. In: Limnology in North America, D. G. Frey, Ed. (Univ. of Wisconsin Press, Madison, 1963), pp. 117-162.

Web 20 Antarctic pack ice zone: G. A. Knox, Antarctic marine ecosystems. In: Antarctic Ecology, M. W. Holdgate, Ed. (Academic Press, New York, 1970) 1:69-96.

Web 21 Ross Sea: B. C. Patten and J. T. Finn, Systems approach to continental shelf ecosystems. In: Theoretical Systems Ecology, E. Halfon, Ed. (Academic Press, New York, 1979) pp. 184-212.

Web 22 Bear Island, Spitsbergen: V. S. Summerhayes and C. S. Elton, Contributions to the ecology of Spitsbergen and Bear Island, J. Ecol. 11:214-286 (1923). Prior number: Cohen (1978) 15

Web 23 Prairie, Manitoba: R. D. Bird, Biotic communities of the Aspen Parkland of central Canada, Ecology, 11:356-442 (1930). Prior number: Cohen (1978) 1.1

Web 24 Willow forest, Manitoba: R. D. Bird, Biotic communities of the Aspen Parkland of central Canada, Ecology, 11:356-442 (1930). Prior number: Cohen (1978) 1.2

Web 25 Aspen communities, Manitoba: R. D. Bird, Biotic communities of the Aspen Parkland of central Canada, Ecology, 11:356-442 (1930). Prior number: Cohen (1978) 1.3

Web 26 Aspen forest, Manitoba: R. D. Bird, Biotic communities of the Aspen Parkland of central Canada, Ecology, 11:356-442 (1930). Prior number: Cohen (1978) 1.4

Web 27 Wytham Wood, England: G. C. Varley, The concept of energy applied to a woodland community. In: Animal Populations in Relation to Their Food Resources, A. Watson, Ed. (Blackwell Scientific, Oxford, England, 1970), pp. 389-401.

Web 28 Salt meadow, New Zealand: K. Paviour-Smith, The biotic community of a salt meadow in New Zealand, Trans. R. Soc. N.Z. 83:525-554 (1956).

Web 29 Arctic seas: M. J. Dunbar, Arctic and subarctic marine ecology: immediate problems, Arctic 7:213-228 (1954).

Web 30 Antarctic seas: N. A. Mackintosh, A survey of antarctic biology up to 1945. In: Biologie antarctique, R. Carrick, M. Holdgate, J. Prevost, Eds. (Hermann, Paris, 1964), pp. 3-38.

Web 31 Epiplankton communities, Black Sea: T. S. Petipa, E. V. Pavlova, G. N. Mironov, The food web structure, utilization transport of energy by trophic levels in the planktonic communities. In: Marine Food Chains, J. H. Steele, Ed. (Oliver and Boyd, Edinburgh, 1970), 142-167.

Web 32 Bathyplankton communities, Black Sea: T. S. Petipa, E. V. Pavlova, G. N. Mironov, The food web structure, utilization transport of energy by trophic levels in the planktonic communities. In: Marine Food Chains, J. H. Steele, Ed. (Oliver and Boyd, Edinburgh, 1970), 142-167.

Web 33 Crocodile Creek, Malawi: G. Fryer, The trophic interrelationships and ecology of some littoral communities of Lake Nyasa, Proc. London Zool. Soc. 132:153-281 (1959). Prior number: Cohen (1978) 28.3

Web 34 River Clydach, Wales: J. R. Jones, A further ecological study of calcareous streams in the "Black Mountain" district of South Wales, J. Anim. Ecol. 18:142-159 (1949).

Web 35 Morgan's Creek, Kentucky: G. W. Minshall, Role of allochthonous detritus in the trophic structure of a woodland springbrook community, Ecology 48:139-149 (1967). Prior number: Cohen (1978) 18

Web 36 Mangrove swamp 6, Hawaii: G. E. Walsh, An ecological study of a Hawaiian mangrove swamp. In: Estuaries, G. H. Lauff, Ed. (AAAS Publication 83, Washington, DC, 1967), pp. 420-431.

Web 37 Marine sublittoral, southern California: T. A. Clarke, A. O. Flechsig, R. W. Grigg, Ecological studies during Project Sealab II, Science 157:1381-1389 (1967). Prior number: Cohen (1978) 2

Web 38 Lake Nyasa, rocky shore, Malawi: G. Fryer, The trophic interrelationships and ecology of some littoral communities of Lake Nyasa, Proc. London Zool. Soc. 132:153-281 (1959). Prior number: Cohen (1978) 28.1

Web 39 Lake Nyasa, sandy shore, Malawi: G. Fryer, The trophic interrelationships and ecology of some littoral communities of Lake Nyasa, Proc. London Zool. Soc. 132:153-281 (1959). Prior number: Cohen (1978) 28.2

Web 40 Rain forest, Malaysia: J. L. Harrison, The distribution of feeding habits among animals in a tropical rain forest, J. Anim. Ecol. 31:53-63 (1962). Prior number: Cohen (1978) 25

Web 41 Tropical seas, epipelagic zone: N. V. Parin, Ichthyofauna of the Epipelagic Zone (Israel Program for Scientific Translations, Jerusalem, 1970).

Web 42 Upwelling areas, Pacific Ocean: M. E. Vinogradov and E. A. Shushkina, Some development patterns of plankton communities in the upwelling areas of the Pacific Ocean. Mar. Biol. 48:357-366 (1978).

Web 43 Kelp bed community, South California: R. J. Rosenthal, W. D. Clarke, P. K. Dayton, Ecology and natural history of a stand of giant kelp, Macrocystis pyrifera, off Del Mar, California. Fish. Bull. (Dublin) 72:670-684 (1974).

Web 44 Marine coastal lagoons, Guerrero, Mexico: A. Yanez-Arancibia, Taxonomia, ecologia y estructura de las comunidades de peces en lagunas costeras con bocas efimeras del Pacifico de Mexico. Cent. Cienc. del Mar y Limnol. Univ. Nal. Auton. Mex. Publ. Espec. 2:1-306 (1978).

Web 45 Cone Spring, Iowa: L. J. Tilly, The structure and dynamics of Cone Spring. Ecol. Monogr. 38:169-197 (1968).

Web 46 Lake Texoma, Texas: B. C. Patten and 40 co-authors, Total ecosystem model for a cove in Lake Texoma. In: Systems Analysis and Simulation in Ecology, B. C. Patten, Ed. (Academic Press, New York, 1975) 3:205-421.

Web 47 Swamps, south Florida: L. D. Harris and G. B. Bowman, Vertebrate predator subsystem. In: Grasslands, Systems Analysis and Man, A. I. Breymeyer and G. M. Van Dyne, Eds. (International Biological Programme Series, no. 19, Cambridge Univ. Press, Cambridge, England, 1980), pp. 591-607.

Web 48 Nearshore marine 1, Aleutian Islands: C. A. Simenstad, J. A. Estes, K. W. Kenyon, Aleuts, sea otters, and alternate stable-state communities, Science 200:403-411 (1978).

Web 49 Nearshore marine 2, Aleutian Islands: C. A. Simenstad, J. A. Estes, K. W. Kenyon, Aleuts, sea otters, and alternate stable-state communities, Science 200:403-411 (1978).

Web 50 Sand beach, California: J. W. Nybakken, Marine Biology: An Ecological Approach (Harper and Row, New York, 1982).

Web 51 Shallow sublittoral, Cape Ann, Massachusetts: R. W. Dexter, The marine communities of a tidal inlet at Cape Ann, Massachusetts: a study in bio-ecology, Ecol. Monogr. 17:263-294 (1947).

Web 52 Rocky shore, Torch Bay, Alaska: R. T. Paine, Food webs: linkage, interaction strength and community infrastructure, J. Anim. Ecol. 49:667-685 (1980).

Web 53 Rocky shore, Cape Flattery, Washington: R. T. Paine, Food webs: linkage, interaction strength and community infrastructure, J. Anim. Ecol. 49:667-685 (1980).

Web 54 Western rocky shore, Barbados: F. Briand, unpublished observations

Web 55 Mudflat, Ythan estuary, Scotland: H. Milne and G. M. Dunnet, Standing crop, productivity and trophic relations of the fauna of the Ythan estuary. In: The Estuarine Environment, R. S. K. Barnes and J. Green, Eds. (Applied Science Publications, Edinburgh, Scotland, 1972), pp. 86-106.

Web 56 Mussel bed, Ythan estuary, Scotland: H. Milne and G. M. Dunnet, Standing crop, productivity and trophic relations of the fauna of the Ythan estuary. In: The Estuarine Environment, R. S. K. Barnes and J. Green, Eds. (Applied Science Publications, Edinburgh, Scotland, 1972), pp. 86-106.

Web 57 Brackish lagoons, Guerrero, Mexico: A. Yanez-Arancibia, Taxonomia, ecologia y estructura de las comunidades de peces en lagunas costeras con bocas efimeras del Pacifico de Mexico. Cent. Cienc. del Mar y Limnol. Univ. Nal. Auton. Mex. Publ. Espec. 2:1-306 (1978).

Web 58 Sphagnum bog, Russia, USSR: N. N. Smirnov, Food cycles in sphagnous bogs, Hydrobiologia 17:175-182 (1961).

Web 59 Trelease woods, Illinois: A. C. Twomey, The bird population of an elm-maple forest with special reference to aspection, territorialism, and coactions, Ecol. Monogr. 15:175-205 (1945).

Web 60 Montane forest, Arizona: D. I. Rasmussen, Biotic communities of Kaibab Plateau, Arizona, Ecol. Monogr. 11:228-275 (1941).

Web 61 Barren regions, Spitsbergen: V. S. Summerhayes and C. S. Elton, Further contributions to the ecology of Spitzbergen, J. Ecol. 16:193-268 (1928).

Web 62 Reindeer pasture, Spitsbergen: V. S. Summerhayes and C. S. Elton, Further contributions to the ecology of Spitzbergen, J. Ecol. 16:193-268 (1928).

Web 63 River Rheidol, Wales: J. R. Jones, A further ecological study of the river Rheidol: the food of the common insects of the main-stream, J. Anim. Ecol. 19:159-174 (1950).

Web 64 Linesville Creek, Pennsylvania: K. W. Cummins, W. P. Coffman, P. A. Roff, Trophic relationships in a small woodland stream, Verh. Int. Ver. Theor. Angew. Limnol. 16:627-638 (1966).

Web 65 Yoshino River rapids, Japan: M. Tsuda, Interim results of the Yoshino River productivity survey, especially on benthic animals. In: Productivity Problems of Freshwaters, Z. Kajak and A. Hillbricht-Ilkowska, Eds. (Polish Scientific, Warsaw, 1972), pp. 829-841.

Web 66 River Thames, England: K. H. Mann, R. H. Britton, A. Kowalczewski, T. J. Lack, C. P. Mathews and Ian McDonald, Productivity and energy flow at all trophic levels in the River Thames, England. In: Productivity Problems of Freshwaters, Z. Kajak and A. Hillbricht-Ilkowska, Eds. (Polish Scientific, Warsaw, 1972), pp. 579-596.

Web 67 Mudflats, Mississippi River, Iowa: C. A. Carlson, Summer bottom fauna of the Mississippi River above Dam 19, Keokuk, Iowa, Ecology 49:162-168 (1968).

Web 68 Loch Leven, Scotland: N. C. Morgan and D. S. McLusky, A summary of the Loch Leven IBP results in relation to lake management and future research, Proc. R. Soc. Edinburgh Series B 74:407-416 (1972).

Web 69 Tagus estuary, Portugal: L. Saldanha, Estudio Ambiental do Estuario do Tejo, Publ. no. 5(4) (CNA/Tejo, Lisbon, 1980).

Web 70 Crystal River estuary, Florida: W. M. Kemp, W. H. Smith, H. N. McKellar, M. B. Debman, M. Homer, D. L. Young and H. T. Odum, Energy cost-benefit analysis applied to power plants near Crystal River, Florida. In: Ecosystem Modeling in Theory and Practice: An Introduction with Case Histories, C. A. Hall and J. W. Day, Jr., Eds. (Wiley, New York, 1977), pp. 507-543.

Web 71 Lake Rybinsk, Russia, USSR: Y. I. Sorokin, Biological productivity of the Rybinsk reservoir. In: Productivity Problems of Freshwaters, Z. Kajak and A. Hillbricht-Ilkowska, Eds. (Polish Scientific, Warsaw, 1972), pp. 493-503.

Web 72 Heney Lake, pelagic zone, Quebec: A. Baril, Effect of the water mite Piona constricta on planktonic community structure, M.Sc. Thesis, University of Ottawa, Canada (1983).

Web 73 Hafner Lake, Austria: F. Schiemer, M. Bobek, P. Gludovatz, A. Ioschenkohl, I. Zweimuller and M. Martinetz, Trophische Interaktionen im Pelagial des Hafnersees, Sitzungsber. Akad. Wiss. Wien Math. Naturwiss. Kl. Abt. 1:191-209 (1982).

Web 74 Sand beach, South Africa: A. C. Brown, Food relationships on the intertidal sandy beaches of the Cape Peninsula, S. Afr. J. Sci. 60:35-41 (1964).

Web 75 Vorderer Finstertaler Lake, Austria: R. Pechlaner, G. Bretschko, P. Gollmann, H. Pfeifer, M. Tilzer and H. P. Weissenbach, Ein Hochgebirgssee (Vorderer Finstertaler See, Kubtai, Tirol) als Modell des Energietransportes durch ein limnisches Oekosystem, Verh. Dtsch. Zool. Ges. 65:47-56 (1972).

Web 76 Neusiedler Lake, Austria: F. Schiemer, The benthic communities of the open lake. In: Neusiedlersee: The Limnology of a Shallow Lake in Central Europe, H. Loeffler, Ed. (Junk, The Hague, Netherlands, 1979), pp. 337-384.

Web 77 Lake Abaya, Ethiopia: D. Riedel, Der Margheritensee (Südabessinien) – Zugleich ein Beitrag zur Kenntnis der Abessinischen Graben-Seen, Arch. Hydrobiol. 58:435-466 (1962).

Web 78 Lake George, Uganda: M. J. Burgis, I. G. Dunn, G. G. Ganf, L. M. McGowan and A. B. Viner, Lake George, Uganda: Studies on a tropical freshwater ecosystem. In: Productivity Problems of Freshwaters, Z. Kajak and A. Hillbricht-Ilkowska, Eds. (Polish Scientific, Warsaw, 1972), pp. 301-309.

Web 79 Lake Paajarvi, offshore, Finland: J. Sarvala, Paarjarven energiatalous, Luonnon Tutkija 78:181-190 (1974).

Web 80 Lake Paajarvi, littoral zone, Finland: J. Sarvala, Paarjarven energiatalous, Luonnon Tutkija 78:181-190 (1974).

Web 81 Sendai Bay, mesopelagic zone, Japan: M. A. Hatanaka, Sendai Bay. In: Productivity of Biocenoses in Coastal Regions of Japan, K. Hogetsu, M. Horanaka, T. Hatanaka,
T. Kawamura, Eds. (Japanese Committee for the International Biological Program Synthesis, Tokyo, 1977), 14:173-221.

Web 82 Permanent freshwater rockpool, France: P. Ohm and H. Remmert, Etudes sur
les rockpools des Pyrenees-Orientales, Vie et Milieu 6:194-209 (1955).

Web 83 Lake Pyhajarvi, littoral zone, Finland: K. Aulio, K. Jumppanen, H. Molsa,
J. Nevalainen, M. Rajasilta, I. Vuorinen, Litoraalin merkitys Pyhajarven kalatuotannolle,
Sakylan Pyhajarven Tila Ja Biologinen Tuotanto (Lounais-Suomen Vesiensuojeluyhdistys
R. Y., Turku, Finland, 1981) 47:173-176.

Web 84 Temporary pond, Michigan: H. M. Wilbur, Competition, predation, and the
structure of the Ambystoma-Rana sylvatica community, Ecology 53:3-21 (1972).

Web 85 Tasek Bera swamp, Malaysia: T. Mizuno and J. I. Furtado. In: Tasek Bera, J. I.
Furtado and S. Mori, Eds. (Junk, The Hague, Netherlands, 1982), pp. 357-359.

Web 86 Suruga Bay, epipelagic zone, Japan: K. Hogetsu, Biological productivity of
some coastal regions of Japan. In: Marine Production Mechanisms, M. J. Dunbar, Ed.
(International Biological Programme Series, no. 20, Cambridge Univ. Press, Cambridge,
England, 1979), pp. 71-87.

Web 87 Ice edge community, High Arctic, Canada: S. W. Bradstreet and W. E. Cross,
Trophic relationships at High Arctic ice edges, Arctic 35:1-12 (1982).

Web 88 Lestijoki River rapids, Finland: K. Kuusela, Early summer ecology and community structure of the macrozoobenthos on stones in the Javajankoski rapids on the river
Lestijoki, Finland, Acta Universitatis Ouluensis (Ser. A, no. 87, Oulu, Finland, 1979).

Web 89 River Cam, England: P. H. T. Hartley, Food and feeding relationships in a community of freshwater fishes, J. Anim. Ecol. 17:1-14 (1948).

Web 90 Old field, New Jersey: D. J. Shure, Radionuclide tracer analysis of trophic relationships in an old-field ecosystem, Ecol. Monogr. 43:1-19 (1973).

Web 91 Shigayama coniferous forest, Japan: Y. Kitazawa, Ecosystem metabolism of
the subalpine coniferous forest of the Shigayama IBP area. In: Ecosystem Analysis of the
Subalpine Coniferous Forest of Shigayama IBP Area, Central Japan, Y. Kitazawa, Ed.
(Japanese Committee for the International Biological Program Synthesis, Tokyo, 1977)
15:181-196.

Web 92 High Himalayas community, Tibet: L. W. Swan, The ecology of the high Himalayas, Sci. Am. 205:68-78 (October 1961).

Web 93 Alpine tundra, Montana: D. L. Pattie and N. A. M. Verbeek, Alpine birds of
the Beartooth Mountains, Condor 68:167-176 (1966); Alpine mammals of the Beartooth
Mountains, Northwest Sci. 41:110-117 (1967).

Web 94 Wet coastal tundra, Barrow, Alaska: J. Brown, The structure and function of
the tundra ecosystem, U.S. Tundra Biome 1971 Progress Rept. 1 (1971).

Web 95 Tundra, Prudhoe, Alaska: J. Brown, Ecological investigations of the Tundra
biome in the Prudhoe Bay Region, Alaska, Special Report, no. 2, Biol. Pap. Univ. Alaska
(1975).

Web 96 Tundra, Yamal Peninsula, Siberia: V. I. Osmolovskaya, Geographical distribution of raptors in Kazakhstan plains and their importance for pest control, Tr. Acad. Sci.
USSR Inst. Geogr. 41:5-77 (1948). (In Russian.)

Web 97 Tundra, South Yamal, Siberia: T. Dunaeva and V. Kucheruk, Material on the
ecology of the terrestrial vertebrates of the tundra of south Yamal, Bull. Soc. Nat. Moscou
(N.S., Zool. Sect.) 4(19):1-80 (1941).

Web 98 Sand dunes, Namib Desert, Namibia: E. Holm and C. H. Scholtz, Structure
and pattern of the Namib Desert dune ecosystem at Gobabed, Madoqua 12:3-39 (1980).

Web 99 Sonora Desert, Arizona: P. G. Howes, The Giant Cactus Forest and Its World
(Duell, Sloan, and Pearce, New York, 1954).

Web 100 Rajasthan Desert, India: I. K. Sharma, A study of ecosystems of the Indian
desert, Trans. Indian Soc. Desert Technol. and Univ. Center Desert Stud. 5:51-55 and A
study of agro-ecosystems in the Indian desert, ibid. 5:77-82 (1980).

Web 101 Temporary freshwater rockpool, France: P. Ohm and H. Remmert, Etudes
sur les rockpools des Pyrenees-Orientales, Vie et Milieu 6:194-209 (1955).

Web 102 Plankton, oligotrophic tropical Pacific: E. A. Shushkina and M. E. Vinogradov,
Trophic relationships in communities and the functioning of marine ecosystems: II. In:
Marine Production Mechanisms, M. J. Dunbar, Ed. (Cambridge Univ. Press, Cambridge,
England, 1979), pp. 251-268.

Web 103 Tropical plankton community, Pacific: T. S. Petipa, Trophic relationships in communities and the functioning of marine ecosystems: I. In: Marine Production Mechanisms, M. J. Dunbar, Ed. (Cambridge Univ. Press, Cambridge, England, 1979), pp. 233-250.

Web 104 Rocky shore, Bay of Panama: B. A. Menge, J. Lubchenco, S. D. Gaines and L. R. Ashkenas, A test of the Menge-Sutherland model of community organization in a tropical rocky intertidal food web, Oecologia (Berlin) 71:75-89 (1986).

Web 105 Rocky shore, Gulf of Maine, USA: D. C. Edwards, D. O. Conover, F. Sutter, Mobile predators and the structure of marine intertidal communities, Ecology 63:1175-1180 (1982).

Web 106 Rocky shore, Monterey Bay, California: P. W. Glynn, Community composition, structure, and interrelationships in the marine intertidal Endocladia muricata - Balanus glandula association in Monterey Bay, California, Beaufortia 148(12):1-198 (1965).

Web 107 Bay pilings community, New Jersey: C. H. Peterson, The importance of predation and competition in organizing the intertidal epifaunal communities of Barnegat Inlet, New Jersey, Oecologia (Berlin) 39:1-24 (1979).

Web 108 Rocky shore, Cabrillo Point, California: W. G. Hewatt, Ecological studies on selected marine intertidal communities of Monterey Bay, California, Am. Midl. Nat. 18:161-206 (1937).

Web 109 Rocky shore, central Chile: J. C. Castilla, Perspectivas de investigacion en estructura y dinamica de communidades intermareales rocosas de Chile Central. II. Depredadores de alto nivel trofico, Medio Ambiente 5:190-215 (1981).

Web 110 Rocky shore, Cape Ann, Massachussetts: R. W. Dexter, The marine communities of a tidal inlet at Cape Ann, Massachusetts: a study in bio-ecology, Ecol. Monogr. 17:263-294 (1947).

Web 111 Mudflat, Cape Ann, Massachussetts: R. W. Dexter, The marine communities of a tidal inlet at Cape Ann, Massachusetts: a study in bio-ecology, Ecol. Monogr. 17:263-294 (1947).

Web 112 Low salt marsh, Cape Ann, Massachussetts: R. W. Dexter, The marine communities of a tidal inlet at Cape Ann, Massachusetts: a study in bio-ecology, Ecol. Monogr. 17:263-294 (1947).

Web 113 High salt marsh, Cape Ann, Massachussetts: R. W. Dexter, The marine communities of a tidal inlet at Cape Ann, Massachusetts: a study in bio-ecology, Ecol. Monogr. 17:263-294 (1947).

1 Cochin backwater, India

1	3	4	5	6	7	8	9
1	1	1	1	1	0	0	0
2	1	1	1	1	0	0	0
3	0	0	0	0	1	1	1
4	0	0	0	0	1	0	0
5	0	0	0	0	1	1	0
6	0	0	0	0	0	1	1
7	0	0	0	0	0	1	0
8	0	0	0	0	0	0	1

1 basic food
2 detritus
3 prawns and shrimps
4 benthos(micro, meio and macro)
5 zooplankton herbivores
6 fish herbivores
7 other carnivores
8 fish carnivores
9 man

2 Knysna estuary, South Africa

2	4	5	6	7	8	9	10	11	12	13	14	15
1	1	0	0	0	1	1	0	0	0	0	0	0
2	0	0	0	0	0	0	0	0	0	0	1	0
3	1	0	1	1	1	1	1	1	0	0	0	0
4	0	1	0	0	0	0	0	0	0	0	0	0
5	0	0	0	0	0	0	0	0	0	0	0	1
6	0	0	0	0	0	0	0	0	1	0	0	1
7	0	0	0	0	0	0	0	0	1	1	0	0
8	0	0	0	0	0	0	0	0	0	1	1	0
9	0	0	0	0	0	0	0	0	0	1	1	0
10	0	0	0	0	0	0	0	0	0	1	0	0
11	0	0	0	0	0	0	0	0	0	1	1	0
13	0	0	0	0	0	0	0	0	0	0	0	1
14	0	0	0	0	0	0	0	0	0	0	0	1

1 phytoplankton
2 attached plants
3 detritus
4 zooplankton
5 *Hyporhamphus*
6 *Mugil*
7 *Upogebia*
8 *Lamya*
9 *Solen*
10 *Arenicola*
11 *Hymenosoma*
12 *Johnius*
13 *Lithognathus*
14 *Rhabdosargus*
15 *Hypacanthus*

3 Salt marsh, Long Island, USA

3	5	6	7	8	9	10	11	12	13	14	15	16	17	18	19	20	21	22	23	24
1	1	0	0	0	0	0	0	1	0	0	0	0	0	0	0	0	0	0	0	0
2	0	0	1	0	0	0	0	1	1	1	1	0	0	0	0	0	0	0	0	0
3	0	1	0	1	0	0	0	0	0	0	1	1	0	0	0	0	0	0	0	0
4	0	0	0	0	1	1	0	0	0	0	0	0	0	0	0	0	0	0	0	0
5	0	0	0	0	0	0	1	0	0	0	0	0	0	0	0	0	0	0	0	0
6	0	0	0	0	0	0	1	0	0	0	0	0	1	0	0	0	0	0	0	0
7	0	0	0	0	0	0	0	0	0	1	0	0	0	0	0	0	0	0	0	0
8	0	0	0	0	0	0	0	0	0	1	0	0	0	0	0	0	0	1	0	0
10	0	0	0	0	0	0	0	0	0	0	0	0	0	0	0	0	0	0	0	1
11	0	0	0	0	0	0	0	0	0	0	0	0	1	1	1	0	0	0	0	0
12	0	0	0	0	0	0	0	0	0	0	0	0	1	0	1	1	0	0	0	0
13	0	0	0	0	0	0	0	0	0	0	0	0	0	0	1	1	0	0	0	0
14	0	0	0	0	0	0	0	0	0	0	0	0	1	0	1	0	0	0	0	0
15	0	0	0	0	0	0	0	0	0	0	0	0	0	0	1	0	0	0	1	0
16	0	0	0	0	0	0	0	0	0	0	0	0	0	0	1	0	0	0	1	0

1 organic debris
2 water plant
3 plankton
4 marsh plants
5 bay shrimp
6 silversides
7 mud snail
8 clam

9 mosquito
10 cricket
11 billfish
12 eel
13 fluke
14 blowfish
15 minnow 1
16 minnow 2

17 tern
18 Osprey
19 Green Heron
20 merganser
21 cormorant
22 gull
23 kingfisher
24 Red-winged Blackbird

4 Salt marsh, California

4	3	4	5	6	7	8	9	10	11	12	13
1	1	0	0	0	0	1	0	1	1	1	0
2	0	1	1	0	0	0	1	0	0	0	0
3	0	0	0	1	0	1	0	1	1	1	0
4	0	0	1	0	1	1	1	1	0	0	0
5	0	0	0	0	1	0	0	0	0	0	0
6	0	0	0	0	0	0	0	0	0	0	1
8	0	0	0	0	0	0	0	0	0	1	1
9	0	0	0	0	0	0	0	0	0	0	1
10	0	0	0	0	0	0	0	0	0	0	1
11	0	0	0	0	0	0	0	0	0	0	1
12	0	0	0	0	0	0	0	0	0	0	1

1 terrestrial plants
2 marine plants
3 terrestrial invertebrates
4 intertidal and marine invertebrates
5 fishes
6 *Sorex*
7 herons

8 *Rallus, Anas*
9 migrant shorebirds and waterfowl
10 passerines
11 *Microtus, Reithrodontomys, Mus*
12 *Rattus*
13 *Circus, Asio*

5 Salt marsh, Georgia

5	2	3	6	7
1	1	0	1	0
2	0	1	0	0
4	0	0	1	0
5	0	0	1	0
6	0	0	0	1

1 *Spartina*
2 *Prokelisia, Orchelimum,* other herbivorous insects
3 spiders, passerines, dragonflies
4 algae
5 bacteria
6 *Uca* and *Sesarma, Modiolus, Littorina, Oligochaete, Streblospio, Capitella, Manayunkia*
7 *Eurytium,* clapper rail, raccoon

6 Tidal flat, California

	3	4	5	6	7	8	9	10	11	12	13	14	15	16	17	18	19	20	21	22	23	24	25
1	1	1	0	0	0	0	1	0	0	0	0	0	0	0	0	0	0	0	0	0	0	0	0
2	1	0	1	1	1	1	1	1	0	0	0	0	0	0	0	0	0	0	0	0	0	0	0
3	0	1	0	1	0	0	0	0	0	0	0	0	0	0	0	0	0	0	0	0	0	0	0
4	0	0	0	0	0	0	0	0	0	1	1	1	1	1	0	0	0	0	0	0	0	0	0
5	0	0	0	0	0	0	0	0	0	1	0	1	0	0	0	0	0	0	0	0	0	0	0
6	0	0	0	0	0	0	0	0	0	0	1	0	0	0	0	0	0	0	0	0	0	0	1
8	0	0	0	0	0	0	0	0	0	0	0	1	0	0	0	0	0	0	0	0	0	0	0
9	0	0	0	0	0	0	0	0	0	0	0	1	0	0	0	0	1	1	0	0	1	1	1
10	0	0	0	0	0	0	0	0	0	1	1	1	1	1	0	0	0	0	0	0	0	0	0
12	0	0	0	0	0	0	0	0	1	0	1	0	0	0	1	1	1	0	0	1	0	1	0
14	0	0	0	0	0	0	0	0	0	0	0	0	0	0	0	0	0	0	0	0	0	0	0
15	0	0	0	0	0	0	0	0	0	1	0	1	0	1	0	0	0	0	0	0	0	0	0
17	0	0	0	0	0	0	0	0	0	0	0	0	0	0	0	0	0	0	0	0	0	0	0
18	0	0	0	0	0	0	0	0	0	0	0	0	0	0	0	0	0	0	1	0	0	0	1
19	0	0	0	0	0	0	0	0	0	0	0	0	0	0	0	0	0	0	1	1	1	0	1
22	0	0	0	0	0	0	0	0	0	0	0	0	0	0	0	0	0	0	0	0	0	0	1

1 primary producers
2 detritus
3 protozoa
4 crustacea
5 Cailianassa
6 shore crabs
7 Phoronopsis
8 Urechis
9 clams

10 Lumbrinereis, Notomastus
11 heron
12 Clevelandia
13 wading birds
14 stingray
15 Nereis, Glycera
16 nemertceans
17 terns, shorebirds
18 cabezon

19 flounder
20 Cerebratulus
21 loon, cormorant
22 striped bass
23 scoter
24 gulls
25 man

7 Narragansett Bay, Rhode Island

7	4	5	6	7	8	9	10	11	12	13	14	15	16	17	18	19	20
1	1	1	0	1	0	0	1	1	0	0	0	0	0	0	0	0	0
2	1	1	1	1	0	0	1	0	0	0	0	0	0	0	1	0	0
3	0	0	1	0	0	0	0	0	0	0	0	0	0	0	0	0	0
4	0	0	0	0	1	1	1	0	0	0	0	0	0	0	0	0	0
5	0	0	0	0	0	0	0	0	0	0	0	0	0	0	0	1	0
6	0	0	0	0	0	0	0	0	1	0	0	0	0	1	1	1	0
7	0	0	0	0	0	0	0	0	0	0	0	0	0	0	0	1	0
8	0	0	0	0	0	0	0	0	0	1	0	0	0	0	0	0	0
10	0	0	0	0	0	0	0	0	0	0	1	1	1	0	0	0	0
11	0	0	0	0	0	0	0	0	0	0	0	0	1	1	1	1	0
12	0	0	0	0	0	0	0	0	0	0	0	0	0	0	0	0	1
14	0	0	0	0	0	0	0	0	0	0	0	0	0	0	0	0	1
15	0	0	0	0	0	0	0	0	0	0	0	0	0	0	0	0	1
16	0	0	0	0	0	0	0	0	0	0	0	0	0	0	0	0	1
17	0	0	0	0	0	0	0	0	0	0	0	0	0	0	0	0	1
19	0	0	0	0	0	0	0	0	0	0	0	0	0	0	0	0	1

1 flagellates, diatoms
2 particulate detritus
3 macroalgae, eelgrass
4 *Acartia*, other copepods
5 sponges
6 benthic macrofauna
7 clams
8 ctenophores
9 meroplankton, fish larvae
10 Pacific menhaden
11 bivalves
12 crabs, lobsters
13 butterfish
14 striped bass
15 bluefish
16 mackerel
17 demersal species
18 starfish
19 flounder
20 man

8 Salt marsh, Rhode Island

8	5	6	7	8	9	10	11	12	13	14	15
1	0	0	0	0	0	0	0	0	1	1	0
2	0	0	0	0	0	0	0	0	1	0	0
3	1	1	0	1	1	1	1	0	0	0	0
4	0	0	1	1	0	0	0	0	0	0	0
5	0	0	0	0	0	1	0	0	0	0	0
6	0	0	0	0	1	1	0	1	0	0	0
7	0	0	0	0	0	0	0	1	0	0	0
9	0	0	0	0	0	0	1	1	1	1	0
10	0	0	0	0	0	0	0	0	1	1	1
11	0	0	0	0	0	0	0	0	0	1	0
12	0	0	0	0	0	0	0	0	0	1	0

1 *Ruppia*
2 *Ulva, Enteromorpha*
3 detritus
4 phytoplankton
5 nematodes, copepods, ostracods
6 amphipods
7 zooplankton
8 juvenile menhaden
9 shrimps
10 common mummichog
11 striped mummichog
12 other fish
13 ducks
14 eel
15 man

9 Lough Ine rapids, Ireland

9	3	4	5	6	7	8	9	10
1	0	1	1	0	0	0	0	0
2	1	0	0	0	0	0	0	0
3	0	0	0	0	0	0	1	0
4	0	0	0	1	1	1	1	1
5	0	0	0	1	1	0	1	0
6	0	0	0	0	0	0	0	1
7	0	0	0	0	0	0	0	1
8	0	0	0	0	0	0	0	1
9	0	0	0	0	0	0	0	1

1 attached algae   6 *Cancer*
2 plankton   7 *Portunus puber*
3 *Anomia*   8 *Carcinus*
4 *Paracentrotus*   9 *Marthasterias*
5 *Gibbula eineraria*   10 birds

10 Exposed rocky shore, New England, USA

10	3	4	5
1	1	1	0
2	1	1	0
3	0	0	1
4	0	0	1

1 detritus   4 *Mytilus edulis*
2 plankton   5 *Thais lapillus*
3 *Balanus balanoides*

11 Protected rocky shore, New England, USA

11	4	5	6	7	8
1	0	0	1	1	0
2	0	0	1	1	0
3	1	1	0	0	0
4	0	0	0	0	1
5	0	0	0	0	1
6	0	0	0	0	1
7	0	0	0	0	1

1 detritus   5 *Littorina*
2 plankton   6 *Mytilus edulis*
3 algae   7 *Balanus balanoides*
4 *Acmaea testudinalis*   8 *Thais lapillus*

12 Exposed rocky shore, Washington

12	4	5	6	7	8	9	10	11	12	13
1	1	0	0	1	1	0	0	0	0	0
2	1	0	0	1	1	0	0	0	0	0
3	0	1	1	0	0	1	1	0	0	0
4	0	0	0	0	0	0	0	1	1	1
5	0	0	0	0	0	0	0	0	1	1
6	0	0	0	0	0	0	0	0	1	1
7	0	0	0	0	0	0	0	1	1	1
8	0	0	0	0	0	0	0	0	1	0
9	0	0	0	0	0	0	0	0	1	1
10	0	0	0	0	0	0	0	0	0	1
11	0	0	0	0	0	0	0	0	1	1

1 detritus   8 *Pollicipes*
2 plankton   9 chitons
3 algae   10 *Littorina*
4 acorn barnacles   11 *Thais*
5 limpets   12 *Pisaster*
6 *Tegula*   13 *Leptasterias*
7 *Mytilus californianus*

13 Protected rocky shore, Washington

13	4	5	6	7	8	9	10	11	12	13
1	1	0	0	1	0	0	0	0	0	0
2	1	0	0	1	0	0	0	0	0	0
3	0	1	1	0	1	1	0	0	0	0
4	0	0	0	0	0	0	1	1	1	1
5	0	0	0	0	0	0	1	1	1	1
6	0	0	0	0	0	0	0	0	1	1
7	0	0	0	0	0	0	1	0	1	1
8	0	0	0	0	0	0	0	1	1	1
9	0	0	0	0	0	0	0	0	1	1
10	0	0	0	0	0	0	0	0	1	1
11	0	0	0	0	0	0	0	0	1	1

1 detritus
2 plankton
3 algae
4 acorn barnacles
5 limpets
6 other herbivorous gastropods
7 *Mytilus edulis*

8 chitons
9 *Littorina*
10 *Thais*
11 *Seariesia*
12 *Pisaster*
13 *Leptasterias*

14 Mangrove swamp 1, Hawaii

14	2	3	4	5	6	7	8
1	1	1	1	0	1	0	0
2	0	0	0	1	1	0	0
4	0	0	0	0	0	1	0
5	0	0	0	0	0	0	1
6	0	0	0	0	0	0	1
7	0	0	0	0	0	0	1

1 detritus
2 *Tendipes* larvae
3 *Procambarus clarkii*
4 mosquito larvae

5 *Xiphophorous helleri*
6 *Melania indefinita*
7 *Lebistes reticulatus*
8 *Eleotris sandwicensis*

15 Mangrove swamp 3, Hawaii

15	2	3	4	5	6	7	8	9
1	1	1	1	1	1	0	0	0
2	0	0	0	0	0	1	0	0
3	0	0	0	0	0	0	0	1
4	0	0	0	0	0	0	0	1
5	0	0	0	0	0	0	0	1
6	0	0	0	0	0	0	1	0
8	0	0	0	0	0	0	0	1

1 detritus
2 copepods
3 *Xiphophorous helleri*
4 *Neritina tahitiensis*
5 *Metopograpsis messor*

6 mosquito larvae
7 *Chonophorous genivittatus*
8 *Lebistes reticulatus*
9 *Kuhlia sandvicensis*

16 Pamlico estuary, North Carolina

16	5	6	7	8	9	10	11	12	13	14
1	0	1	0	1	0	0	0	0	0	0
2	1	1	0	1	0	0	0	0	0	0
3	0	1	1	1	0	0	0	0	0	0
4	0	0	0	0	0	0	0	1	0	0
5	0	0	0	0	1	1	1	0	1	0
6	0	0	0	0	0	0	0	1	0	0
7	0	0	0	0	1	0	1	1	0	0
8	0	0	0	0	0	0	0	0	0	1
10	0	0	0	0	0	0	0	0	0	1
11	0	0	0	0	0	0	0	0	0	1

1 detritus	8 *Mugil*
2 dinoflagellates	9 small *Paralichthys*
3 benthic diatoms	10 *Brevoortia*
4 *Ruppia*	11 small *Leiostomus, Micropogon*
5 *Acartia tonsa,* harpacticoids	12 *Callinectes*
6 *Rangia*	13 *Ctenophora*
7 *Gammarus, Paleomonetes*	14 *Roccus, Cynoscion*

17 Coral reefs, Marshall Islands

17	4	5	6	7	8	9	10	11	12	13	14
1	0	1	0	0	0	1	0	0	0	0	0
2	1	0	0	0	0	0	0	0	0	0	0
3	0	0	1	0	0	1	0	0	0	0	0
4	0	0	0	1	1	1	0	0	0	0	0
5	0	0	0	0	0	1	1	0	0	0	0
6	0	0	0	0	0	1	1	0	0	0	0
7	0	0	0	0	0	0	0	1	0	0	1
8	0	0	0	0	0	1	0	0	1	0	0
9	0	0	0	0	0	0	0	0	0	1	1
10	0	0	0	0	0	1	0	0	0	1	1
11	0	0	0	0	0	0	0	0	0	0	1
13	0	0	0	0	0	0	0	0	0	0	1

1 detritus	8 corals
2 phytoplankton	9 omnivores
3 benthic algae	10 small benthic carnivores
4 zooplankton	11 large midwater carnivores
5 detritus feeders	12 coral feeders
6 algal feeders	13 large benthic carnivores
7 small midwater plankton feeders	14 transient carnivores

18 Kapingamarangi atoll, Polynesia

18	3	4	5	6	7	8	9	11	12	14	15	16	21	22	23	24	25	26	27
1	0	0	1	0	0	0	0	0	0	0	0	0	0	0	0	0	0	0	0
2	1	0	1	0	0	0	0	0	0	0	0	0	0	0	0	0	0	0	0
3	0	1	1	0	0	0	0	0	0	0	0	0	0	0	0	0	0	0	0
4	0	0	1	0	0	0	0	0	0	0	0	0	0	0	0	0	0	0	0
5	0	0	0	1	1	1	1	0	0	0	0	0	0	0	-1	0	0	0	0
10	0	0	0	0	0	0	0	1	0	0	0	0	0	0	0	0	0	0	0
11	0	0	0	0	0	0	1	0	0	0	0	0	0	0	0	0	0	0	0
12	0	0	0	0	0	0	1	0	0	0	0	0	0	0	0	0	0	0	0
13	0	0	0	0	0	0	1	0	1	1	1	0	0	0	0	0	1	0	0
15	0	0	0	0	0	0	1	0	0	0	0	0	0	0	0	0	0	0	0
16	0	0	0	0	0	0	1	0	0	0	0	0	0	0	0	0	0	0	0
17	0	0	0	0	0	0	1	0	0	0	0	1	1	0	0	0	1	1	0
18	0	0	0	0	0	0	1	0	0	0	0	0	0	0	0	0	1	1	0
19	0	0	0	0	0	0	1	0	0	0	0	0	0	0	0	0	1	1	0
20	0	0	0	0	0	0	1	0	0	0	0	0	0	0	0	1	1	1	0
21	0	0	0	0	0	0	0	0	0	0	0	0	0	1	1	0	0	0	1
22	0	0	0	0	0	0	0	0	0	0	0	0	0	0	1	0	0	0	0
27	0	0	0	0	0	0	0	0	0	0	0	0	0	0	0	1	0	0	0

1 algae
2 phytoplankton
3 zooplankton
4 invertebrates
5 fish
6 terns
7 frigate birds
8 boobies
9 man
10 turtle grass
11 sea turtles
12 pig
13 coconut
14 rat
15 coconut crabs
16 fowl
17 land vegetation
18 *Cyrtosperma*
19 *Pandanus*
20 breadfruit
21 insects
22 skinks
23 reef heron
24 starlings
25 land crustacea
26 fungi, snails, annelids
27 geckos

19 Moosehead Lake, Maine

19	3	4	5	6	7	8	9	10	11	12	13	14	15	16	17
1	1	0	1	0	0	0	0	0	0	0	0	0	0	0	0
2	0	1	0	0	0	0	0	0	0	0	0	0	0	0	0
3	0	0	0	1	1	1	1	0	0	0	0	0	0	0	0
4	0	0	0	0	0	0	1	1	0	1	1	0	1	0	0
5	0	0	0	0	0	0	0	1	1	1	1	0	1	1	0
7	0	0	0	0	0	0	0	0	0	0	0	0	0	0	1
8	0	0	0	0	0	0	0	0	0	0	0	1	1	1	1
9	0	0	0	0	0	0	0	0	0	0	0	0	0	1	1
10	0	0	0	0	0	0	0	0	0	0	0	0	1	0	1
11	0	0	0	0	0	0	0	0	0	0	0	0	1	1	0
12	0	0	0	0	0	0	0	0	0	0	0	0	0	1	1
15	0	0	0	0	0	0	0	0	0	0	0	0	0	0	1

1 phytoplankton
2 ooze, bacteria
3 zooplankton
4 chironomids
5 littoral browsers
6 *Chaoborus*
7 *Leptodora*
8 *Osmerus*
9 *Catostomus*
10 *Gasterosteus*
11 cyprinids
12 *Cottus*
13 *Prosopium*
14 *Salmo*
15 *Salvelinus*
16 *Lota*
17 *Cristivomer*

20 Antarctic pack ice zone

20	4	5	6	7	8	9	10	11	12	13	14	15	16	17	18	19
1	1	1	0	0	0	0	0	0	0	0	0	0	0	0	0	0
2	0	0	1	1	1	0	0	0	0	0	0	0	0	0	0	0
3	0	0	0	0	1	0	0	0	0	0	0	0	0	0	0	0
4	0	1	0	0	0	1	0	0	0	0	0	0	0	0	0	0
5	0	0	0	0	0	0	1	0	0	0	0	0	0	0	0	0
6	0	1	0	0	1	0	1	1	0	0	0	0	0	0	0	0
7	0	0	0	0	0	1	0	1	1	1	0	0	1	0	1	0
8	0	0	0	0	0	0	0	0	0	0	1	0	0	0	0	1
9	0	0	0	0	0	0	0	0	0	0	0	1	1	0	0	0
10	0	0	0	0	0	0	0	1	0	0	0	0	0	0	0	0
11	0	0	0	0	0	0	0	0	0	0	0	0	1	0	0	0
13	0	0	0	0	0	0	0	0	0	0	0	0	1	1	1	0
14	0	0	0	0	0	0	0	0	0	0	0	0	0	0	0	1
15	0	0	0	0	0	0	0	0	0	0	0	0	1	0	1	0

1 epontic microalgae
2 planktonic microalgae
3 detritus
4 copepods, *Orchomonopsis*
5 bottom notothenids, fry of *Trematomus*
6 phytoplankton filter feeders
7 *Euphausia superba, E. crystallorophias*
8 benthic filter feeding invertebrates
9 *Trematomus borchgrevinki, Pleurogramma antarctica*
10 predatory planktonic invertebrates

11 predatory cephalopods
12 whales, crabeater seals
13 penguins, petrels
14 pycnogonids, polychaetes, *Glyptonotus, Lineus, Odonaster*
15 predatory fishes
16 Weddell seals
17 toothed whales
18 leopard seals
19 predatory benthic fishes

21 Ross Sea

21	4	5	6	7	8	9	10
1	1	1	0	0	0	0	0
2	0	1	0	0	0	0	0
3	0	1	0	0	0	0	0
4	0	1	1	0	0	0	0
5	1	0	1	1	1	1	1
6	0	0	0	1	1	1	1
7	0	0	0	0	1	1	1
8	0	0	0	0	0	1	0
9	0	0	0	0	0	0	1

1 ice algae
2 phytoplankton
3 detritus
4 ice invertebrates
5 zooplankton

6 fish
7 cephalopods
8 penguins
9 seals
10 whales

22 Bear Island, Spitsbergen

	2	6	7	8	9	10	11	12	13	14	16	17	18	19	20	21	22	24	25	26	27	28
1	0	0	0	0	0	0	0	0	0	0	0	-1	0	0	0	0	0	0	0	0	0	0
2	0	0	0	0	0	0	0	0	0	0	0	0	1	0	0	0	0	0	0	0	-1	-1
3	0	-1	-1	-1	-1	0	1	1	1	1	0	0	0	0	0	0	0	0	0	-1	0	0
4	0	0	0	1	0	1	0	0	0	0	0	0	0	0	0	0	0	0	0	-1	-1	0
5	1	0	0	0	0	0	0	0	0	0	0	0	1	0	0	0	0	0	0	-1	-1	-1
8	1	0	0	0	0	0	0	0	0	0	0	0	0	0	0	0	1	0	0	0	0	0
9	0	0	0	0	1	0	0	0	0	0	0	0	0	0	0	0	1	0	0	0	0	0
10	0	0	0	0	1	0	0	0	0	0	0	0	0	0	0	0	1	0	0	0	-1	0
11	0	0	0	1	1	0	0	0	0	0	1	0	0	0	0	0	0	0	0	-1	0	0
12	0	0	0	1	1	0	0	0	0	0	1	0	0	0	0	0	0	0	0	0	0	0
13	0	0	1	1	1	0	0	0	0	0	1	0	0	0	0	0	0	0	0	-1	0	0
14	0	0	0	1	1	0	0	0	0	0	0	0	0	0	0	0	0	0	0	0	0	0
15	0	0	0	0	0	0	0	0	0	0	0	0	0	0	0	0	0	0	0	0	-1	0
16	0	0	0	1	1	0	0	0	0	0	0	0	0	-2	0	0	0	0	0	0	0	0
18	0	0	0	0	1	0	0	0	0	0	0	0	0	0	-2	0	0	0	0	0	-1	-1
19	0	0	0	0	0	0	0	0	0	0	0	0	0	0	-2	0	0	0	0	0	0	0
20	0	0	0	0	0	0	0	0	0	0	0	0	0	0	0	0	0	0	0	0	0	0
21	0	0	0	0	0	0	0	0	0	0	0	0	0	1	1	1	1	1	0	0	1	1
23	0	0	0	0	0	0	0	0	0	0	0	0	0	0	0	0	1	0	1	0	0	0
24	0	0	0	1	-1	0	0	0	-1	0	1	0	0	0	0	1	1	0	0	0	0	0
26	0	0	0	0	0	0	0	0	0	0	0	0	0	1	0	1	0	0	0	0	0	0
27	0	0	0	0	0	0	0	0	0	0	0	0	0	1	0	1	0	0	0	0	0	0
28	0	0	0	0	1	0	0	0	0	0	0	0	0	0	0	0	0	0	0	0	0	0

1 bacteria
2 protozoa (fresh water plankton)
3 plants, dead plants
4 algae {fresh water bottom and littoral}
5 algae {fresh water plankton}
6 worms
7 geese
8 snow bunting, ptarmigan
9 purple sandpiper
10 protozoa (fresh water bottom and littoral)
11 Hymenoptera
12 mites
13 Diptera (adult)
14 Collembola
15 moss
16 spiders
17 protozoa (land)
18 Entomostraca, Rotifera (fresh water plankton)
19 skua, glaucous gull
20 kittiwake, guillemots, fulmar petrel, little auk, puffin
21 northern eider, long-tailed duck, red-throated diver
22 arctic fox
23 marine animals
24 seals
25 polar bear
26 Diptera (immature)
27 Entomostraca, Rotifera, Tardigrada, Oligochaeta, Nematoda (fresh water bottom and littoral)
28 *Lepidurus*

23 Prairie, Manitoba

23	1	2	3	4	5	6	7	9	10	11	12	13	14	15
1	0	1	1	0	0	0	0	0	0	0	0	0	0	0
4	0	1	0	0	1	0	1	0	0	0	0	0	0	0
5	0	1	1	0	0	0	0	0	0	0	0	0	0	0
6	0	1	0	0	0	0	0	0	0	0	0	0	0	0
8	1	0	0	1	1	1	0	1	0	1	-1	-1	0	0
9	0	0	0	0	1	0	0	0	1	0	1	1	1	1
10	0	0	0	0	0	0	0	0	0	0	0	0	1	0
11	0	0	0	0	0	0	0	0	0	0	1	1	0	0
13	0	0	0	0	0	0	1	0	0	0	0	0	0	0
14	0	0	0	0	0	0	0	0	0	0	0	0	0	1

1 Richardson spermophile (ground squirrel)
2 marsh hawk, coyote, red-tailed hawk, weasel
3 badger
4 vole (*Microtus*)
5 13-striped spermophile (ground squirrel)
6 pocket gopher (*Thomomys*)
7 great horned owl
8 *Agropyron, Stipa, Helianthus*
9 insects in herb and surface stratum, Diptera, Hemiptera, grasshoppers, etc.
10 spiders
11 insects in soil stratum, wire worms, cutworms, white grubs, etc.
12 meadow lark, chipping sparrow, clay-colored sparrow, vesper sparrow, horned lark, upland plover
13 crow
14 frog
15 garter snake

24 Willow forest, Manitoba

24	2	3	4	7	8	9	10	12
1	1	0	0	0	0	-1	0	0
5	1	0	0	0	1	-1	0	0
6	1	0	0	0	0	1	0	0
7	0	1	1	0	0	0	-1	0
8	0	-1	0	1	0	0	1	0
9	0	0	1	1	0	0	1	0
10	0	0	0	0	0	0	0	1
11	0	0	0	0	0	0	1	0

1 *Salix discolor*
2 *Galerucella decora*
3 redwinged blackbird, bronze grackle, song sparrow
4 Maryland yellowthroat, yellow warbler, song sparrow
5 *Salix petiolaris*
6 *Salix longifolia*
7 spiders
8 insects, *Pontania petiolaridis*, Collembola
9 insects, *Disyonicha quinquevitata*, Collembola
10 *Rana pipiens*
11 snails
12 garter snake

25 Aspen communities, Manitoba

25	1	2	3	4	5	6	7	9	10	11	12	13	15	16	17	18	19	21	22	23	24	25
1	0	0	0	0	0	0	0	0	1	0	0	0	0	0	0	0	0	0	0	0	0	0
4	1	0	0	0	0	0	0	0	0	0	0	0	0	0	0	0	0	0	0	0	0	1
5	1	0	0	1	0	0	0	0	0	0	0	0	0	1	0	0	0	0	0	0	0	1
6	0	0	1	0	0	0	0	0	0	0	0	0	0	0	0	0	0	0	0	0	0	0
7	0	0	0	0	0	0	0	0	0	1	0	0	0	0	0	0	0	0	0	0	0	0
8	0	1	0	0	1	1	1	0	1	0	0	1	1	0	0	0	0	0	0	0	1	0
10	0	0	0	0	0	0	0	0	0	1	0	0	0	0	0	0	0	0	0	0	0	0
13	0	0	0	0	0	0	0	1	0	0	1	0	0	0	0	0	0	0	0	0	0	0
14	0	0	0	0	0	0	0	0	0	0	1	0	0	0	0	0	0	0	0	0	0	0
15	0	0	0	0	0	0	0	0	0	0	1	0	0	0	0	0	0	0	0	0	0	0
16	0	0	0	0	0	0	0	0	0	1	0	0	0	0	0	0	0	0	0	0	0	0
18	0	0	0	0	0	0	0	1	0	0	1	0	0	0	1	0	0	0	0	0	0	0
19	0	0	0	0	0	0	0	0	0	1	0	0	0	0	0	0	0	0	0	0	0	0
20	0	1	0	0	0	1	0	0	0	0	0	1	1	0	0	1	0	1	1	0	0	0
21	0	0	0	0	0	0	0	0	0	0	1	0	0	0	0	0	0	0	0	0	0	0
22	0	0	0	0	0	0	0	0	0	0	0	0	0	1	0	0	1	0	0	1	0	0
24	0	0	0	0	0	0	0	0	0	0	0	0	0	0	0	0	0	0	0	0	0	1

1 Baltimore oriole, chickadee, least flycatcher, warbling vireo, rosebreasted grossbeak, willow thrush
2 canker, fomes
3 hairy and downy woodpeckers
4 spiders (mature forest)
5 insects (mature forest)
6 *Dicera, Saperda*
7 red squirrel
8 *Populus, Cornus, Corylus, Pyrola, Aralia*
9 goshawk
10 redbacked vole (*Evolomys*)
11 Cooper's and sharpshinned hawks
12 great horned owl
13 ruffed grouse
14 flicker
15 crow
16 house wren
17 ticks
18 snowshoe rabbit
19 red-eyed vireo, yellow warbler, gold finch, catbird, brown thrasher, towhee, robin
20 *Populus, Symphoricarpos, Corylus, Prunus, Amelanchier*
21 redbacked vole, Franklin ground squirrel
22 insects (forest edge)
23 spiders (forest edge)
24 snails
25 frogs

26 Aspen forest, Manitoba

26	1	2	3	4	5	6	8	9	10	11	12	13	14	15	16	18	20	22	23	24	25	28	29	30	33
1	0	0	0	0	0	0	1	0	0	0	0	0	1	0	0	0	0	0	0	0	1	0	0	0	0
4	0	0	1	0	0	0	0	0	0	0	0	0	0	0	0	0	0	0	0	0	0	0	0	0	0
5	1	1	0	1	0	0	0	-1	0	1	0	1	1	1	1	1	1	1	0	0	0	0	0	0	0
6	1	0	0	0	0	0	1	0	0	0	0	0	1	0	1	1	0	0	0	0	0	0	0	0	0
7	0	0	0	0	1	0	0	1	1	0	0	0	0	0	0	0	0	0	0	0	0	0	0	0	0
9	0	1	0	0	0	0	0	0	0	0	0	0	1	0	0	0	0	0	0	0	0	0	0	0	0
10	0	0	0	0	0	0	0	0	0	1	0	1	1	0	0	0	0	0	0	0	0	0	0	0	0
13	0	0	0	0	0	0	0	-1	0	1	1	1	1	0	0	0	0	0	0	0	0	0	0	0	0
15	0	0	0	0	0	0	0	0	0	1	0	0	0	0	0	0	0	0	0	0	0	0	0	0	0
16	0	0	0	0	0	0	1	0	0	0	1	0	1	0	0	0	0	0	0	0	0	0	0	0	0
17	0	0	0	0	1	0	0	0	0	0	0	0	0	0	0	0	0	0	0	0	0	0	0	0	0
19	0	0	0	0	1	0	0	0	0	0	0	0	0	0	0	0	0	0	0	0	0	0	0	0	0
20	0	0	0	0	0	0	1	0	0	0	1	1	1	0	0	0	0	0	0	0	0	0	0	0	0
21	0	0	0	0	1	0	0	0	0	0	0	0	0	0	0	0	0	0	0	0	0	0	0	0	0
22	0	0	0	0	0	0	0	0	0	0	0	0	1	0	0	0	0	0	0	0	0	0	0	0	0
23	0	0	0	0	0	0	0	0	0	0	0	0	0	0	0	0	0	0	0	1	0	0	0	0	0
24	0	0	0	0	0	0	0	0	0	0	0	0	0	0	0	0	0	0	1	0	0	0	0	0	0
25	0	0	0	0	0	0	0	0	0	0	0	0	0	0	0	0	0	0	0	0	0	1	0	0	0
26	0	0	0	0	0	0	0	0	0	0	0	0	0	0	0	0	0	0	0	0	0	1	0	0	0
27	0	0	0	0	0	0	0	-1	0	0	0	0	0	0	0	0	0	0	0	0	0	1	0	0	0
28	0	0	0	0	0	0	0	0	0	1	0	0	0	0	0	0	0	0	0	0	0	0	0	0	0
29	0	0	0	0	0	0	0	0	0	0	0	0	1	0	0	0	0	0	0	1	0	1	0	0	0
30	0	0	0	0	0	0	0	0	0	0	0	0	1	1	0	0	0	0	0	0	0	1	0	0	0
31	0	0	0	0	0	0	0	0	0	0	0	0	0	0	0	0	0	0	0	0	0	0	1	1	0
32	0	0	0	0	0	0	0	0	0	0	0	0	0	0	0	0	0	0	0	1	0	0	0	0	0
33	0	0	0	0	0	0	0	0	0	0	0	0	0	1	0	0	0	0	0	0	0	0	0	0	0
34	0	0	0	0	0	0	0	0	0	0	0	0	0	0	0	0	0	0	0	0	0	0	0	0	0

1 Baltimore oriole, chickadee, least flycatcher, rosebreasted grosbeak, willow thrush
2 canker, fomes
3 hairy and downy woodpeckers
4 Dicera, Saperda
5 insects
6 spiders
7 Populus, Cornus, Corylus, Pryola, Aralia
8 Cooper's and sharpshinned hawks
9 crow
10 ruffed grouse
11 man
12 goshawk
13 rabbit
14 great horned owl
15 flicker
16 Maryland yellowthroat, yellow warbler, songsparrow
17 Salix longifolia
18 Galerucella decora
19 Salix discolor
20 red squirrel
21 Salix petiolaris
22 yellow warbler, redwinged blackbird, bronze grackle
23 Rana pipiens
24 garter snake
25 fish
26 coots
27 ducks
28 coyote, weasel, skunk
29 prairie vole
30 pocket gophers, ground squirrels
31 Agrostis, Agropyron, Stipa, Helianthus
32 mice
33 cutworms, grasshoppers, clickbeetles
34 ants

27 Wytham Wood, England

27	5	6	7	8	9	10	11	12	13	14	15	16	17	18	19	20	21	22
1	1	0	0	0	0	0	1	0	0	0	0	0	0	0	0	0	0	0
2	0	1	0	1	0	0	1	0	1	0	0	0	0	0	0	0	0	0
3	0	1	1	1	0	0	0	0	0	0	0	0	0	0	0	0	0	0
4	0	0	0	0	1	1	0	0	0	0	0	1	0	0	0	0	0	0
5	0	0	0	0	0	0	1	1	1	0	0	0	0	0	0	0	0	0
6	0	0	0	0	0	0	1	0	1	1	1	0	0	0	0	0	1	1
7	0	0	0	0	0	0	0	0	1	0	0	0	0	0	0	0	0	0
8	0	0	0	0	0	0	0	0	1	0	1	0	0	0	0	0	0	0
9	0	0	0	0	0	0	0	0	0	0	1	0	0	1	0	0	0	1
10	0	0	0	0	0	0	0	0	0	0	0	1	0	0	0	0	0	0
11	0	0	0	0	0	0	0	0	0	0	0	0	0	1	1	0	0	0
12	0	0	0	0	0	0	0	0	1	0	0	0	1	0	0	0	0	0
13	0	0	0	0	0	0	0	0	0	0	0	0	0	0	1	0	0	0
14	0	0	0	0	0	0	0	0	0	0	1	0	0	0	0	1	0	0
15	0	0	0	0	0	0	0	0	0	0	0	0	0	0	0	0	1	1
16	0	0	0	0	0	0	0	0	0	0	1	0	0	0	0	0	1	0

1 herbs
2 trees and bushes
3 oak trees
4 total litter
5 insects
6 winter moth
7 *Tartrix*
8 leaf feeders

9 earthworms
10 fungi
11 voles, mice
12 spiders
13 titmice
14 *Cyzenis*
15 *Philanthus, Abax, Feronia*
16 soil insects, mites

17 parasites
18 owls
19 weasels
20 hyperparasites
21 shrews
22 moles

28 Salt meadow, New Zealand

28	8	9	10	11	12	13	14	15	16	17	18	19	20	21	22	23	24	25	26	27
1	0	0	0	0	0	0	0	0	1	1	1	1	1	1	1	1	1	0	0	0
2	0	0	1	0	1	1	1	1	0	0	0	0	0	0	0	0	0	0	0	0
3	0	0	0	0	1	0	0	0	0	0	0	0	0	0	0	0	0	1	1	0
4	0	0	0	0	0	0	0	0	0	0	0	0	0	0	0	0	0	0	0	1
5	0	0	1	1	1	1	0	0	0	0	0	0	0	0	0	0	0	0	0	0
6	0	1	0	0	0	0	0	0	0	0	0	0	0	0	0	0	0	0	0	0
7	1	1	0	0	0	0	0	0	0	0	0	0	0	0	0	0	0	0	0	0
8	0	0	0	0	0	0	0	0	0	0	0	0	0	0	0	0	0	0	0	0
9	0	0	0	0	0	0	0	0	0	0	0	0	0	0	0	0	0	0	0	0
11	0	0	0	0	0	0	0	0	0	0	0	0	0	0	0	0	0	0	0	0
12	0	0	0	0	0	0	0	0	0	0	0	0	0	0	0	0	0	0	0	0
14	0	0	0	0	0	0	0	0	0	0	0	0	0	0	0	0	0	0	0	0
16	0	0	0	0	0	0	0	0	0	1	0	0	0	0	0	0	0	0	0	0
17	0	0	0	0	0	0	0	0	0	0	0	0	0	0	0	0	0	0	0	0
23	0	0	0	0	0	0	0	0	0	0	0	0	0	0	0	0	0	0	0	0
24	0	0	0	0	0	0	0	0	0	0	0	0	0	0	0	0	0	0	0	0
40	0	0	0	0	0	0	0	0	0	0	0	0	0	0	0	0	0	0	0	0

1 humus
2 leaves
3 flowers
4 seeds
5 roots
6 algae
7 bacteria
8 protozoa, rotifers
9 phytophagous nematodes
10 weevil larvae
11 coccids
12 lepidopterous larvae
13 other hemiptera
14 rabbits
15 *Uromyces scaevolae*
16 fungi
17 collembola
18 harpacticoids
19 staphylinids
20 dipterous larvae
21 haplotaxid worms
22 oribatids
23 mites
24 amphipods
25 bumblebees
26 adult hymenopterans
27 redpolls
28 carnivorous nematodes
29 tartigrades
30 parasitic hymenopterous larvae
31 *Trichostrongylus retortaeformis*
32 *Graphidium strigosum*
33 *Passalurus ambiguus*
34 stoats
35 harrier hawks
36 trombidiform mites

29 Arctic seas

29	3	4	5	6	7	8	9	10	11	12	13	14	15	16	17	18	19	20	21	22
1	1	1	0	0	0	0	0	0	0	0	0	0	0	0	0	0	0	0	0	0
2	0	0	1	1	0	0	0	0	0	0	0	0	0	0	0	0	0	0	0	0
3	0	1	0	0	1	1	0	0	0	0	0	0	0	0	0	0	0	0	0	0
4	0	0	0	0	0	0	1	1	1	1	0	0	0	0	0	0	0	0	0	0
6	0	0	0	0	0	0	0	0	0	0	1	1	1	1	1	0	0	0	0	1
11	0	0	0	0	0	0	0	0	0	1	0	0	0	0	1	1	0	0	0	0
12	0	0	0	0	0	0	0	0	0	0	0	0	0	0	0	0	0	0	1	1
13	0	0	0	0	0	0	0	0	0	0	1	1	0	0	0	0	0	0	1	1
18	0	0	0	0	0	0	0	0	0	0	0	0	0	0	0	0	0	0	1	1
19	0	0	0	0	0	0	0	0	0	0	0	0	0	0	0	0	0	0	1	1
20	0	0	0	0	0	0	0	0	0	0	0	0	0	0	0	0	0	0	1	1

1 phytoplankton
2 detritus
3 smaller zooplankton
4 larger zooplankton
5 bacteria
6 benthonic invertebrates
7 right whales
8 clupeid fishes
9 rorquals
10 arctic char
11 caplin
12 cod
13 benthonic vertebrates
14 bearded seals
15 beluga
16 narwhal
17 walrus
18 harp seal
19 harbour seal
20 ringed seal
21 killer whale
22 Greenland shark

(Web 28 cont.)

28	29	30	31	32	33	34	35	36	37	38	39	40	41	42	43	44	45
0	0	0	0	0	0	0	0	0	0	0	0	0	0	0	0	0	0
0	0	0	0	0	0	0	0	0	0	0	0	0	0	0	0	0	0
0	0	0	0	0	0	0	0	0	0	0	0	0	0	0	0	0	0
0	0	0	0	0	0	0	0	0	0	0	0	0	0	0	0	0	0
0	0	0	0	0	0	0	0	0	0	0	0	0	0	0	0	0	0
0	0	0	0	0	0	0	0	0	0	0	0	0	0	0	0	0	0
0	0	0	0	0	0	0	0	0	0	0	0	0	0	0	0	0	0
1	0	0	0	0	0	0	0	0	0	0	0	0	0	0	0	0	0
0	1	0	0	0	0	0	0	0	0	0	0	0	0	0	0	0	0
0	0	1	0	0	0	0	0	0	0	0	0	0	0	0	0	0	0
0	0	1	0	0	0	0	0	0	0	0	0	0	0	0	0	0	0
0	0	0	1	1	1	1	1	0	0	0	0	0	0	0	0	0	0
0	0	0	0	0	0	0	0	0	0	0	0	0	0	0	0	0	0
1	0	0	0	0	0	0	0	1	1	0	0	0	0	0	0	0	0
0	0	0	0	0	0	0	0	0	0	1	0	0	0	0	0	0	0
0	0	0	0	0	0	0	0	0	0	1	1	1	1	0	0	0	0
0	0	0	0	0	0	0	1	0	0	0	0	0	1	1	1	1	1

37 spiders
38 ants
39 starlings
40 Dotterel
41 hymenolepidid cestodes
42 nematode *Echinuria*
43 analgesid mites
44 other mites
45 lice

30 Antarctic seas

30	2	3	4	5	6	7	8	9	10	11	12	13	14
1	1	1	0	0	0	0	0	0	0	0	0	0	0
2	0	0	1	1	1	1	0	1	0	0	0	0	0
3	0	0	1	0	1	1	1	0	0	0	0	0	0
4	0	0	0	0	1	1	1	0	0	0	0	0	0
5	0	0	0	0	0	0	0	0	0	0	0	0	1
6	0	0	0	0	0	0	0	0	1	0	1	0	0
7	0	0	0	0	1	0	1	0	1	1	0	0	0
8	0	0	0	0	1	1	0	0	1	1	1	1	0
9	0	0	0	0	0	0	0	0	0	1	0	0	
10	0	0	0	0	0	0	0	0	0	1	0	0	
11	0	0	0	0	0	0	0	0	0	1	0	1	
13	0	0	0	0	0	0	0	0	0	0	0	1	

1 phytoplankton
2 *Euphausia superba*
3 other herbivorous plankton
4 carnivorous plankton
5 baleen whales
6 birds
7 fish
8 squid
9 crabeater seal
10 leopard seal
11 elephant seal
12 smaller toothed whales
13 sperm whale
14 man

31 Epiplankton communities, Black Sea

31	3	4	5	6	7	8	9	10	11	12	13	14
1	1	1	1	1	1	1	1	1	1	1	0	0
2	0	0	1	1	1	1	1	1	1	0	0	0
3	0	0	1	1	0	1	1	1	1	0	0	0
4	0	0	0	1	0	1	1	1	1	0	0	0
5	0	0	0	0	0	0	0	0	1	1	1	0
6	0	0	0	0	0	0	0	0	1	1	1	1
7	0	0	0	0	0	0	0	0	0	1	1	1
8	0	0	0	0	0	0	0	0	1	0	1	0
9	0	0	0	0	0	0	0	0	1	1	1	0
10	0	0	0	0	0	0	0	0	0	0	1	1
11	0	0	0	0	0	0	0	0	0	1	1	1
12	0	0	0	0	0	0	0	0	0	0	1	1
13	0	0	0	0	0	0	0	0	0	0	0	1

1 detritus
2 phytoplankton
3 saprophagous plankton
4 large-sized phytoplankton
5 *Oikopleura*
6 II-III copepodites
7 IV-VI copepodites of *Paracalanus*

8 larvae of molluscs and polychaetes
9 naupliuses
10 IV-VI copepodites of *Calanus* and *Pseudocalanus*
11 mixed-food consumers
12 primary carnivores
13 secondary carnivores
14 tertiary carnivores

32 Bathyplankton communities, Black Sea

32	3	4	5	6	7	8	9	10	11	12	13	14
1	1	1	1	1	1	1	1	1	1	1	0	0
2	0	0	1	1	1	1	1	1	1	0	0	0
3	0	0	1	1	0	1	1	1	1	0	0	0
4	0	0	0	1	0	1	1	1	1	0	0	0
5	0	0	0	0	0	0	0	0	1	1	1	1
6	0	0	0	0	0	0	0	0	1	1	1	1
7	0	0	0	0	0	0	0	0	0	1	1	1
8	0	0	0	0	0	0	0	0	1	0	1	0
9	0	0	0	0	0	0	0	0	1	1	1	0
10	0	0	0	0	0	0	0	0	0	0	1	1
11	0	0	0	0	0	0	0	0	0	1	1	1
12	0	0	0	0	0	0	0	0	0	0	1	1
13	0	0	0	0	0	0	0	0	0	0	0	1

1 detritus
2 phytoplankton
3 saprophagous plankton
4 large-sized phytoplankton
5 *Oikopleura*
6 II-III copepodites
7 IV-VI copepodites of *Paracalanus*

8 larvae of molluscs and polychaetes
9 naupliuses
10 IV-VI copepodites of *Calanus* and *Pseudocalanus*
11 mixed-food consumers
12 primary carnivores
13 secondary carnivores
14 tertiary carnivores

33 Crocodile Creek, Malawi

33	1	2	3	4	5	6	7	9	10	11	12	13	14	15	16	17	18	19	20	21	22	23	24	25	26	28	30	32
7	1	0	0	0	0	0	0	0	0	0	0	0	0	0	0	0	0	0	0	0	0	0	0	0	0	0	0	0
8	1	0	0	0	0	1	0	0	0	0	0	0	0	0	0	0	0	0	0	0	0	0	0	0	0	0	0	0
9	0	0	0	1	1	0	0	0	0	0	0	0	0	0	0	0	0	0	0	0	0	0	0	0	0	0	0	0
12	0	0	0	1	0	0	0	1	0	0	0	0	0	0	0	1	0	0	0	0	0	0	0	0	0	0	0	0
13	0	1	1	1	0	0	0	0	0	0	0	0	0	0	0	0	0	0	0	0	0	0	0	0	0	0	0	0
15	0	0	0	0	1	0	0	1	0	0	0	0	0	0	0	1	0	0	0	0	0	0	0	0	0	0	0	0
18	0	1	1	1	0	0	0	0	0	0	0	0	0	0	0	0	0	0	0	0	0	0	0	0	0	0	0	0
19	0	1	1	0	0	0	0	1	1	0	0	1	0	0	0	0	0	0	0	0	0	0	0	0	0	0	0	0
20	0	0	1	1	1	0	1	0	0	0	0	0	1	1	0	1	1	1	1	1	1	1	1	1	1	1	1	0
21	0	0	0	1	0	0	0	0	1	0	0	0	1	1	0	0	1	1	1	1	1	1	1	1	1	1	1	0
22	0	1	0	1	0	0	0	0	0	0	1	1	0	1	1	1	1	1	1	1	1	1	1	1	1	1	1	1
27	0	0	0	0	0	0	0	0	0	0	0	0	0	0	0	0	0	0	0	0	0	0	0	0	0	0	0	0
29	0	0	0	0	0	0	0	0	0	0	0	0	0	0	0	0	0	0	0	0	0	0	0	0	0	0	0	0
31	0	0	0	0	0	0	0	0	0	0	0	0	0	0	0	0	0	0	0	0	0	0	0	0	0	1	1	0
33	0	0	0	0	0	0	0	0	0	1	0	0	0	0	1	0	0	0	0	0	0	0	1	0	0	0	1	1

1 Crocodilus niloticus
2 juvenile Cichlidae
3 Barbus innocens
4 Serranochromis robustus
5 Naucoris sp.
6 Clarias mellandi
7 frogs
8 Dytiscid beetles
9 Anisopterid larvae
10 Barbus paludinosus
11 Alestes imberi
12 mosquito larvae
13 Cyclopoid copepods
14 Zygopterid larvae
15 Caridina nilotica
16 Barilius microcephalus
17 Tilapia shirana, T. saka-squamipinnis
18 Gyraulus costulatus
19 Cladocera
20 Chironomid larvae
21 caddis larvae
22 Baetid nymphs
23 Micronecta
24 Barbus sp.
25 Segmentorbis angustus
26 Limnaea sp.
27 bottom algae and detritus
28 Haplochromis similis
29 higher plants
30 Clarias mossambicus
31 indet. fishes
32 Gerrids
33 terrestrial insects

34 River Clydach, Wales

34	5	6	7	8	9	10	11	12
1	1	1	0	1	1	0	0	0
2	0	1	0	0	1	0	1	0
3	0	1	1	1	0	0	0	0
4	0	1	1	1	1	0	0	0
5	0	0	0	0	0	0	1	0
6	0	0	0	0	1	1	1	1
7	0	0	0	0	0	0	1	1
8	0	0	0	0	0	1	0	1
9	0	0	0	0	0	1	1	1
10	0	0	0	0	0	0	0	1

1 leaf fragments
2 *Ulothrix* and other green algae
3 detritus
4 diatoms
5 *Protonemura*
6 *Leuctra, Baetis, Ephemerella,*
   Simulium chironomids

7 *Philopotamus*
8 *Ecdyonurus*
9 *Hydropsyche*
10 *Rhyacophila*
11 *Perla*
12 *Dinocras*

35 Morgan's Creek, Kentucky

35	1	2	3	4	5	6	7	8	9	10	11
6	1	0	0	1	1	0	0	0	0	0	0
7	0	0	1	1	1	1	0	0	0	0	0
8	1	0	1	1	1	1	0	0	0	0	0
9	0	1	1	0	0	1	1	0	0	0	0
10	0	1	1	1	0	0	0	0	0	0	0
11	0	1	1	0	1	1	1	0	0	0	0
12	0	1	0	0	0	1	1	1	1	1	1
13	0	0	0	0	0	0	1	1	1	1	1

1 Phagocata
2 Decapoda - *Orconectes, Cambarus*
3 Plecoptera - *Isoperia, Isogenus*
4 Megaloptera - *Nigronia, Sialis*
5 Pisces - *Rhinichthys, Semotilus*
6 *Gammarus*
7 Trichoptera - *Diplectrona, Rhyacophila*
8 *Asellus*
9 Ephemeroptera - *Baetis, Centroptilum, Epeorus, Paraleptophlebia, Pseudocloeon*
10 Trichoptera - *Neophylax, Glossosoma*
11 Diptera - Tendipedidae, *Simulium*
12 detritus
13 diatoms

36 Mangrove swamp 6, Hawaii

36	9	10	11	12	13	14	15	16	17	18	19	20	21	22
1	1	0	1	0	0	0	0	0	1	0	1	0	0	0
2	1	1	1	1	1	1	1	0	1	0	0	0	1	1
3	0	0	1	1	0	1	0	0	0	0	0	0	0	0
4	0	0	0	1	1	1	1	0	1	0	0	0	0	0
5	0	0	0	0	0	1	0	0	0	0	0	0	0	0
6	0	0	0	0	0	1	1	0	0	0	1	0	0	0
7	0	0	0	0	0	0	0	0	0	0	1	0	0	0
8	0	0	0	0	0	0	0	0	0	0	1	0	0	0
11	0	0	0	0	0	1	1	1	0	0	0	0	0	0
13	0	0	0	0	0	0	0	0	1	1	1	0	0	0
15	0	0	0	0	0	0	0	0	1	1	0	0	0	0
16	0	0	0	0	0	0	0	0	0	0	0	1	0	0
17	0	0	0	0	0	0	0	0	0	0	0	1	0	0

1 detritus
2 diatoms
3 *Chlamydomonas*
4 *Ulothrix*
5 nematodes
6 ostracods
7 *Melampus parvulus*
8 *Littorina scabra*

9 *Tilapia mossambica*
10 *Podophthalmus vigil*
11 copepods
12 *Metopograpsis messor*
13 *Macrobrachium*
14 *Oxyurichthyes lonchotus*
15 *Palaemonetes*
16 *Kuhlia sandvicensis*

17 *Charybdis orientalis*
18 *Scylla serrata*
19 *Eleotris sandwicensis*
20 *Conger marginatus*
21 *Mugil cephalus*
22 *Chonophorous genivittatus*

37 Marine sublittoral, southern California

37	1	2	3	4	5	6	7	8	9	10	11	12	13	14	15	16	17	18	19	20	21	22	23	24
9	0	1	1	1	1	-1	0	0	0	0	0	0	0	0	0	0	0	0	0	0	0	0	0	0
10	0	0	1	1	1	-1	-1	-1	0	0	0	0	0	0	0	0	0	0	0	0	0	0	0	0
11	0	0	1	0	0	0	0	-1	0	0	0	0	0	0	0	0	0	0	0	0	0	0	0	0
16	0	0	0	0	0	0	0	0	0	0	0	0	0	1	1	0	0	0	0	0	0	0	0	0
17	0	0	0	0	0	0	0	0	0	0	0	1	-1	1	1	0	0	0	0	0	0	0	0	0
18	0	0	0	0	0	0	0	0	0	0	0	1	0	1	1	0	0	0	0	0	0	0	0	0
19	0	1	1	1	0	0	0	0	0	0	0	1	0	1	0	0	0	0	0	0	0	0	0	0
20	0	1	1	1	1	0	0	0	1	0	0	0	0	0	0	1	1	1	1	1	1	1	1	1
21	-1	1	1	1	0	0	0	0	0	-1	0	0	0	0	0	1	1	1	1	1	1	1	1	0
22	-1	1	1	1	1	0	0	0	0	0	0	0	0	0	0	1	1	1	1	0	1	-1	1	0
23	0	1	0	0	0	0	0	0	0	-1	0	0	0	0	0	1	1	1	1	1	1	1	1	1
25	0	0	0	0	0	0	0	0	0	0	1	0	0	0	0	1	1	1	1	1	1	1	1	1

1 Pacific bonito
2 vermilion rockfish
3 California scorpion fish
4 California sea lion
5 cabezon
6 sand bass
7 Pacific angel shark
8 California halibut
9 squid
10 octopus
11 small fishes and invertebrates
12 Pacific sand dab
13 pink sea perch
14 hornyhead turbot
15 longfin sand dab
16 ophiuroids
17 polychaetes
18 benthic crustacea
19 hypoplanktonic crustacea
20 white croaker
21 jack mackerel
22 northern anchovy
23 zooplankton
24 blacksmith, black perch, rubber lip sea perch, sharpnose sea perch, shiner perch
25 detritus

38 Lake Nyasa, rocky shore, Malawi

38	1	2	3	4	5	6	7	8	9	10	11	12	13	14	15	16	17	18	19	20	21	22	23	24	25	26	30	31
4	1	1	1	0	0	0	0	0	0	0	0	0	0	0	0	0	0	0	0	0	0	0	0	0	0	0	0	0
5	0	1	1	0	0	0	0	0	0	0	0	0	0	0	0	0	0	0	0	0	0	0	0	0	0	0	0	0
15	0	0	0	0	0	0	0	1	1	1	1	1	1	1	0	0	0	0	0	0	0	0	0	0	0	0	1	0
16	0	0	1	0	0	0	0	1	1	0	0	1	1	0	0	0	0	0	0	0	0	0	0	0	0	0	0	0
18	0	0	0	0	0	0	0	1	1	0	0	1	1	0	0	0	0	0	0	0	0	0	0	0	0	0	0	0
19	0	0	0	0	0	0	1	1	1	1	0	0	0	0	1	1	0	0	0	0	0	1	0	0	0	0	0	0
21	0	0	0	0	0	0	1	0	0	1	1	1	1	0	0	1	1	0	0	0	0	0	0	0	0	0	1	1
23	0	0	0	0	0	0	0	0	1	1	1	1	1	0	0	0	0	0	0	0	0	0	0	0	0	0	0	0
24	0	0	1	0	0	0	0	0	0	1	1	1	1	0	1	0	1	0	0	1	0	1	0	0	0	0	0	0
25	0	0	1	0	0	0	1	0	0	0	0	0	1	1	0	1	0	1	1	0	0	0	0	0	0	0	1	0
26	0	0	1	0	0	0	0	1	1	1	1	1	1	1	1	1	1	0	0	0	0	0	0	0	0	0	1	0
27	0	0	0	1	1	1	1	1	1	1	1	1	1	1	0	0	0	0	0	0	1	1	1	1	1	1	1	1
28	0	0	1	0	0	0	0	0	0	0	0	0	1	-1	0	0	0	0	0	0	0	1	0	0	0	0	0	1
29	0	0	1	0	0	0	0	0	1	1	0	1	0	0	0	0	1	0	0	0	0	1	0	0	0	0	0	1

1 *Haplochromis pardalis*, other predatory fishes
2 *Haplochromis polyodon*
3 *Haplochromis kiwinge*
4 juvenile cichlids
5 *Pseudotropheus zebra*
6 *Pseudotropheus elongatus, P. tropheops, P. minutus, P. auratus, P. fuscus, Labeotropheus fuelleborni, L. trewavasae*
7 *Haplochromis guentheri, H. fenestratus*
8 *Mastacembelus shiranus*
9 *Bathyclarias worthingoni*
10 *Labidochromis vellicans*
11 *Labidochromis caeruleus*
12 *Haplochromis euchilus*
13 *Haplochromis ornatus*
14 *Pseudotropheus fuscoides*
15 Hydropsychid larvae
16 *Neoperla spio* nymphs
17 *Cynotilapia afra*
18 *Potamonautes lirrangensis*
19 *Schizopera consimilis*
20 leech
21 ostracods
22 *Barilius microcephalus*
23 *Eubrianax* larvae
24 Elmid larvae
25 Chironomid larvae
26 *Afronurus* nymphs
27 Aufwuchs
28 terrestrial insects
29 plankton
30 *Melanochromis melanopterus*
31 *Varicorhinus nyasensis*

39 Lake Nyasa, sandy shore, Malawi

39	1	2	3	4	5	6	7	8	9	10	11	12	13	14	15	16	17	18	19	20	21	22	23	24	25	26	27	28	29	30	36	37
7	0	0	1	0	0	0	0	0	0	0	0	0	0	0	0	0	0	0	0	0	0	0	0	0	0	0	0	0	0	0	0	0
8	0	1	1	1	1	1	0	0	0	0	0	0	0	0	0	0	1	0	0	0	0	0	0	0	0	0	0	0	0	0	0	0
10	0	1	0	1	0	0	0	1	0	0	0	0	0	0	0	0	0	0	0	0	0	0	0	0	0	0	0	0	0	0	0	0
14	0	0	0	0	0	0	1	1	1	0	1	1	1	0	1	0	1	0	0	0	0	0	0	0	0	0	0	0	0	0	0	0
20	1	0	1	0	0	0	0	0	1	0	1	1	1	1	0	1	1	0	1	0	0	0	0	0	0	0	0	0	0	0	0	0
21	1	0	1	0	0	0	0	0	0	0	1	0	0	0	0	0	1	1	0	0	0	0	0	0	0	0	0	0	0	0	0	0
22	1	0	1	0	0	0	0	0	0	0	1	1	0	0	0	0	1	0	0	0	0	0	0	0	0	0	0	0	0	0	0	0
23	1	0	0	0	0	0	0	0	0	0	0	0	1	0	0	1	1	0	1	0	0	0	0	0	0	0	0	0	0	0	0	0
24	0	0	0	0	0	0	0	0	0	0	0	0	0	0	1	0	1	0	0	0	0	0	0	0	0	0	0	0	0	0	0	0
26	0	0	0	0	0	0	0	0	0	0	0	0	0	0	0	0	1	1	0	0	0	0	0	0	0	0	0	0	0	0	0	0
27	1	0	1	0	0	0	0	0	0	1	0	0	0	1	0	1	1	0	0	0	0	0	0	0	0	0	0	0	0	0	1	0
31	1	0	0	0	0	0	1	0	0	0	0	0	0	0	0	0	0	1	1	1	1	1	1	1	0	0	0	0	0	0	0	0
32	0	0	1	0	0	0	0	0	0	0	0	0	0	0	0	0	0	0	1	0	0	0	0	0	1	1	0	0	0	0	1	0
33	1	0	1	0	0	0	0	0	0	1	0	0	0	0	0	0	0	0	1	1	0	0	0	0	1	0	0	1	1	1	1	0
34	0	0	0	0	0	0	0	0	0	0	0	0	0	0	0	0	0	0	1	0	0	0	0	0	0	0	0	0	0	0	0	0
35	0	0	0	0	0	0	0	0	0	0	0	0	0	0	0	0	0	1	0	0	0	0	0	0	0	0	0	0	0	0	0	0
37	1	0	1	0	0	1	0	0	0	0	0	0	0	0	0	0	1	0	1	0	0	0	0	0	0	0	0	0	0	0	0	0

1 Haplochromis johnstoni
2 Barbus rhoadesii
3 Haplochromis dimidiatus
4 Haplochromis rostratus
5 Haplochromis sp.
6 Rhamphochromis spp.
7 Engraulicypris sardella
8 various immature fishes
9 Lethrinops brevis
10 Tilapia saka-squamipinnis
11 Synodontis zambesensis
12 Lethrinops
13 Alestes imberi

14 Microcyclops nyasae
15 Lethrinops furcifer
16 Barbus eurystomus
17 Haplochromis mola
18 Barilius microcephalus
19 Barbus johnstoni
20 Baëtid nymphs
21 Ostracods
22 Chironomid larvae
23 Caridina nilotica
24 Corbicula africana
25 Barbus innocens, Haplochromis similis, Haplochromis moori
26 Lanistes procerus

27 Malanoides tuberculata, other gastropods
28 Tilapia melanopleura
29 Tilapia shirana
30 Labeo mesops
31 terrestrial insects
32 Aufwuchs on Vallisneria
33 Vallisneria
34 bottom detritus, algae
35 plankton
36 Haplochromis chrysonotus
37 Cyclopoid copepods

40 Rain forest, Malaysia

40	2	3	4	5	7	8	9	10	11
1	1	0	1	1	0	0	0	0	0
4	1	1	0	0	1	1	0	1	0
6	0	0	0	0	1	0	0	0	0
7	0	0	0	0	0	1	0	0	0
9	0	0	1	1	0	0	0	1	1
11	0	0	0	0	0	0	0	1	0

1 canopy - leaves, fruits, flowers
2 canopy animals - birds, fruit bats and other mammals
3 upper air animals - birds and bats, insectivorous, carnivorous
4 insects
5 large ground animals - large mammals and birds
6 trunk, fruit, flowers
7 middle zone scansorial animals - mammals in both canopy and ground zones
8 middle zone flying animals - birds and insectivorous bats
9 ground - roots, fallen fruit, leaves and trunks
10 small ground animals - birds and small mammals
11 fungi

41 Tropical seas, epipelagic zone

41	3	4	5	6	7	8	9	10	11	12	13	14	15	16	17	18	19
1	1	1	1	0	0	0	0	0	0	0	0	0	0	0	0	0	0
2	1	1	1	0	0	0	0	0	0	0	0	0	0	0	0	0	0
3	0	0	0	1	1	0	0	0	1	0	0	0	1	0	0	0	0
4	1	0	1	1	1	1	1	1	0	0	0	0	0	0	0	0	0
5	0	0	0	0	0	0	0	0	1	0	0	0	1	0	0	0	0
6	0	0	0	0	0	0	0	0	1	0	0	0	1	1	0	0	0
7	0	0	0	0	0	0	0	0	0	1	1	1	0	0	0	0	0
8	0	0	0	0	1	0	1	0	1	0	0	0	1	1	0	0	0
9	0	0	0	0	0	0	0	0	0	1	1	1	1	0	0	0	0
10	0	0	0	0	0	0	0	0	0	0	0	0	0	0	0	1	0
11	0	0	0	0	0	0	0	0	0	0	0	0	1	1	1	0	0
12	0	0	0	0	0	0	0	0	0	0	0	0	1	1	0	0	0
13	0	0	0	0	0	0	0	0	0	0	0	0	1	1	1	1	0
15	0	0	0	0	0	0	0	0	0	0	0	0	0	0	1	1	1
16	0	0	0	0	0	0	0	0	0	0	0	0	1	0	1	1	0
17	0	0	0	0	0	0	0	0	0	0	0	0	0	0	0	0	1
18	0	0	0	0	0	0	0	0	0	0	0	0	0	0	0	0	1

1 coccolithophores
2 dinoflagellates
3 euphausiids
4 copepods
5 shrimps
6 vertically migrating mesopelagic fishes
7 flying fishes
8 hyperiid amphipods
9 lanternfish
10 ocean sunfish

11 other mesopelagic fishes
12 snake mackerel
13 squid
14 dolphin *Coryphaena*
15 tuna
16 lancetfish
17 marlin
18 medium-sized sharks
19 large sharks

42 Upwelling areas, Pacific Ocean

42	6	7	8	9	10	11	12	13	14	15	16
1	0	0	1	1	1	1	1	0	0	0	0
2	0	0	0	0	0	1	1	0	0	0	1
3	0	0	0	0	0	0	1	0	0	0	1
4	1	0	0	0	0	0	0	0	0	0	0
5	1	1	0	0	0	0	0	0	0	0	0
6	0	1	1	1	1	0	0	0	0	0	0
7	0	0	1	1	1	0	0	0	0	0	0
8	0	0	0	1	1	1	1	1	1	0	0
9	0	0	0	0	0	0	0	1	1	1	1
10	0	0	0	0	0	0	0	1	1	1	1
11	0	0	0	0	0	0	0	0	1	1	1
12	0	0	0	0	0	0	0	0	0	1	0
13	0	0	0	0	0	0	0	0	1	1	1
14	0	0	0	0	0	0	0	0	0	1	1
15	0	0	0	0	0	0	0	0	0	0	1

1 small phytoplankton
2 medium phytoplankton
3 large phytoplankton
4 detritus
5 dissolved organic matter
6 bacteria
7 zooflagellates
8 ciliates
9 meroplankton, appendicularians, doliolids
10 small calanoids
11 medium-sized calanoids
12 juvenile euphausiids
13 cyclopoids
14 calanoids, small tomopterids, small coelenterates
15 chaetognaths, polychaetes
16 anchovy

43 Kelp bed community, south California

43	5	6	7	8	9	10	11	12	13	14	15	16	17	18	19	20	21
1	1	0	1	1	0	0	1	1	0	1	0	0	0	0	0	0	0
2	1	0	0	0	1	0	0	1	0	0	0	0	0	0	0	0	0
3	0	1	1	1	0	0	0	0	1	0	0	0	0	0	0	0	0
4	0	0	1	1	0	0	0	0	0	0	0	0	0	0	0	0	0
5	0	0	0	0	1	1	1	0	0	0	0	0	0	0	0	0	0
6	0	0	0	0	0	0	0	0	0	0	1	0	0	1	0	0	0
7	0	0	0	0	0	1	0	0	0	0	1	1	0	0	1	0	1
8	0	0	0	0	0	0	0	0	0	0	0	0	0	1	0	0	0
9	0	0	0	0	0	0	0	0	0	0	1	1	0	1	0	0	0
10	0	0	0	0	0	0	0	0	0	0	0	0	0	0	0	0	1
11	0	0	0	0	0	0	0	0	0	0	1	0	0	1	0	1	0
12	0	0	0	0	0	0	0	0	0	0	0	0	0	0	0	0	1
13	0	0	0	0	0	0	0	0	0	0	1	0	0	1	0	0	0
14	0	0	0	0	0	0	0	0	0	0	0	1	0	0	0	0	0
15	0	0	0	0	0	0	0	0	0	0	0	1	1	0	0	1	0
16	0	0	0	0	0	0	0	0	0	0	0	0	0	0	0	0	1

1 detritus
2 phytoplankton
3 *Macrocystis pyrifera*
4 *Pterygophora californica*
5 zooplankton
6 *Astraea undosa*
7 *Strongylocentrotus purpuratus*
8 *Strongylocentrotus franciscanus*
9 *Styela montereyensis*
10 *Tealia coriacea*
11 *Parapholas californica*
12 *Tethya aurantia*
13 *Diopatra ornata*
14 *Paguristes ulreyi*
15 *Kelletia kelletii*
16 *Astrometis sertulifera*
17 *Octopus bimaculatus*
18 *Pisaster giganteus*
19 *Pimelometopon pulchrum*
20 *Pisaster brevispinus*
21 *Dermasterias imbricata*

44 Marine coastal lagoons, Guerrero, Mexico

44	4	5	6	7	8	9	10	11	12
1	1	1	0	0	1	1	0	0	0
2	1	1	1	1	0	1	1	1	0
3	0	1	1	1	0	0	1	0	0
4	0	0	0	0	1	1	1	0	0
5	0	0	0	0	1	0	1	1	0
6	0	0	0	0	0	0	1	0	0
7	0	0	0	0	0	0	1	1	0
8	0	0	0	0	0	0	1	0	0
9	0	0	0	0	0	0	1	0	0
10	0	0	0	0	0	0	0	1	1
11	0	0	0	0	0	0	0	0	1

1 benthic algae
2 detritus
3 phytoplankton
4 micro- and meio-benthos
5 mollusks
6 zooplankton

7 insects
8 *Callinectes*
9 *Penaeus*
10 fishes (type 1)
11 fishes (type 2)
12 fishes (type 3)

45 Cone Spring, Iowa

45	2	3	4	5	6	7	8	9	10	11
1	1	1	1	1	1	1	0	0	1	0
2	0	0	0	0	0	0	1	1	0	0
3	0	0	0	0	0	0	1	1	1	1
4	0	0	0	0	0	0	0	0	1	1
6	0	0	0	0	0	0	1	0	0	1
7	0	0	0	0	0	0	0	0	1	1
10	0	0	0	0	0	0	0	0	0	1

1 detritus
2 *Frenesia*
3 *Gammarus*
4 tendipedid
5 *Physa*
6 *Cardiocladius*

7 *Tubifex*
8 *Rhantus*
9 *Chauliodes*
10 *Pentaneura*
11 *Phagocata*

46 Lake Texoma, Texas

46	8	9	10	11	12	13	14	15	16	17	18	19
1	1	0	0	0	1	1	0	1	1	1	0	0
2	0	0	0	1	0	1	0	1	1	1	0	0
3	0	0	0	0	0	1	0	1	1	1	0	0
4	0	1	1	0	0	1	0	1	1	1	1	0
5	0	1	1	0	0	1	0	1	0	1	1	0
6	0	1	1	0	0	0	0	0	0	0	0	0
7	0	0	1	0	0	0	0	0	0	0	0	0
8	0	0	0	1	1	1	1	1	1	1	1	0
10	0	0	0	0	0	0	0	0	0	0	1	0
11	0	0	0	0	0	1	1	1	1	1	0	0
12	0	0	0	0	0	0	1	1	1	1	1	1
13	0	0	0	0	0	0	1	1	1	1	1	0
14	0	0	0	0	0	0	0	1	1	1	1	1
15	0	0	0	0	0	0	0	0	1	1	1	0
16	0	0	0	0	0	0	0	0	0	0	1	1
17	0	0	0	0	0	0	0	0	0	0	1	1
18	0	0	0	0	0	0	0	0	0	0	0	1

1 small phytoplankton
2 medium phytoplankton
3 large phytoplankton
4 algal mats, crusts
5 submergent vascular plants
6 emergent vascular plants
7 animal carcasses
8 small zooplankton
9 herbivorous vertebrate harvesters
10 turtles

11 medium zooplankton
12 suspension-feeding invertebrates
13 fish larvae
14 invertebrate predators
15 fingerling fishes
16 filter-feeding, minnow-like fishes
17 bottom-feeding fishes
18 carnivorous fishes
19 carnivorous vertebrate harvesters

47 Swamp, south Florida

47	4	5	6	7	8	9	10	11	12	13	14	15	16	17	18	19	20	21	22	23	24	25	26	27
1	1	1	1	1	0	0	1	0	0	0	0	0	0	0	0	0	0	0	0	0	0	0	0	0
2	1	1	1	1	0	0	0	0	0	0	0	0	0	0	0	0	0	0	0	0	0	0	0	0
3	1	1	1	1	1	0	1	0	0	0	0	0	0	0	0	0	0	0	0	0	0	0	0	0
4	0	0	0	0	0	1	1	0	0	0	0	0	0	0	0	0	0	0	0	0	0	0	0	0
5	0	0	0	0	0	1	1	1	0	0	0	0	0	0	0	0	0	0	0	0	0	0	0	0
6	0	0	0	0	0	1	1	0	0	0	0	0	0	0	0	0	0	0	0	0	0	0	0	0
7	0	0	0	0	0	1	0	0	1	0	0	0	0	0	0	0	0	0	0	0	0	0	0	0
8	0	0	0	0	0	0	0	0	0	0	0	0	0	0	0	0	0	0	0	0	0	0	0	1
9	0	0	0	0	0	0	0	0	1	0	0	0	0	0	0	0	0	1	1	1	0	0	0	0
10	0	0	0	0	0	0	0	0	0	1	1	0	0	0	0	0	0	0	0	0	0	0	0	0
11	0	0	0	0	0	0	0	0	0	0	0	1	0	0	0	0	0	0	0	0	0	0	0	0
12	0	0	0	0	0	0	0	0	0	1	1	0	1	0	1	0	0	0	0	0	0	0	0	0
13	0	0	0	0	0	0	0	0	0	0	0	1	0	0	0	1	1	1	0	0	0	0	0	0
14	0	0	0	0	0	0	0	0	0	0	0	0	1	1	0	1	1	1	1	0	0	0	0	0
15	0	0	0	0	0	0	0	0	0	0	0	0	0	1	1	0	1	0	0	1	0	0	0	0
16	0	0	0	0	0	0	0	0	0	0	0	0	0	0	1	0	0	0	0	1	0	0	0	0
17	0	0	0	0	0	0	0	0	0	0	0	0	0	0	1	0	0	0	0	0	0	0	0	0
18	0	0	0	0	0	0	0	0	0	0	0	0	0	0	0	0	0	0	0	0	1	1	1	0
19	0	0	0	0	0	0	0	0	0	0	0	0	0	0	0	0	0	0	0	0	0	1	0	0
20	0	0	0	0	0	0	0	0	0	0	0	0	0	0	0	0	0	0	0	0	0	1	0	0
21	0	0	0	0	0	0	0	0	0	0	0	0	0	0	0	0	0	0	0	0	0	1	1	1
22	0	0	0	0	0	0	0	0	0	0	0	0	0	0	0	0	0	0	0	0	0	1	0	1
23	0	0	0	0	0	0	0	0	0	0	0	0	0	0	0	0	0	0	0	0	0	0	0	1
24	0	0	0	0	0	0	0	0	0	0	0	0	0	0	0	0	0	0	0	0	0	0	1	1

1 phytoplankton, periphyton
2 detritus
3 vascular plants
4 copepods
5 cladocerans
6 amphipods
7 insect larvae
8 waterfowl, marsh rabbits, deer, water rat
9 plecopterans, odonates, hemipterans
10 crayfish
11 cyprinodontids
12 mosquitofish
13 coleopterans
14 dipterans
15 hemipterans
16 centrarchids
17 snakes, turtles
18 bowfin
19 gar
20 pickerel
21 herons, ibises
22 egrets
23 raccoons
24 opossums
25 alligators
26 raptors
27 bobcats

48 Nearshore marine 1, Aleutian Islands

48	4	5	6	7	8	9	10	11	12	13
1	0	0	1	1	0	1	0	0	0	0
2	0	1	1	0	0	0	0	0	0	0
3	1	0	1	0	0	0	0	0	0	0
4	0	0	0	0	1	1	0	0	0	0
5	0	0	0	0	1	0	1	0	0	0
6	0	0	0	0	0	0	0	0	1	0
7	0	0	0	0	0	0	1	0	0	0
9	0	0	0	0	0	0	1	0	1	0
10	0	0	0	0	0	0	0	1	1	1
11	0	0	0	0	0	0	0	0	0	1
12	0	0	0	0	0	0	0	0	0	1

1 detritus
2 macroalgae
3 nearshore phytoplankton
4 mussels
5 chitons, gastropods, limpets
6 sea urchins
7 epibenthic crustacea

8 asteroids
9 decapods
10 nearshore fishes
11 harbor seals
12 sea otters
13 prehistoric Aleut man

49 Nearshore marine 2, Aleutian Islands

49	4	5	6	7	8	9	10	11	12
1	0	0	1	1	0	1	0	0	0
2	0	1	1	0	0	0	0	0	0
3	1	0	1	0	0	0	0	0	0
4	0	0	0	0	1	1	0	0	1
5	0	0	0	0	1	0	1	0	1
6	0	0	0	0	0	0	1	0	1
7	0	0	0	0	0	0	1	0	0
9	0	0	0	0	0	0	0	1	1
10	0	0	0	0	0	0	0	0	1
11	0	0	0	0	0	0	0	0	1

1 detritus
2 macroalgae
3 nearshore phytoplankton
4 mussels
5 chitons, gastropods, limpets
6 sea urchins

7 epibenthic crustacea
8 asteroids
9 decapods
10 nearshore fishes
11 octopuses
12 prehistoric Aleut man

50 Sand beach, California

50	3	4	5	6	7	8	9	10	11	12	13	14
1	1	1	1	0	0	0	1	1	0	0	0	0
2	0	1	1	1	1	1	0	0	0	0	0	0
3	0	0	0	0	0	0	0	0	0	0	1	0
4	0	0	0	0	0	0	0	1	0	0	0	0
5	0	0	0	0	0	0	0	0	0	0	1	1
6	0	0	0	0	0	0	0	0	1	1	0	0
7	0	0	0	0	0	0	0	0	1	0	0	0
8	0	0	0	0	0	0	0	0	0	0	0	1
9	0	0	0	0	0	0	0	0	0	0	0	1
10	0	0	0	0	0	0	0	0	0	0	1	1
11	0	0	0	0	0	0	0	0	0	1	0	1

1 debris
2 plankton
3 amphipods
4 *Blepharipoda*
5 *Emerita analoga*
6 *Tivela stultorum*
7 *Donax*

8 *Olivella*
9 *Thoracophelia*
10 *Nepthys*
11 *Policines*
12 sea otter
13 birds
14 fishes

51 Shallow sublittoral, Cape Ann, Massachusetts

51	5	6	7	8	9	10	11	12	13	14	15	16	17	18	19	20	21	22	23	24	25
1	1	1	1	1	1	0	0	0	0	0	1	0	0	0	0	0	0	0	0	0	0
2	0	0	0	0	0	1	0	0	0	0	0	0	1	0	0	0	0	0	0	0	0
3	0	0	0	0	0	1	1	1	0	0	0	0	0	0	0	0	0	0	0	0	0
4	0	0	0	0	0	0	0	0	1	1	0	0	0	0	0	0	0	0	0	0	0
5	0	0	0	0	0	0	0	0	0	0	1	1	0	0	0	1	0	1	0	0	1
6	0	0	0	0	0	0	0	0	0	0	0	0	1	0	0	0	0	0	1	0	0
7	0	0	0	0	0	0	0	0	0	0	0	0	1	1	0	0	1	0	1	0	0
8	0	0	0	0	0	0	0	0	0	0	0	0	1	0	0	0	0	0	1	0	0
9	0	0	0	0	0	0	0	0	0	0	0	0	1	0	0	0	0	0	0	0	0
10	0	0	0	0	0	0	0	0	0	0	0	0	0	0	0	0	0	1	1	0	0
11	0	0	0	0	0	0	0	0	0	0	0	0	0	1	0	0	0	0	1	0	0
12	0	0	0	0	0	0	0	0	0	0	0	0	0	0	0	0	0	1	0	0	0
13	0	0	0	0	0	0	0	0	0	0	0	0	0	0	0	0	0	1	1	0	0
14	0	0	0	0	0	0	0	0	0	0	0	0	0	0	1	0	0	0	1	0	0
15	0	0	0	0	0	0	0	0	0	0	0	0	0	0	0	0	0	1	0	1	0
16	0	0	0	0	0	0	0	0	0	0	0	0	0	0	0	1	0	1	0	0	0
17	0	0	0	0	0	0	0	0	0	0	0	0	0	0	0	0	0	0	1	0	0
18	0	0	0	0	0	0	0	0	0	0	0	0	0	0	0	1	0	0	0	0	0
21	0	0	0	0	0	0	0	0	0	0	0	0	0	0	0	0	0	1	0	0	0
22	0	0	0	0	0	0	0	0	0	0	0	0	0	0	0	0	0	0	0	1	0
23	0	0	0	0	0	0	0	0	0	0	0	0	0	0	0	0	0	0	0	0	1

1 plankton and detritus
2 organic debris
3 macroalgae
4 organic matter in mud
5 *Fundulus*, fish fry
6 *Chalina*
7 *Mytilus, Gemma*
8 *Abietinaria, Sertularia, Metridium*
9 *Lichenophora*

10 isopods, *Gammarus, Caprella*
11 *Littorina littorea*
12 *Strongylocentrotus*
13 *Crago*
14 annelids
15 *Scomber, Clupea*
16 *Loligo*
17 *Pagurus, Cancer*
18 *Polinices*

19 *Limulus*
20 *Raja*
21 *Asterias*
22 *Pomatomus, Poronatus*
23 *Myoxocephalus*
24 *Phoca*
25 *Sterna*

52 Rocky shore, Torch Bay, Alaska

52	3	4	5	6	7	8	9	10	11	12	13	14	15	16	17	18	19	20	21	22
1	1	1	1	1	0	0	0	0	0	1	0	0	0	0	0	0	0	0	0	0
2	0	0	0	0	1	1	1	1	1	0	0	0	0	0	0	0	0	0	0	0
5	0	0	0	0	0	0	0	0	0	0	0	0	1	0	0	0	0	0	1	1
6	0	0	0	0	0	0	0	0	0	0	0	1	0	0	0	0	0	0	0	1
7	0	0	0	0	0	0	0	0	0	0	0	1	1	1	0	0	1	1	1	1
8	0	0	0	0	0	0	0	0	0	0	0	1	1	0	0	1	1	0	1	0
9	0	0	0	0	0	0	0	0	0	0	0	0	0	1	0	0	0	0	0	0
10	0	0	0	0	0	0	0	0	0	0	0	0	0	1	0	0	0	0	1	0
11	0	0	0	0	0	0	0	0	0	1	0	0	0	1	0	0	0	0	0	0
14	0	0	0	0	0	0	0	0	0	0	0	0	1	0	0	0	0	0	0	1

1 benthic algae
2 plankton, detritus
3 *Tonicella, Acmaea mitra*
4 urchins
5 limpets
6 *Littorina*
7 barnacles
8 *Mytilus edulis*
9 anemones, tunicates
10 bryozoa
11 *Halichondria*
12 *Katharina*
13 *Searlesia*
14 *Thais* sp.
15 *Pycnopodia*
16 *Henricia, Archidoris*
17 *Dermasterias*
18 *Pisaster*
19 *Thais canaliculata*
20 *Emplectonema*
21 *Leptasterias*
22 *Thais lima*

53 Rocky shore, Cape Flattery, Washington

53	3	4	5	6	7	8	9	10	11	12	13	14	15	16	17	18	19	20	21	22
1	1	1	1	1	0	0	0	0	0	1	1	0	0	0	0	0	0	0	0	0
2	0	0	0	0	1	1	1	1	1	0	0	1	0	0	0	0	0	0	0	0
4	0	0	0	0	0	0	0	0	0	0	0	1	0	0	0	0	0	0	0	0
5	0	0	0	0	0	0	0	0	0	0	0	0	1	0	1	0	0	0	0	1
6	0	0	0	0	0	0	0	0	0	0	0	0	1	0	1	0	0	0	0	0
7	0	0	0	0	0	0	0	0	0	0	0	0	0	1	1	1	0	0	0	1
8	0	0	0	0	0	0	0	0	0	0	0	0	0	0	0	1	0	0	0	1
9	0	0	0	0	0	0	0	0	0	0	0	0	0	0	0	0	1	0	0	0
10	0	0	0	0	0	0	0	0	0	0	0	0	0	0	0	0	0	0	0	1
11	0	0	0	0	0	0	0	0	0	0	0	0	0	0	0	0	0	0	1	0
12	0	0	0	0	0	0	0	0	0	0	0	0	1	0	0	0	0	0	0	1
13	0	0	0	0	0	0	0	0	0	0	0	0	0	0	0	0	0	0	0	1
14	0	0	0	0	0	0	0	0	0	0	0	0	0	0	0	1	0	0	0	0

1 benthic algae
2 plankton, detritus
3 *Tonicella, Acmaea mitra*
4 urchins
5 limpets
6 *Littorina*
7 barnacles
8 *Mytilus edulis*
9 anemones
10 *Mytilus californianus*
11 *Halichondria*
12 *Katharina*
13 *Tegula*
14 *Spirorbis*
15 *Pycnopodia*
16 *Searlesia*
17 *Thais* sp, *Emplectonema, Ceratostoma*
18 *Leptasterias*
19 *Thais canaliculata*
20 *Dermasterias*
21 *Henricia, Archidoris*
22 *Pisaster*

54 Western rocky shore, Barbados

54	5	6	7	8	9	10	11	12	13	14	15
1	0	0	0	0	0	0	0	1	0	0	0
2	1	1	1	0	0	0	0	1	0	0	0
3	0	0	1	0	0	0	0	0	1	0	0
4	0	0	0	1	1	1	1	0	0	0	0
6	0	0	0	0	0	0	0	0	0	1	0
7	0	0	0	0	0	0	0	0	0	0	1
8	0	0	0	0	0	0	0	1	1	0	1
9	0	0	0	0	0	0	0	0	1	0	0
10	0	0	0	0	0	0	0	0	0	1	0
11	0	0	0	0	0	0	0	0	1	0	0
12	0	0	0	0	0	0	0	0	0	1	0
13	0	0	0	0	0	0	0	0	0	0	1
14	0	0	0	0	0	0	0	0	0	0	1

1 *Porolithon, Lithophyllum*
2 *Chaetomorpha, Enteromorpha, Cladophora*, diatoms
3 organic debris
4 plankton, detritus
5 *Nodolittorina tuberculata, Littorina* sp.
6 *Fissurella barbadensis, Acmaea jamaicensis*
7 *Enchinometra lucunter*
8 sponges
9 *Spirobranchus giganteus*
10 *Tetraclita squamosa*
11 bryozoans
12 *Acanthopleura granulata*
13 *Grapsus grapsus*
14 *Purpura patula*
15 reef fishes

55 Mudflat, Ythan estuary, Scotland

55	2	3	4	5	6	7	8	9	10	11	12
1	1	1	1	1	0	0	0	0	0	0	0
2	0	0	0	0	1	1	1	1	0	0	0
3	0	0	0	0	0	0	0	0	1	0	0
4	0	0	0	0	0	0	0	1	1	0	0
5	0	0	0	0	0	1	1	1	1	0	0
7	0	0	0	0	0	0	0	0	0	1	1
8	0	0	0	0	0	0	0	0	0	0	1

1 detritus
2 *Corophium, Hydrobia, Littorina*
3 *Arenicola*
4 *Macoma, Cardium*
5 nereids
6 goby
7 flounder
8 shelduck
9 redshank, dunlin, knot
10 oystercatcher
11 cormorant, heron, merganser
12 gulls

56 Mussel bed, Ythan estuary, Scotland

56	2	3	4	5	6	7	8	9	10	11	12
1	1	1	1	1	0	0	0	0	0	0	0
2	0	0	0	0	1	1	0	0	0	0	1
3	0	0	0	0	0	1	1	0	0	0	1
4	0	0	0	0	0	0	1	0	0	0	0
5	0	0	0	0	0	0	1	1	1	1	0
7	0	0	0	0	0	0	0	0	0	0	1

1 detritus　　　　　　　　7 eider
2 *Mytilus*　　　　　　　　8 turnstone
3 *Carcinus*　　　　　　　9 butterfish
4 *Corophium*　　　　　　10 blenny
5 gammarids　　　　　　　11 goby
6 oystercatcher　　　　　12 gulls

57 Brackish lagoons, Guerrero, Mexico

57	4	5	6	7	8	9
1	1	0	1	0	0	0
2	1	1	1	1	1	0
3	0	1	0	1	0	0
4	0	0	1	1	0	0
5	0	0	0	1	1	0
6	0	0	0	1	1	1
7	0	0	0	0	1	1
8	0	0	0	0	0	1

1 benthic algae　　　　　　6 *Macrobranchium*
2 detritus　　　　　　　　7 fishes (type 1)
3 phytoplankton　　　　　8 fishes (type 2)
4 micro- and meio-benthos　9 fishes (type 3)
5 zooplankton, insects

58 Sphagnum bog, Russia, USSR

58	6	7	8	9	10	11	12	13	14	15	16	17	18
1	1	0	0	0	0	0	0	0	0	0	0	0	0
2	1	0	0	0	0	0	0	0	0	0	0	0	0
3	0	1	0	0	0	0	0	0	0	0	0	0	0
4	0	0	1	0	0	1	0	0	0	0	0	0	0
5	0	0	0	1	0	0	0	0	0	0	0	1	0
6	0	0	0	0	0	0	1	1	0	0	0	0	0
7	0	0	0	0	1	0	1	1	0	0	0	0	0
8	0	0	0	0	0	1	0	0	0	1	0	0	0
9	0	0	0	0	0	0	0	0	0	0	0	1	0
12	0	0	0	0	0	0	0	1	0	0	0	0	0
13	0	0	0	0	0	0	0	0	1	1	1	0	0
15	0	0	0	0	0	0	0	0	0	0	1	0	0
16	0	0	0	0	0	0	0	0	0	0	0	1	0
17	0	0	0	0	0	0	0	0	0	0	0	0	1

1 *Sphagnum riparium*
2 algae
3 detritus
4 decomposing sphagnum
5 angiosperms
6 *Psectrocladius* larvae
7 chironomid larvae, cladocerans, rotifers
8 fungi
9 caterpillars

10 *Utricularia*
11 collembola
12 *Chaoborus,* Odonata larvae, Hemiptera, Coleoptera
13 *Nematocera* imagines
14 *Drosera*
15 ants
16 spiders
17 frogs, lizards, birds
18 birds of prey

59 Trelease Woods, Illinois

59	7	8	9	10	11	12	13	14	15	16	17	18	19	20	21	22	23	24	25	26	27	28	29	30
1	0	0	0	0	0	0	0	0	0	0	0	0	0	0	0	1	0	0	0	0	0	0	0	0
2	0	1	1	1	0	0	0	0	0	0	0	0	0	0	0	0	0	0	0	0	0	0	0	0
3	1	1	0	1	1	1	0	1	1	1	0	0	0	0	0	1	0	0	0	0	0	0	0	0
4	0	0	0	1	0	0	1	0	0	1	0	0	0	0	0	0	0	0	0	0	0	0	0	0
5	0	0	0	0	0	0	0	0	0	0	0	0	0	0	0	0	0	0	0	0	0	0	0	0
6	0	0	0	0	0	1	0	1	1	0	0	0	0	0	0	0	0	0	0	0	0	0	0	0
7	0	0	0	0	0	0	0	0	0	0	0	0	0	0	0	0	0	0	0	0	0	0	0	0
8	0	0	0	0	0	0	0	0	0	0	0	0	0	0	0	0	0	0	0	0	1	1	0	0
9	0	0	0	0	0	0	0	0	0	0	0	0	0	0	0	0	0	0	0	0	1	0	0	0
10	0	0	0	0	0	0	0	0	0	0	1	0	1	0	1	0	1	1	1	1	1	1	0	1
11	0	0	0	0	0	0	0	0	0	0	0	0	1	0	1	0	0	1	1	0	1	0	1	0
12	0	0	0	0	0	0	0	0	0	0	1	0	1	0	1	0	1	1	1	0	0	1	0	1
13	0	0	0	0	0	0	0	0	0	0	0	0	0	0	0	0	0	0	0	0	1	0	0	0
14	0	0	0	0	0	0	0	0	0	0	0	1	1	0	1	0	1	0	0	0	0	1	1	0
15	0	0	0	0	0	0	0	0	0	0	1	0	0	0	0	0	0	1	1	1	1	0	0	1
16	0	0	0	0	0	0	0	0	0	0	0	0	1	0	1	0	1	1	0	0	0	1	0	1
17	0	0	0	0	0	0	0	0	0	0	0	0	0	0	0	0	0	0	0	0	0	1	0	0
18	0	0	0	0	0	0	0	0	0	0	0	0	0	0	0	0	0	0	1	1	0	0	0	1
19	0	0	0	0	0	0	0	0	0	0	0	0	0	0	0	0	0	1	1	0	1	0	0	1
20	0	0	0	0	0	0	0	0	0	0	0	0	0	0	0	0	0	0	0	0	0	0	0	0
22	0	0	0	0	0	0	0	0	0	0	0	0	0	0	0	0	0	0	0	0	0	0	0	1
23	0	0	0	0	0	0	0	0	0	0	0	0	0	0	0	0	0	0	0	0	0	0	0	1
24	0	0	0	0	0	0	0	0	0	0	0	0	0	0	0	0	0	0	0	0	0	0	0	0

1 acorns
2 sap and plant juices
3 leaves
4 flowers
5 dead organic material
6 roots, bark, wood
7 Hemiptera
8 Orthoptera
9 Diptera
10 Homoptera
11 Lepidoptera
12 rabbit
13 Coleoptera
14 deermice
15 wood-borers, weevils
16 Hymenoptera
17 indigo bunting
18 shrew
19 wood thrush
20 arachnids

21 Maryland yellow-throat
22 fox squirrel
23 downy woodpecker
24 crested flycatcher
25 tufted titmouse
26 white-breasted nuthatch
27 red-eyed vireo
28 red-headed woodpecker
29 cat
30 barred owl

60 Montane forest, Arizona

60	6	7	8	9	10	11	12	13	14	15	16	17	18	19	20	21	22	23	24	25	26	27	28	29	30	31	32	33
1	1	1	1	1	1	0	0	0	1	0	0	0	1	0	0	0	0	0	0	0	0	1	0	0	0	0	0	0
2	1	1	1	1	1	1	0	0	0	0	0	0	1	0	0	0	0	0	0	0	0	0	0	0	0	0	0	0
3	1	0	1	1	0	0	1	1	1	1	0	0	1	1	0	0	0	0	0	0	0	0	0	0	0	0	0	0
4	0	0	1	0	0	1	0	0	0	0	1	1	0	0	0	0	0	0	0	0	0	0	0	0	0	0	0	0
5	0	1	0	0	0	0	0	0	1	0	0	1	0	0	0	0	0	0	0	0	0	0	0	0	0	0	1	0
6	0	0	0	0	0	0	0	0	0	0	0	0	0	0	0	1	1	0	0	0	1	0	1	1	1	0	0	0
7	0	0	0	0	0	0	0	0	0	0	0	0	0	0	0	0	1	0	0	0	0	0	1	0	0	0	0	0
8	0	0	0	0	0	0	0	0	0	0	0	0	0	0	1	0	0	1	0	0	0	0	0	0	0	0	0	0
9	0	0	0	0	0	0	0	0	0	0	0	0	0	0	1	0	0	0	0	1	0	0	0	0	0	0	0	0
10	0	0	0	0	0	0	0	0	0	0	0	0	0	0	0	0	0	0	0	0	0	0	0	0	0	0	0	1
11	0	0	0	0	0	0	0	0	0	0	0	0	0	0	0	0	0	1	1	1	0	1	0	0	0	0	0	0
12	0	0	0	0	0	0	0	0	0	0	0	0	0	0	0	0	0	0	0	0	0	0	0	0	0	0	0	0
13	0	0	0	0	0	0	0	0	0	0	0	0	0	0	1	0	0	1	0	1	0	0	0	0	0	0	0	0
14	0	0	0	0	0	0	0	0	0	0	0	0	0	0	0	1	1	0	0	0	1	0	1	1	1	0	1	0
15	0	0	0	0	0	0	0	0	0	0	0	0	0	0	1	0	0	0	0	1	0	0	0	0	0	0	0	0
16	0	0	0	0	0	0	0	0	0	0	0	0	0	0	0	0	0	1	1	1	0	1	0	0	0	0	0	0
17	0	0	0	0	0	0	0	0	0	0	0	0	0	0	0	0	0	0	0	0	0	0	0	0	0	0	0	0
18	0	0	0	0	0	0	0	0	0	0	0	0	0	0	0	0	0	0	0	0	1	0	1	0	1	0	0	1
19	0	0	0	0	0	0	0	0	0	0	0	0	0	0	0	0	0	1	1	1	0	0	0	0	0	0	0	0
20	0	0	0	0	0	0	0	0	0	0	0	0	0	0	0	0	0	0	0	0	0	0	0	0	0	1	1	0
23	0	0	0	0	0	0	0	0	0	0	0	0	0	0	0	0	0	0	0	0	0	0	0	0	0	1	0	0
25	0	0	0	0	0	0	0	0	0	0	0	0	0	0	0	0	0	0	0	0	0	0	0	0	0	1	1	0

1 yellow pine
2 aspen
3 herbs, grass, shrubs
4 decayed vegetation
5 fungi
6 Peromyscus, Microtus
7 Kaibab squirrel
8 aphids
9 leafhoppers
10 porcupines
11 beetles

12 horses
13 Orthoptera, Hemiptera
14 chipmunks, ground squirrels
15 meadow insects
16 ground invertebrates
18 deer
19 ants
20 flycatcher
21 badger
22 horned owl

23 pygmy nuthatch, Audubon warbler
24 horned toad
25 chipping sparrow, robin, red-backed junco, bluebird
26 red-tailed hawk
27 long-crested jay
28 bobcat
29 weasel
30 coyote
31 sharp-shinned hawk
32 goshawk
33 cougar

61 Barren regions, Spitsbergen

61	3	4	5	6	7	8	9
1	1	1	1	0	0	1	0
2	0	0	0	1	0	0	0
5	0	0	0	0	1	1	1
6	0	0	0	0	0	0	1
7	0	0	0	0	0	1	1

1 lichens, mosses
2 detritus
3 scarlet mite
4 reindeer
5 landflies, adult
  chironomids,
  Achorutes

6 worm (*Lumbricillus*)
7 spider
8 snow bunting
9 purple sandpiper

62 Reindeer pasture, Spitsbergen

62	4	5	6	7	8	9	10	11	12
1	1	1	1	1	0	0	0	0	0
2	0	0	0	0	1	0	0	0	0
3	0	0	0	0	0	1	0	0	0
6	0	0	0	0	0	0	0	1	0
7	0	0	0	0	0	0	1	1	0
8	0	0	0	0	0	0	1	0	0
9	0	0	0	0	0	0	0	1	0
10	0	0	0	0	0	0	0	1	0
11	0	0	0	0	0	0	0	0	1

1 lichens, mosses, phanerogams
2 reindeer dung
3 algae, detritus
4 reindeer
5 pink-footed goose
6 worm (*Henlea*)
7 mites, landflies, adult chironomids, springtails
8 dung fly
9 bog and intertidal invertebrates
10 spiders
11 purple sandpiper
12 arctic fox

63 River Rheidol, Wales

63	6	7	8	9	10	11	12	13	14	15	16	17	18
1	0	0	1	0	1	1	1	1	1	0	0	0	0
2	1	0	1	1	0	1	1	1	1	0	0	0	0
3	1	1	1	1	1	1	1	1	1	0	0	0	0
4	1	1	1	1	0	0	0	1	1	0	0	1	0
5	1	1	1	1	1	0	1	1	1	0	0	1	0
6	0	0	0	0	0	0	0	1	0	1	1	1	1
7	0	0	0	0	0	0	0	1	1	1	1	1	1
8	0	0	0	0	0	0	0	1	0	0	1	1	1
9	0	0	0	0	0	0	0	1	0	0	1	1	0
10	0	0	0	0	0	0	0	1	1	0	0	0	0
11	0	0	0	0	0	0	0	0	1	1	1	0	1
12	0	0	0	0	0	0	0	1	1	1	1	0	1
13	0	0	0	0	0	0	0	0	0	0	1	1	1
14	0	0	0	0	0	0	0	0	0	0	1	0	1
15	0	0	0	0	0	0	0	0	0	0	1	0	1
16	0	0	0	0	0	0	0	0	0	0	0	1	1

1 green algae
2 fragmented leaf, stem tissue, moss
3 diatoms
4 *Batrachospermum, Lemanea*
5 detritus
6 *Rhithrogena*
7 *Baetis*
8 *Leuctra, Protonemeura, Amphinemura*
9 Oligochaeta

10 Copepoda, Cladocera
11 *Simulium*
12 Chironomidae
13 *Chloroperla*
14 *Hydropsyche*
15 Dystiscidae (*Deronectes, Oreonectes*)
16 *Polycentropus*
17 *Isoperla*
18 *Perlodes*

64 Linesville Creek, Pennsylvania

64	3	4	5	6	7	8	9	10	11	12	13	14	15	16	17	18	19
1	1	1	1	1	1	1	1	0	0	0	1	0	0	0	0	1	0
2	0	0	0	0	1	0	1	1	1	1	1	0	0	0	0	0	0
3	0	0	0	0	0	0	0	0	0	0	0	1	0	0	0	0	0
4	0	0	0	0	0	0	0	0	0	0	0	1	1	0	0	0	0
5	0	0	0	0	0	0	0	0	0	0	0	0	0	0	0	0	1
6	0	0	0	0	0	0	0	0	0	0	0	1	0	0	0	0	1
9	0	0	0	0	0	0	0	0	0	0	0	1	0	0	0	0	0
11	0	0	0	0	0	0	0	0	0	0	0	0	0	0	1	0	1
12	0	0	0	0	0	0	0	0	0	0	1	0	0	1	0	1	1

1 diatoms
2 detritus
3 *Centroptilum, Stenonema, Chimarra, Tipula* sp. 1
4 *Habrophleboides, Prosimulium*
5 *Corynoneura, Psychomyia*
6 *Polypedilum*
7 *Ephemera*
8 *Agapetus, Psilotreta, Stenelmis, Isonychia, Geora, Antocha, Helicopsyche, Psephenus*
9 *Eukiefferiella*

10 *Pycnopsyche, Tipula* sp. 2
11 *Brillia, Microtendipes*
12 *Tanytorsus*
13 *Hydropsyche*
14 *Cottus*
15 *Phasganophora, Nigronia*
16 *Antherix, Eucalia*
17 *Cyrnellus*
18 *Chenmatopsyche*
19 *Pentaneura*

65 Yoshino River rapids, Japan

65	3	4	5	6	7	8	9	10	11	12	13	14	15	16
1	1	1	1	1	1	1	1	1	1	0	0	0	0	0
2	0	0	0	0	0	0	0	0	0	0	0	0	1	1
3	0	0	0	0	0	0	0	0	0	1	0	1	1	1
4	0	0	0	0	0	0	0	0	0	1	0	0	0	0
5	0	0	0	0	0	0	0	0	0	1	1	1	1	1
6	0	0	0	0	0	0	0	0	0	1	1	1	1	1
7	0	0	0	0	0	0	0	0	0	1	1	1	1	1
8	0	0	0	0	0	0	0	0	0	1	1	1	1	1
9	0	0	0	0	0	0	0	0	0	1	0	1	0	1
10	0	0	0	0	0	0	0	0	0	1	0	1	1	1

1 algae
2 surface bait
3 Diptera
4 *Ephemerella*
5 Baetidae
6 Ecdyonuridae
7 Plecoptera
8 *Stenopsyche*

9 *Hydropsyche*
10 *Micrasema*
11 *Plecoglossus altivelis*
12 *Cobitis biwae, Rhinogobius flumineus*
13 *Liobagrus reini*
14 *Cottus pollux*
15 *Maroco jouyi*
16 *Oncorhynchus rhodurus*

66 River Thames, England

66	4	5	6	7	8	9	10
1	1	1	0	1	0	0	0
2	0	0	1	0	1	1	0
3	0	0	0	0	1	1	0
4	0	0	0	1	1	1	0
5	0	0	0	0	1	1	0
6	0	0	0	0	0	1	0
7	0	0	0	0	0	1	0
8	0	0	0	0	0	1	1
9	0	0	0	0	0	0	1

1 phytoplankton, suspended detritus
2 periphyton, benthic algae
3 allochthonous matter
4 zooplankton
5 chironomids

6 gastropods, crustacea, tubificids
7 mollusks, sponges
8 roach
9 bleak
10 pike, large perch

67 Mudflats, Mississippi River, Iowa

67	3	4	5	6	7	8	9	10	11	12	13	14	15	16	17	18	19	20	21
1	1	1	1	1	0	0	0	1	0	0	0	0	0	0	0	0	0	1	0
2	0	0	0	1	1	1	1	1	1	0	0	0	0	0	0	1	0	0	0
3	0	0	0	0	0	0	0	0	0	0	0	1	0	1	0	1	0	1	0
4	0	0	0	0	0	0	0	0	0	0	0	0	1	1	0	0	0	1	0
5	0	0	0	0	0	0	0	0	0	0	1	1	1	1	1	1	0	1	0
6	0	0	0	0	0	0	0	0	0	0	0	0	1	1	1	0	1	0	0
7	0	0	0	0	0	0	0	1	1	1	0	0	0	0	0	1	0	0	0
8	0	0	0	0	0	0	0	0	0	0	0	0	1	1	1	0	1	0	0
9	0	0	0	0	0	0	0	0	0	0	0	0	1	0	1	0	0	0	0
10	0	0	0	0	0	0	0	0	0	0	0	0	1	1	1	0	0	0	0
11	0	0	0	0	0	0	0	0	0	0	0	0	1	1	1	0	0	0	0
12	0	0	0	0	0	0	0	0	0	0	0	0	1	0	1	0	1	1	0
13	0	0	0	0	0	0	0	0	0	0	0	0	1	0	1	0	0	0	0
14	0	0	0	0	0	0	0	0	0	0	0	0	0	0	1	0	0	0	0
15	0	0	0	0	0	0	0	0	0	0	0	0	1	0	1	0	0	0	0
16	0	0	0	0	0	0	0	0	0	0	0	0	0	0	1	0	0	0	0
17	0	0	0	0	0	0	0	0	0	0	0	0	0	0	0	0	0	0	1
18	0	0	0	0	0	0	0	0	0	0	0	0	0	0	0	0	1	1	1
19	0	0	0	0	0	0	0	0	0	0	0	0	0	0	0	0	0	0	1

1 detritus
2 phytoplankton
3 burrowing ephemerids
4 Oligochaeta
5 Chironomidae
6 Gastropoda
7 zooplankton

8 Amnicolidae
9 herbivorous insects
10 Pelecypoda
11 *Potamya*
12 macrocrustacea
13 *Oecetis*
14 *Sialis*

15 carnivorous Diptera
16 *Gomphus*
17 ducks (Lesser Scaup)
18 benthos-eating fish
19 piscivorous fish
20 Hirudinea
21 mammals

68 Loch Leven, Scotland

68	5	6	7	8	9	10	11	12	13	14	15	16	17	18	19	20	21	22
1	1	1	1	0	0	0	0	0	1	0	0	0	0	0	0	0	0	0
2	0	1	1	1	0	0	0	0	0	0	0	0	0	0	0	0	0	0
3	0	0	0	1	1	1	0	0	0	0	0	0	0	1	0	0	0	0
4	0	0	0	0	0	0	1	0	0	0	0	0	0	0	0	0	0	0
5	0	0	0	0	0	0	0	1	1	1	0	0	0	0	0	0	0	0
6	0	0	0	0	0	0	0	0	0	0	0	0	0	1	1	0	0	0
7	0	0	0	0	0	0	0	0	0	0	0	1	0	1	0	0	0	0
8	0	0	0	0	0	0	0	0	0	0	0	1	1	0	1	1	0	0
10	0	0	0	0	0	0	0	0	0	0	1	0	0	0	1	1	0	0
12	0	0	0	0	0	0	0	0	0	0	0	0	0	1	0	0	0	0
15	0	0	0	0	0	0	0	0	0	0	0	0	0	0	1	1	0	0
19	0	0	0	0	0	0	0	0	0	0	0	0	0	0	0	0	1	1
20	0	0	0	0	0	0	0	0	0	0	0	0	0	0	0	0	1	0

1 detritus
2 benthic algae
3 phytoplankton
4 *Phragmites, Potamogeton*
5 bacteria
6 *Asellus*
7 *Valvata*
8 herb, chironomids

9 *Anadonta*
10 *Daphnia*
11 pochard, swan, coot
12 rotifers
13 oligochaetes
14 nematodes, ciliates
15 fry
16 tufted duck

17 insect emergence
18 *Cyclops*
19 trout
20 perch
21 pike
22 man

69 Tagus estuary, Portugal

69	5	6	7	8	9	10	11	12	13	14	15	16	17	18	19	20	21	22	23	24	25	26	27	28	29
1	1	1	0	0	0	0	0	0	0	0	0	0	0	0	0	0	0	0	0	0	0	0	0	0	0
2	1	1	1	0	0	0	0	0	0	0	0	0	0	0	0	0	0	0	0	0	0	0	0	0	0
3	0	0	1	1	1	1	1	1	1	1	1	0	0	0	0	0	0	0	0	0	0	0	0	0	0
4	0	0	1	1	1	0	1	1	1	1	1	1	1	1	1	1	0	0	0	0	0	0	0	0	0
5	0	0	1	0	0	0	0	0	0	0	0	1	1	1	0	0	1	0	1	0	0	0	0	1	1
6	0	0	1	0	0	0	0	0	0	0	0	1	1	1	0	0	1	0	1	0	0	0	0	1	0
9	0	0	0	0	0	0	0	0	0	0	0	0	0	0	1	0	0	0	0	0	0	0	1	1	1
10	0	0	0	0	0	0	0	0	0	0	0	0	0	0	1	0	0	0	0	0	0	0	0	0	0
12	0	0	0	0	0	0	0	0	0	0	0	0	0	0	0	1	0	1	1	1	1	0	1	1	1
13	0	0	0	0	0	0	0	0	0	0	0	0	0	0	0	0	0	0	1	1	1	0	1	1	1
14	0	0	0	0	0	0	0	0	0	0	0	0	0	0	1	1	1	1	0	1	1	0	1	1	1
15	0	0	0	0	0	0	0	0	0	0	0	0	0	0	0	1	0	0	1	1	1	0	1	1	0
16	0	0	0	0	0	0	0	0	0	0	0	0	0	0	0	0	0	0	1	0	0	0	1	0	0
17	0	0	0	0	0	0	0	0	0	0	0	0	0	0	0	0	0	0	1	0	0	0	1	0	0
18	0	0	0	0	0	0	0	0	0	0	0	0	0	0	0	0	0	0	1	0	0	0	1	0	0
19	0	0	0	0	0	0	0	0	0	0	0	0	0	0	0	0	0	0	0	0	0	0	0	1	1
20	0	0	0	0	0	0	0	0	0	0	0	0	0	0	1	0	1	1	1	1	1	0	0	1	0
21	0	0	0	0	0	0	0	0	0	0	0	0	0	0	1	0	1	1	1	1	1	0	1	1	1
22	0	0	0	0	0	0	0	0	0	0	0	0	0	0	0	0	0	0	1	0	0	0	1	1	1
23	0	0	0	0	0	0	0	0	0	0	0	0	0	0	0	0	0	0	0	0	0	0	0	0	0
24	0	0	0	0	0	0	0	0	0	0	0	0	0	0	0	0	0	0	0	0	0	0	1	1	1
25	0	0	0	0	0	0	0	0	0	0	0	0	0	0	0	0	0	0	0	0	0	0	0	0	0
26	0	0	0	0	0	0	0	0	0	0	0	0	0	0	0	0	0	0	0	0	0	0	1	1	1
28	0	0	0	0	0	0	0	0	0	0	0	0	0	0	0	0	0	0	0	0	0	0	0	0	1

1 phytoplankton
2 Ulva, Enteromorpha
3 suspended detritus
4 sedimented detritus
5 Mysidacea
6 Copepoda
7 Chelon labrosus, Liza ramada, L. aurata
8 Crassostrea angulata
9 Scrobicularia plana, Cerastoderma
10 Mytilus galloprovincialis

11 bacteria
12 Gammaridae
13 Corophium
14 Cirratulidae, Capitellidae, Maldanidae
15 meiofauna
16 Engraulis encrasicolus
17 Sardina pilchardus
18 Clupeidae
19 Carcinus maenas
20 Crangon crangon

21 Nereis diversicolor
22 Pomatoschistus minutus
23 Conger conger
24 Trigla lucerna
25 Solea vulgaris
26 Nereis succinea
27 Dicentrarchus labrax
28 Ciliata mustela
29 birds

70 Crystal River estuary, Florida

70	3	4	5	6	7	8	9	10	11	12	13	14
1	1	0	1	1	1	0	0	0	1	0	0	0
2	1	1	1	1	1	0	0	0	0	0	0	0
3	0	0	0	0	1	0	0	0	0	0	0	0
4	0	0	0	0	0	0	0	0	0	1	0	1
5	0	0	0	0	0	0	0	0	0	1	1	1
6	0	0	0	0	1	1	1	1	1	0	1	0
7	0	0	0	0	0	0	0	0	0	1	1	1
8	0	0	0	0	0	0	0	0	0	0	1	0
9	0	0	0	0	0	0	0	0	0	0	0	1
11	0	0	0	0	0	0	0	0	0	0	0	1

1 producers
2 detritus
3 zooplankton
4 juvenile penaeid shrimp
5 mullet
6 benthic invertebrates
7 small fish
8 blue crabs
9 rays
10 black drum
11 sheepshead
12 trout, jack
13 red drum
14 porpoise shark

71 Lake Rybinsk, Russia, USSR

71	4	5	6	7	8	9	10	11	12	13	14	15	16
1	1	1	0	0	0	0	0	0	0	0	0	0	0
2	1	1	1	1	1	0	0	1	0	1	0	0	0
3	0	1	0	0	0	0	0	0	0	0	0	0	0
4	0	0	1	1	0	0	0	0	0	0	0	0	0
5	0	0	0	0	1	0	0	0	0	1	0	0	0
6	0	0	0	0	0	1	0	1	0	0	0	0	0
7	0	0	0	0	0	1	0	1	1	0	0	0	0
8	0	0	0	0	0	0	1	0	0	1	0	0	0
9	0	0	0	0	0	0	0	1	1	0	0	0	0
10	0	0	0	0	0	0	0	0	0	1	0	0	0
11	0	0	0	0	0	0	0	0	0	0	1	0	0
12	0	0	0	0	0	0	0	0	0	0	1	0	1
13	0	0	0	0	0	0	0	0	0	0	1	0	1
14	0	0	0	0	0	0	0	0	0	0	0	1	1
15	0	0	0	0	0	0	0	0	0	0	0	0	1

1 allochthonous organic matter
2 phytoplankton
3 macrophytes
4 bacterioplankton
5 benthic bacteria
6 planktonic protozoa
7 Rotatoria, Calanoida, Cladocera
8 macrobenthos
9 predatory zooplankton
10 predatory benthos
11 fish larvae
12 planktivorous fishes
13 benthos-feeding fishes
14 predatory fishes
15 other predatory fishes
16 man

72 Heney Lake, pelagic zone, Quebec

72	6	7	8	9	10	11	12	13	14	15	16	17
1	0	0	0	1	0	0	0	0	0	0	0	0
2	0	1	0	1	1	0	0	0	0	0	0	0
3	0	0	1	1	1	1	0	0	0	0	0	0
4	1	0	1	0	0	1	0	0	0	0	0	0
5	1	0	0	0	0	0	0	0	0	0	0	0
6	0	0	1	0	0	1	0	0	0	0	0	0
7	0	0	0	0	0	0	1	1	0	1	1	0
8	0	0	0	0	0	0	1	1	1	1	0	0
9	0	0	0	0	0	0	0	1	0	0	1	0
10	0	0	0	0	0	0	0	1	1	0	0	0
11	0	0	0	0	0	0	0	1	1	0	0	0
12	0	0	0	0	0	0	0	0	0	0	1	0
13	0	0	0	0	0	0	0	0	0	0	1	0
14	0	0	0	0	0	0	0	0	0	0	1	0
16	0	0	0	0	0	0	0	0	0	0	0	1

1 phytoplankton cells above 30 micrometer
2 phytoplankton cells 10-30 micrometer
3 phytoplankton cells below 10 micrometer
4 particulate organic matter
5 dissolved organic matter
6 bacteria
7 *Daphnia, Diaphanosoma, Ceriodaphnia*
8 *Bosmina, Chydorus, Tropocyclops*
9 *Diaptomus*
10 copepodites
11 rotifers and copepod nauplii
12 *Leptodora kindtii*
13 *Chaoborus*
14 *Epischura, Mesocyclops, Acanthocyclops*
15 *Piona constricta*
16 *Osmerus eperlanus mordax*
17 *Salvelinus namaycush, Esox lucius*

73 Hafner Lake, Austria

73	5	6	7	8	9	10
1	1	1	1	0	0	0
2	0	0	1	0	0	0
3	0	0	0	0	1	0
4	0	0	0	0	0	1
5	0	0	0	1	0	0
6	0	0	0	1	1	1
7	0	0	0	1	1	1
8	0	0	0	0	1	1

1 nannoplankton
2 phytoplankton
3 terrestrial insects
4 littoral food items
5 rotifers

6 *Bosmina longirostris*
7 copepods
8 *Chaoborus flavicans*
9 *Alburnus alburnus* (bleak)
10 *Blica bjorkna* (silver bream)

74 Sand beach, South Africa

74	3	4	5	6	7	8	9	10	11	12	13	14	15	16	17	18	19	20	21
1	1	1	1	1	1	1	0	0	0	0	0	1	0	0	0	1	1	0	0
2	0	1	0	0	1	0	1	1	1	1	1	0	0	0	1	0	0	0	0
3	0	0	0	0	0	0	0	0	0	0	0	0	0	0	0	0	0	0	1
4	0	0	0	0	0	0	0	0	0	0	0	1	0	0	0	0	0	0	0
5	0	0	0	0	0	0	0	0	0	0	0	0	0	0	0	1	1	0	0
6	0	0	0	0	0	0	0	0	0	0	0	0	0	0	0	1	0	0	0
7	0	0	0	0	0	0	0	0	0	0	0	0	1	0	0	0	0	0	0
8	0	0	0	0	0	0	0	0	0	0	1	0	1	0	0	0	0	0	0
9	0	0	0	0	0	0	0	0	0	0	0	0	0	0	0	0	0	1	0
10	0	0	0	0	0	0	0	0	0	0	0	0	0	0	0	0	0	0	1
11	0	0	0	0	0	0	0	0	0	0	0	0	1	1	1	0	0	0	0
13	0	0	0	0	0	0	0	0	0	0	0	0	1	1	0	0	0	0	0
14	0	0	0	0	0	0	0	0	0	0	0	0	0	0	0	1	1	0	0
17	0	0	0	0	0	0	0	0	0	0	0	0	0	0	0	0	0	1	0
18	0	0	0	0	0	0	0	0	0	0	0	0	0	0	0	0	0	0	1

1 detritus
2 organic macrodebris
3 *Gastrosaccus, Callianassa, Donax*
4 bacteria
5 Cumacea
6 Turbellaria
7 Nemertea

8 *Bathyporeia*
9 *Bullia*
10 *Talorchestia*
11 errant polychaeta
12 *Larus*
13 isopods
14 protozoa

15 Curlew sandpiper
16 Sanderling
17 *Ovalipes*
18 sedentary polychaeta
19 nematode worms
20 elasmobranch fishes
21 predatory fishes

75 Vorderer Finstertaler Lake, Austria

75	4	5	6	7	8	9
1	1	1	1	1	0	0
2	0	0	0	1	0	0
3	0	0	0	1	0	1
4	0	0	1	1	0	0
5	0	0	0	0	1	0
6	0	0	0	0	1	0
7	0	0	0	0	1	1
8	0	0	0	0	0	1

1 phytoplankton
2 phytobenthos
3 terrestrial insects
4 bacterioplankton
5 *Polyathra, Synchaeta*
6 *Keratella*, young *Cyclops*
7 zoobenthos
8 adult *Cyclops*
9 fishes (*Salmo trutta, Salvelinus alpinus*)

76 Neusiedler Lake, Austria

76	4	5	6	7	8	9	10	11	12	13	14
1	1	0	0	0	0	0	0	0	0	0	0
2	0	1	0	0	0	1	0	0	0	0	0
3	0	0	1	1	1	0	0	0	0	0	0
4	0	0	0	0	0	0	0	1	0	0	0
5	0	0	0	0	0	1	1	0	0	1	0
7	0	0	0	0	0	0	0	0	1	0	1
8	0	0	0	0	0	0	0	0	0	0	1
9	0	0	0	0	0	0	0	0	1	0	0
10	0	0	0	0	0	0	0	0	0	1	0
12	0	0	0	0	0	0	0	0	0	1	1

1 organic matter
2 phytoplankton
3 benthic algae
4 bacteria
5 zooplankton
6 *Hypsibius augusti* (tartigrade)
7 *Tanypus punctipennis*
8 *Blicca bjorkna*
9 *Procladius*
10 *Pelecus cultratus*
11 *Limnocythere, Paraplectonema,* tubificidae
12 *Acerina cernua* (pope)
13 *Lucioperca*
14 *Anguilla anguilla*

77 Lake Abaya, Ethiopia

77	4	5	6	7	8	9	10	11	12	13
1	1	0	0	0	1	0	0	0	0	0
2	0	1	0	0	1	0	0	0	0	1
3	0	0	1	0	0	0	0	0	0	0
4	0	0	0	1	1	0	0	0	0	0
5	0	0	0	0	0	0	0	0	0	1
6	0	0	0	0	1	1	0	0	0	0
7	0	0	0	0	0	0	1	0	0	0
8	0	0	0	0	0	0	1	1	1	1
9	0	0	0	0	0	0	0	1	1	1
10	0	0	0	0	0	0	0	0	1	1
11	0	0	0	0	0	0	0	0	1	1
12	0	0	0	0	0	0	0	0	0	1

1 phytoplankton   8 *Tilapia zilli*
2 aquatic plants   9 *Barbus, Mormyrus*
3 ooze   10 carnivorous fishes
4 zooplankton   11 piscivorous birds
5 hippopotamus   12 crocodile
6 zoobenthos   13 man
7 fish fry

78 Lake George, Uganda

78	3	4	5	6	7	8	9	10	11	12	13	14	15	16
1	1	1	1	1	1	0	0	0	0	0	0	0	0	0
2	0	0	0	0	0	1	0	1	0	0	0	0	0	0
3	0	0	0	0	0	0	1	0	0	1	1	0	0	0
4	0	0	0	0	0	0	0	1	0	0	0	1	0	0
5	0	0	0	0	0	0	0	0	0	0	0	0	1	1
6	0	0	0	0	0	0	0	0	1	0	0	0	1	1
7	0	0	0	0	0	0	0	1	0	0	0	0	0	0
8	0	0	0	0	0	0	0	1	0	0	0	0	0	0
9	0	0	0	0	0	0	0	0	0	1	1	0	0	0
10	0	0	0	0	0	0	0	0	0	0	0	0	1	1
12	0	0	0	0	0	0	0	0	0	0	0	1	0	0
13	0	0	0	0	0	0	0	0	0	0	0	0	1	1
15	0	0	0	0	0	0	0	0	0	0	0	0	0	1

1 algae
2 detritus
3 zooplankton
4 herbivorous Diptera
5 mid-water feeding fishes
6 *Tilapia leucosticta, T. nilotica*
7 mollusks
8 worms

9 *Mesocyclops, Asplanchna*
10 bottom-feeding fishes
11 man
12 carnivorous Diptera
13 surface feeding fishes
14 insectivorous birds
15 piscivorous fishes
16 piscivorous birds

79 Lake Paajarvi, offshore, Finland

79	4	5	6	7	8	9	10	11	12	13	14	15	16	17	18	19	20	21
1	0	0	0	0	0	1	0	0	0	0	0	1	0	0	0	0	0	0
2	1	0	0	0	0	0	0	0	0	0	0	0	0	0	0	0	0	0
3	0	1	1	1	1	0	0	0	0	0	0	0	0	0	0	0	0	0
4	0	0	0	0	0	1	0	0	0	0	0	0	0	0	0	0	0	0
5	0	0	0	0	0	0	1	0	0	0	0	0	0	0	0	0	0	0
6	0	0	0	0	0	0	0	1	0	0	0	0	0	0	0	0	0	0
7	0	0	0	0	0	0	0	0	0	0	0	0	0	0	1	0	0	0
8	0	0	0	0	0	0	0	0	0	0	0	0	0	0	1	1	0	0
9	0	0	0	0	0	0	0	0	1	1	1	1	1	1	0	0	1	0
10	0	0	0	0	0	0	0	0	0	0	0	0	0	0	1	0	0	0
11	0	0	0	0	0	0	0	0	0	0	0	0	0	0	1	0	0	0
12	0	0	0	0	0	0	0	0	0	0	0	0	1	0	0	0	0	0
13	0	0	0	0	0	0	0	0	0	0	0	0	0	0	1	1	0	0
15	0	0	0	0	0	0	0	0	0	0	0	0	0	0	0	0	1	0
16	0	0	0	0	0	0	0	0	0	0	0	0	0	0	0	0	0	1
17	0	0	0	0	0	0	0	0	0	0	0	0	0	0	0	1	0	0
20	0	0	0	0	0	0	0	0	0	0	0	0	0	0	0	0	0	1

1 phytoplankton
2 allochthonous organic matter
3 bottom organic matter
4 bacteria, protozoa, fungi
5 Turbellaria, Halacaridae, Tartigrada, Copepoda (*Diacyclops, Bryocamptus*),
   Cladocera (*Alona, Ilyocryptus*), Chironomidae (*Mesocricotopus, Micropsectra*), *Cytheridea*
6 Nematoda (*Ironus, Tobrilus*), Oligochaeta (*Psammoryctes, Tubifex, Peloscolex, Stylodrilus*)
7 *Pisidium*
8 *Pallasea*
9 *Daphnia, Bosmina, Polyarthra, Conochilus, Eudiaptomus, Synchaeta*
10 Chironomidae (*Procladius, Pentaneurini, Protanypus*)
11 *Acanthocyclops, Paracladopelma, Monodiamesa*
12 *Asplanchna, Heterocope, Thermocyclops, Leptodora*
13 *Gammaracanthus*
14 *Chaoborus*
15 *Mysis*
16 *Coregonus albula*
17 young smelt
18 ruff
19 burbot
20 smelt
21 walleye

80 Lake Paajarvi, littoral zone, Finland

80	6	7	8	9	10	11	12	13	14	15	16	17	18	19	20	21	22	23	24	25	26	27
1	1	1	0	0	0	0	0	0	1	0	0	0	0	0	0	0	0	0	0	0	0	0
2	1	1	0	0	0	0	0	0	0	0	0	0	0	0	0	0	0	0	0	0	0	0
3	0	0	0	0	0	0	0	0	0	0	0	0	0	0	0	0	0	0	0	0	1	0
4	0	1	1	1	1	1	1	1	0	0	0	0	0	0	0	0	0	0	0	0	0	0
5	0	0	0	0	0	0	0	0	0	0	0	0	0	0	0	0	0	1	0	0	0	0
6	0	0	0	0	0	0	0	0	1	0	0	0	0	0	0	0	0	0	0	0	0	0
8	0	0	0	0	0	0	0	0	0	0	1	1	1	0	1	0	0	0	1	1	0	0
9	0	0	0	0	0	0	0	0	0	0	1	0	1	0	1	1	0	0	0	0	0	0
10	0	0	0	0	0	0	0	0	0	0	1	0	1	0	1	1	0	0	1	1	1	0
11	0	0	0	0	0	0	0	0	0	0	1	0	1	1	1	0	0	0	1	1	1	0
12	0	0	0	0	0	0	0	0	0	0	1	0	1	1	1	1	0	0	0	0	1	0
13	0	0	0	0	0	0	0	0	0	0	0	0	0	0	0	0	0	1	0	1	1	0
14	0	0	0	0	0	0	0	0	0	0	0	0	0	0	0	0	1	1	1	1	1	0
15	0	0	0	0	0	0	0	0	0	0	0	0	0	0	0	0	0	1	1	1	0	0
16	0	0	0	0	0	0	0	0	0	0	0	0	0	0	0	0	0	1	1	1	0	0
19	0	0	0	0	0	0	0	0	0	0	0	0	0	0	0	0	0	0	0	0	1	0
21	0	0	0	0	0	0	0	0	0	0	0	0	0	0	0	0	0	1	1	1	1	0
23	0	0	0	0	0	0	0	0	0	0	0	0	0	0	0	0	0	0	0	0	0	1
24	0	0	0	0	0	0	0	0	0	0	0	0	0	0	0	0	0	0	0	0	0	1
25	0	0	0	0	0	0	0	0	0	0	0	0	0	0	0	0	0	0	0	0	0	1
26	0	0	0	0	0	0	0	0	0	0	0	0	0	0	0	0	0	0	0	0	0	1

1 phytoplankton
2 allochthonous organic matter
3 *Potamogeton, Lobelia, Isoetes, Sparganium*
4 bottom organic matter, benthic macroalgae
5 terrestrial adult insects
6 bacterioplankton
7 *Anodonta*
8 Copepoda (*Attheyella, Paracyclops, Paracamptus,...*), Ostracoda (*Candona, Darwinula, Cyclocypris, Cypridopsis*)
9 Nematoda (*Ironus, Tobrilus*), Oligochaeta (*Limnodrilus, Psammoryctes, Peloscolex*, Lumbriculidae, Naididae)
10 *Asellus, Pallasea*, Cladocera (*Eurycercus, Alona, Alonella*)
11 Chironomidae (*Microtendipes, Tanytarsus, Pseudochironomus, Limnochironomus*)
12 Ephemeroptera (*Caenis, Ephemera, Cloeon,...*)
13 Trichoptera (*Athripsodes, Oxyethira*), Mollusca (*Pisidium, Lymnaea Valvata, Gyraulus, Sphaerium*)
14 *Sida, Eurycercus, Bosmina, Diaphanosoma*, Rotatoria
15 *Acanthocyclops, Macrocyclops, Demicryptochironomus, Monodiameda*
16 Chironomidae (*Procladius, Cryptochironomus, Leptochironomus, Ablabesmyia*)
17 Hirudinea (*Erpobdella, Helobdella*)
18 Mermithidae
19 Trichoptera (*Cyrnus, Oecetis, Molanna*)
20 Turbellaria
21 *Asplanchna, Polyphemus*
22 *Coregonus lavaretus*
23 salakka
24 *Perca fluviatilis*
25 ruff (*Gymnocephalus cernus*)
26 *Rutilus rutilus*
27 pike, adult yellow perch (*Perca*)

81 Sendai Bay, mesopelagic zone, Japan

81	2	3	4	5	6	7	8	9	10	11	12
1	1	1	1	1	1	0	0	0	0	0	0
2	0	0	0	0	0	0	1	0	1	0	0
3	0	0	0	0	0	0	1	1	1	0	0
4	0	0	0	1	1	1	0	0	0	0	0
5	0	0	0	0	0	0	1	0	0	0	0
6	0	0	0	0	0	0	0	1	1	0	0
7	0	0	0	0	0	0	0	0	0	1	0
9	0	0	0	0	0	0	0	0	0	1	1

1 suspended or deposited organic matter
2 bivalves
3 sedentary polychaetes
4 benthic invertebrates
5 *Pinnixa rathbuni*
6 errant polychaetes
7 *Crangon affinis, Metapenaeopsis dalei*
8 *Erynnis japonica*
9 *Limanda herzensteini, L. yokohamae*
10 *Chaeturichthys hexanema*
11 *Liparis tanakai*
12 *Lophius litulon*

82 Permanent freshwater rockpool, France

82	3	4	5	6	7	8	9	10
1	1	1	0	0	0	0	0	0
2	0	0	1	0	0	0	0	0
3	0	0	0	1	0	0	1	0
4	0	0	0	1	1	0	1	0
5	0	0	0	0	0	0	1	0
6	0	0	0	0	0	0	1	0
7	0	0	0	0	0	1	1	0
8	0	0	0	0	0	0	1	0
9	0	0	0	0	0	0	0	1

1 organic material
2 small algae
3 *Culex*
4 *Chironomus*
5 *Sigara, Discoglossus* larvae
6 Tanypodinae
7 *Cloeon*
8 *Plactynemis*
9 *Notonecta, Stictotarsus, Meladema, Sympetrum, Gomphus*
10 *Hydrometra*

83 Lake Pyhajarvi, littoral zone, Finland

83	5	6	7	8	9	10	11	12	13	14	15	16	17	18	19	20	21	22	23	24	25
1	1	0	1	0	0	0	0	0	0	0	0	0	0	0	0	0	0	0	1	0	0
2	1	1	0	0	0	0	0	0	0	0	0	0	0	0	0	0	0	0	0	0	0
3	0	0	0	1	0	0	0	0	0	0	0	0	0	0	0	0	0	0	0	0	0
4	0	0	0	0	1	1	1	1	1	1	1	1	1	1	1	1	0	0	1	0	0
5	0	0	0	0	0	0	0	0	0	0	0	0	0	0	0	0	0	0	1	0	0
6	0	0	0	0	0	0	0	0	0	0	0	1	0	0	0	0	0	0	0	0	0
7	0	0	0	0	0	0	0	0	0	0	0	0	0	0	1	1	0	0	0	1	0
8	0	0	0	0	0	0	0	0	0	0	0	1	1	1	1	1	1	1	0	1	1
9	0	0	0	0	0	0	0	0	0	0	0	0	0	0	1	1	1	0	0	0	0
10	0	0	0	0	0	0	0	0	0	0	0	1	0	0	0	1	1	1	0	0	1
11	0	0	0	0	0	0	0	0	0	0	0	0	1	0	0	0	0	0	0	0	0
12	0	0	0	0	0	0	0	0	0	0	0	1	1	1	1	1	0	1	0	0	1
13	0	0	0	0	0	0	0	0	0	0	0	0	0	0	0	1	1	1	0	0	1
14	0	0	0	0	0	0	0	0	0	0	0	0	0	0	1	1	0	0	0	0	0
15	0	0	0	0	0	0	0	0	0	0	0	1	1	1	1	1	0	0	0	0	0
16	0	0	0	0	0	0	0	0	0	0	0	0	0	0	0	0	0	1	0	0	1
17	0	0	0	0	0	0	0	0	0	0	0	0	0	0	0	0	0	1	0	0	1
18	0	0	0	0	0	0	0	0	0	0	0	0	0	0	0	0	0	1	0	0	0
23	0	0	0	0	0	0	0	0	0	0	0	0	0	0	0	0	0	0	0	1	0
24	0	0	0	0	0	0	0	0	0	0	0	0	0	0	0	0	0	0	0	0	1

1 phytoplankton
2 suspended detritus
3 aquatic plants
4 sedimented detritus
5 *Keratella, Kellicottia*
6 *Codonella, Vorticella*
7 *Daphnia, Bosmina*
8 *Sida, Eurycercus*
9 *Lymnaea, Planorbis, Goera*
10 chironomids
11 *Stylodrilus*
12 *Ephemera, Caenis, Heptagenia*
13 *Pisidium, Sphaerium*

14 Nematoda, Harpacticoida, Cyclopoida, Ostracoda
15 *Asellus*
16 *Polycentropus, Molanna*
17 *Ablabesmyia*
18 *Erpobdella, Helobdella*
19 *Coregonus albula*
20 *Coregonus lavaretus*
21 *Rutilus rutilus*
22 *Gymnocephalus cernus*
23 Cyclopoida
24 young fish
25 *Perca fluviatilis*

84 Temporary pond, Michigan

84	3	4	5	6	7	8	9	10	11	12
1	1	1	1	0	0	0	0	0	0	0
2	0	1	1	0	0	0	0	0	0	0
3	0	0	0	0	0	0	1	1	0	0
4	0	0	0	1	1	1	0	0	1	0
5	0	0	0	0	0	1	0	1	1	0
6	0	0	0	0	0	1	1	0	1	0
7	0	0	0	0	0	0	1	0	0	0
8	0	0	0	0	0	0	0	1	0	1
9	0	0	0	0	0	0	0	1	1	0
10	0	0	0	0	0	0	0	0	0	1

1 detritus
2 periphyton, phytoplankton
3 *Rana sylvatica*
4 *Daphnia pulex, Tendipes*, amphipods, ostracods, corixids
5 gastropods, pelecypods
6 *Chaoborus, Acilius*

7 Hydracarina, Odonata, *Ranatra, Belostoma*
8 *Ambystoma laterale, A. maculatum A. tremblayi*
9 *Dytiscus*
10 *Ambystoma tigrinum*
11 *Notophthalmus viridescens*
12 *Batracobdella picta* (leech)

85 Tasek Bera swamp, Malaysia

85	9	10	11	12	13	14	15	16	17	18	19	20	21	22	23	24	25	26	27
1	1	0	0	0	0	0	0	0	0	0	0	0	0	0	0	0	0	0	0
2	0	1	0	1	0	0	0	0	0	0	0	0	0	0	0	0	0	0	0
3	0	1	0	1	1	0	0	0	0	0	0	0	0	0	0	0	0	0	0
4	0	0	0	1	0	0	0	0	0	0	0	0	0	0	0	0	0	0	0
5	0	0	0	1	1	0	0	1	0	0	0	0	0	0	0	0	0	0	0
6	1	0	0	0	0	0	0	0	1	1	0	0	0	0	0	0	0	0	0
7	0	0	1	0	0	1	1	0	0	0	0	0	0	0	0	0	0	0	0
8	0	0	0	0	0	1	1	0	0	0	0	0	0	0	0	0	0	0	0
9	0	0	0	1	1	0	0	1	1	1	0	0	0	0	0	0	0	0	0
10	0	0	0	0	0	0	0	0	0	0	1	0	0	1	0	0	0	0	0
11	0	0	0	0	0	0	0	0	0	0	0	0	1	0	1	1	1	0	0
12	0	0	0	0	0	0	0	0	0	0	1	0	0	1	0	0	0	0	0
13	0	0	0	0	0	0	0	0	0	0	0	0	0	1	0	0	0	1	0
14	0	0	0	0	0	0	0	0	0	0	0	0	0	0	0	0	1	0	0
16	0	0	0	0	0	0	0	0	0	0	1	0	0	1	0	0	0	0	0
17	0	0	0	0	0	0	0	0	0	0	0	1	0	1	0	0	0	0	0
18	0	0	0	0	0	0	0	0	0	0	0	0	0	1	0	0	0	1	0
19	0	0	0	0	0	0	0	0	0	0	0	0	0	1	0	0	0	0	0
20	0	0	0	0	0	0	0	0	0	0	0	0	0	1	0	0	0	0	0
21	0	0	0	0	0	0	0	0	0	0	0	0	0	0	1	1	1	0	0
22	0	0	0	0	0	0	0	0	0	0	0	0	0	0	0	0	1	1	1
23	0	0	0	0	0	0	0	0	0	0	0	0	0	0	1	1	0	0	0
24	0	0	0	0	0	0	0	0	0	0	0	0	0	0	0	0	1	0	

1 suspended detritus
2 phytoplankton
3 periphyton
4 *Utricularia*
5 submerged macrophytes (*Blyxa, Cryptocoryne, Nitella*)
6 sedimented detritus
7 *Pandanus, Lepironia*
8 swamp forest
9 bacteria, fungi

10 zooplankton
11 emergent herbivorous insects
12 benthic herbivores
13 herbivorous fishes
14 invertebrate defoliators
15 vertebrate herbivores
16 shrimps
17 detritivorous invertebrates
18 detritivorous fishes
19 benthic carnivores

20 carnivorous invertebrates
21 emergent carnivorous insects
22 carnivorous fishes
23 spiders
24 frogs
25 swallows
26 snakes, gharial, turtles
27 Kingfisher, Teal

86 Suruga Bay, epipelagic zone, Japan

86	3	4	5	6	7	8	9	10	11	12	13	14	15	16
1	1	1	0	0	0	0	0	0	0	0	0	0	0	0
2	1	1	1	0	0	0	0	0	0	0	0	0	0	0
3	0	0	0	0	1	0	0	1	0	0	1	0	0	0
4	0	0	0	1	1	1	1	0	1	0	0	0	0	0
5	0	0	0	1	1	1	1	0	0	0	0	0	0	0
6	0	0	0	0	1	0	0	0	0	0	0	0	0	0
7	0	0	0	0	0	0	0	1	1	0	1	1	0	1
8	0	0	0	0	0	0	0	0	0	0	0	1	0	0
9	0	0	0	0	0	0	0	0	1	1	1	1	0	1
10	0	0	0	0	0	0	0	0	0	0	1	0	1	0
11	0	0	0	0	0	0	0	0	0	0	0	0	0	1
12	0	0	0	0	0	0	0	0	0	0	0	1	0	1
13	0	0	0	0	0	0	0	0	0	0	0	0	0	1
14	0	0	0	0	0	0	0	0	0	0	0	0	0	1
15	0	0	0	0	0	0	0	0	0	0	0	0	0	1

1 detritus
2 phytoplankton
3 *Euphausia similis, E. pacifica*
4 *Calanus pacificus*
5 *Paracalanus gracilis*
6 *Sagitta nagae*
7 *Sergia lucens*
8 *Parathemisto gracilis*

9 *Engraulis japonica* (post-larva)
10 *Diaphus coeruleus, D. elucens*, other myctophids
11 *Trachiurus japonica* (adult)
12 *Engraulis japonica* (adult)
13 *Todarodes pacificus*, other cephalopods
14 *Scomber japonicus* (adult)
15 *Stenella* spp.
16 man

87 Ice edge community, high Arctic, Canada

87	3	4	5	6	7	8	9	10	11	12
1	1	1	1	0	0	0	0	0	0	0
2	0	0	0	0	0	0	0	0	1	0
3	0	0	1	1	0	0	0	1	1	0
4	0	0	0	0	1	1	0	0	0	1
5	0	0	0	1	1	1	0	0	0	0
6	0	0	0	0	1	1	1	1	1	1

1 diatoms
2 marine mammal carcasses
3 calanoid copepods
4 epontic copepods
5 *Parathemisto* spp. (amphipods)
6 arctic cod

7 murres
8 ringed seals
9 narwhals
10 kittiwakes
11 fulmars
12 guillemots

88 Lestijoki River rapids, Finland

88	5	6	7	8	9	10	11	12	13	14	15	16
1	1	1	1	1	0	1	0	1	1	0	0	1
2	0	0	1	1	0	1	0	0	0	0	0	0
3	1	1	1	1	1	0	0	0	0	0	0	0
4	0	1	1	1	0	0	1	0	0	0	0	0
5	0	0	0	0	0	0	0	1	1	1	1	1
6	0	0	0	0	0	0	0	1	0	0	1	1
7	0	0	0	0	0	0	0	1	0	0	1	1
8	0	0	0	0	0	0	0	1	1	1	1	1
9	0	0	0	0	0	0	0	0	0	0	1	1
10	0	0	0	0	0	0	0	0	1	0	1	0
11	0	0	0	0	0	0	0	0	0	1	1	0

1 detritus
2 mosses
3 filamentous algae
4 diatoms
5 *Baetis rhodani, B. vernus*
6 *Rheotanytarsus, Eukiefferiella tshernovskii*
7 *Amphinemura, Ephemerella, Heptagenia*
8 Simulidae, *Cricotopus, triannulatus, Orthocladius*

9 *Agraylea multipunctata, Hydroptila tineoides*
10 *Micrasema*
11 *Cricotopus bicinctus*
12 *Isoperla obscura*
13 *Athripsodes*
14 *Erpobdella octolucata*
15 *Rhyacophila nubila*, Pentaneurini spp.
16 *Hydropsyche pellucidula*

89 River Cam, England

89	5	6	7	8	9	10	11	12	13	14	15	16	17	18
1	0	0	0	0	0	0	0	0	0	1	0	0	0	0
2	0	0	0	0	0	0	0	1	0	0	0	0	0	0
3	1	1	1	1	0	0	0	0	0	0	0	1	0	0
4	0	1	1	1	1	1	0	0	0	0	0	0	0	0
5	0	0	0	0	0	0	0	0	0	1	1	1	0	1
6	0	0	0	0	0	0	1	1	1	0	1	0	0	0
7	0	0	0	0	0	0	0	0	0	0	1	0	0	0
8	0	0	0	0	0	0	0	0	0	1	0	0	0	0
9	0	0	0	0	0	0	0	0	1	0	1	0	0	0
10	0	0	0	0	0	0	0	1	0	1	0	0	0	0
12	0	0	0	0	0	0	0	0	0	0	0	1	0	1
13	0	0	0	0	0	0	0	0	0	0	0	0	0	1
14	0	0	0	0	0	0	0	0	0	0	0	0	1	0
15	0	0	0	0	0	0	0	0	0	0	0	0	0	1
16	0	0	0	0	0	0	0	0	0	0	0	0	0	1

1 *Ulothrix*
2 *Spirogyra*
3 plant fragments
4 *Synedra, Coscinodiscus*, other diatoms
5 chironomid larvae
6 *Simulium* larvae (black fly)
7 Trichoptera larvae
8 *Limnophilus*, Ephemeropteran nymph
9 *Baetis*

10 *Ephemera* nymph
11 brown trout, dace
12 gudgeon
13 minnow
14 roach
15 loach
16 three-spined stickleback
17 eel
18 pike

90 Old field, New Jersey

90	5	6	7	8	9	10	11	12	13	14	15	16	17	18	19	20	21	22
1	1	1	1	1	1	0	0	0	0	0	0	0	0	0	0	0	0	0
2	0	0	1	1	1	1	1	1	1	1	1	0	0	0	0	0	0	0
3	0	0	1	0	0	0	0	0	1	1	1	0	0	0	0	0	0	0
4	0	0	0	0	0	0	0	0	0	1	0	1	0	0	0	0	0	0
5	0	0	0	0	0	0	0	0	0	0	0	0	0	1	0	0	0	0
6	0	0	0	0	0	0	0	0	0	0	0	0	1	1	0	0	0	0
7	0	0	0	0	0	0	0	0	0	0	0	0	0	1	1	1	0	0
8	0	0	0	0	0	0	0	0	0	0	0	0	0	1	1	1	0	0
9	0	0	0	0	0	0	0	0	0	0	0	0	0	0	1	1	1	0
10	0	0	0	0	0	0	0	0	0	0	0	0	0	0	1	1	1	0
11	0	0	0	0	0	0	0	0	0	0	0	0	0	0	0	1	1	0
12	0	0	0	0	0	0	0	0	0	0	0	0	0	0	0	0	0	1
13	0	0	0	0	0	0	0	0	0	0	0	0	0	0	0	0	0	1

1 *Raphanus* (wild radish)
2 *Ambrosia* (ragweed)
3 radish debris
4 ragweed debris
5 *Macrosteles, Phyllotreta chalbeipennis*
6 *Hyadaphis, Myzus*
7 *Melanoplus*
8 Lepidoptera larvae
9 *Lygus*
10 *Empoasca, Oecanthus*
11 *Philaenus, Scaphytopius, Reuteroscopus*

12 *Chlamydatus, Plagiognathus, Smicronyx*
13 *Trigonorhinus*
14 *Nemobius*
15 *Gryllus*
16 isopods, millipeds
17 *Chauliognathus, Coccinella*
18 *Harpalus*
19 *Nabis*
20 web spiders
21 ground spiders
22 *Coleomegilla, Sinea*

91 Shigayama coniferous forest, Japan

91	3	4	5	6	7	8	9	10
1	1	0	1	0	0	0	0	0
2	0	1	0	0	0	0	0	0
3	0	0	1	1	0	0	0	0
4	0	0	0	0	1	0	1	0
5	0	0	0	0	0	1	0	1
6	0	0	0	0	0	1	0	1
7	0	0	0	0	0	0	1	0
8	0	0	0	0	0	0	0	1

1 soil organic matter
2 leaves
3 bacteria, fungi
4 defoliating invertebrates, small rodents
5 decomposing soil fauna
6 bacterial and fungal feeders
7 predatory invertebrates
8 predatory soil invertebrates
9 birds
10 insectivores

92 High Himalayas community, Tibet

92	4	5	6	7	8	9	10	11	12	13	14	15	16	17	18
1	1	1	0	0	0	0	0	0	0	0	0	0	0	0	0
2	0	0	1	1	1	1	1	0	0	0	0	0	0	0	0
3	0	0	0	0	0	0	0	1	0	0	0	0	0	0	0
5	0	0	0	0	0	0	0	0	1	1	0	0	0	0	0
6	0	0	0	0	0	0	0	0	0	0	1	0	0	0	0
7	0	0	0	0	0	0	0	0	0	0	0	1	0	0	0
8	0	0	0	0	0	0	0	0	0	0	0	1	1	0	0
9	0	0	0	0	0	0	0	0	0	0	0	0	1	0	0
12	0	0	0	0	0	0	0	0	0	0	0	0	0	0	1	1
13	0	0	0	0	0	0	0	0	0	0	0	0	0	0	0	1

1 wind-blown organic debris, pollen
2 flowering plants
3 dead carcasses
4 *Machilanus* (flea)
5 fungi
6 snow partridge
7 *Ochotona ladacensis* (pika), mice
8 *Pseudois nahura* (bharal sheep)
9 yak

10 bumblebees, butterflies, aphids, weevils
11 lammergeier, yellow-billed chough
12 springtails
13 anthomyiid fly
14 *Mustela altaica* (weasel)
15 wolf, fox
16 snow leopard
17 mites, centipedes
18 salticid spider

93 Alpine tundra, Montana

93	2	3	4	5	6	7	8	9	10	11	12	13	14	15	16	17	18	19	20	21	22	23	24	25	26
1	1	1	1	1	1	1	1	1	1	1	1	1	1	0	0	0	0	0	1	0	0	0	0	0	0
2	0	0	0	0	1	1	1	1	1	0	0	0	0	0	1	0	1	0	1	0	1	1	1	1	0
3	0	0	0	0	0	0	0	0	0	0	0	0	0	0	1	0	1	1	1	0	1	1	0	1	0
4	0	0	0	0	0	0	0	0	0	0	0	0	0	0	0	0	1	1	1	0	0	1	0	0	0
5	0	0	0	0	0	0	0	0	0	0	0	0	0	0	0	0	0	0	0	0	1	0	0	1	0
6	0	0	0	0	0	0	0	0	0	0	0	0	0	0	0	1	0	0	0	1	0	0	0	0	0
7	0	0	0	0	0	0	0	0	0	0	0	0	0	0	0	1	1	0	1	1	1	1	1	1	0
8	0	0	0	0	0	0	0	0	0	0	0	0	0	0	0	1	1	0	0	1	1	1	1	1	0
9	0	0	0	0	0	0	0	0	0	0	0	0	0	0	1	0	0	0	1	0	1	1	1	1	0
10	0	0	0	0	0	0	0	0	0	0	0	1	0	1	1	0	1	0	1	0	1	1	1	1	0
11	0	0	0	0	0	0	0	0	0	0	1	0	0	1	0	0	0	0	0	1	0	0	0	0	0
12	0	0	0	0	0	0	0	0	0	0	0	0	1	0	0	0	0	0	0	1	0	0	0	0	0
13	0	0	0	0	0	0	0	0	0	0	0	0	0	0	0	0	0	0	0	0	1	1	1	1	0
14	0	0	0	0	0	0	0	0	0	0	0	0	0	0	0	0	0	0	0	0	0	0	0	0	1
15	0	0	0	0	0	0	0	0	0	0	0	0	0	0	0	0	0	0	0	0	0	0	0	0	1

1 alpine vegetation
2 pocket gopher
3 marmot
4 bighorn sheep, wapiti, mule deer
5 jack rabbit
6 pika
7 voles (4 species)
8 chipmunk
9 mantled ground squirrel
10 invertebrates
11 Diptera, Hymenoptera, Hemiptera, Coleoptera
12 deer mouse
13 shrews (3 species)

14 Black Rosy finch
15 water pipits, horned larks
16 badger
17 marten
18 wolverine
19 Golden eagle
20 grizzly bear
21 weasels (2 species)
22 hawks and owls (7 species)
23 coyote
24 bobcat
25 red fox
26 prairie falcon

94 Wet coastal tundra, Barrow, Alaska

94	4	5	6	7	8	9	10	11	12
1	1	0	0	0	0	1	0	0	0
2	0	0	0	0	0	1	0	0	0
3	0	1	1	0	0	0	0	0	0
4	0	0	0	0	0	0	1	1	1
6	0	0	0	1	1	1	0	0	0
7	0	0	0	0	1	1	0	0	0
8	0	0	0	0	0	0	1	0	1
9	0	0	0	0	0	0	1	0	1
10	0	0	0	0	0	0	0	1	1

1 monocots
2 dicots
3 detritus, organic matter
4 lemmings
5 microorganisms
6 saprovores
7 carnivorous arthropods
8 shorebirds
9 longspurs
10 weasels
11 owls
12 jaegers

95 Tundra, Prudhoe, Alaska

95	3	4	5	6	7	8	9	10
1	0	1	1	1	0	0	0	0
2	1	0	0	0	1	0	0	0
3	0	0	0	0	1	0	0	0
4	0	0	0	0	0	0	1	1
5	0	0	0	0	0	1	0	0
7	0	0	0	0	0	0	1	1
9	0	0	0	0	0	0	0	1

1 grasses, sedges, willows
2 decaying organic matter
3 bacteria, fungi
4 adult flies
5 waterfowl, lemmings, squirrels
6 caribou
7 fly larvae
8 owls, jaegers, fox, weasels
9 beetles, spiders
10 sandpipers

96 Tundra, Yamal Peninsula, Siberia

96	2	3	4	5	6	7	8	9
1	1	1	1	1	1	0	0	0
2	0	0	0	0	1	1	1	1
3	0	0	0	0	1	1	0	0
4	0	0	0	0	1	1	0	1
5	0	0	0	0	1	0	0	1

1 tundra vegetation, berries
2 *Dicrostonyx*, Passeres
3 insects
4 *Lagopus, Lemnus*
5 Limicolae, Anseres, *Microtus* (2 sp.), *Arvicola*
6 *Stercorarius longicaudus* (jaeger)
7 *Stercorarius parasiticus*
8 *Falco columbarius*
9 *Buteo lagopus, Falco peregrinus*

97 Tundra, South Yamal, Siberia

97	3	4	5	6	7	8	9	10	11
1	1	1	1	0	0	0	0	0	0
2	0	0	1	0	0	0	0	1	0
3	0	0	0	0	0	0	1	0	0
4	0	0	0	1	1	1	1	1	1
5	0	0	0	0	0	1	1	1	1
6	0	0	0	0	0	0	0	0	1

1 tundra vegetation
2 berry crop
3 caribou
4 microtine mammals
5 ptarmigan
6 least weasel
7 arctic fox, red fox, ermine, merlin
8 peregrine
9 white-tailed eagle
10 long-tailed jaeger, parasitic jaeger, herring gull
11 rough-legged hawk

98 Sand dunes, Namib Desert, Namibia

98	5	6	7	8	9	10	11	12	13	14	15	16	17
1	1	1	0	0	0	1	0	0	0	0	0	0	0
2	1	0	1	0	0	1	0	0	0	0	0	0	0
3	0	0	0	1	0	0	0	0	1	0	0	0	0
4	0	0	0	0	1	0	0	0	0	0	0	0	0
5	0	0	0	0	0	1	1	1	0	1	0	1	1
7	0	0	0	0	0	0	1	1	0	0	0	0	0
8	0	0	0	0	0	0	1	1	1	1	0	1	1
9	0	0	0	0	0	0	1	0	0	1	0	0	1
10	0	0	0	0	0	0	0	0	0	0	0	0	1
11	0	0	0	0	0	0	0	1	1	1	0	0	0
12	0	0	0	0	0	0	0	0	0	0	0	1	1
13	0	0	0	0	0	0	0	0	0	0	1	1	1
14	0	0	0	0	0	0	0	0	0	0	1	1	1
15	0	0	0	0	0	0	0	0	0	0	0	0	1

1 annuals (*Monsonia, Stipagrostis, Eragrostis*)
2 perennials
3 wind-blown detritus
4 animal dung
5 Tenebrionidae, Orthoptera, Curculionidae
6 oryx, hare
7 *Aclerda*
8 Thysanura, Isoptera, other Tenebrionidae
9 scarabs
10 *Gerbillus*
11 spiders, solpugids, scorpions
12 mole
13 *Aporosaura*
14 *Typhlosaurus*, lizards
15 snakes
16 jackal, hyena
17 birds

99 Sonora Desert, Arizona

99	10	11	12	13	14	15	16	17	18	19	20	21	22	23	24	25	26	27	28	29
1	1	1	1	1	1	1	1	0	0	0	0	0	0	0	0	0	0	0	0	0
2	0	0	0	0	0	1	1	1	1	1	1	0	0	0	0	0	0	0	0	0
3	1	1	0	1	1	0	1	0	1	0	0	0	0	0	0	0	0	0	0	0
4	0	0	0	0	0	0	1	0	0	0	0	0	0	0	0	0	0	0	0	0
5	0	0	0	0	0	0	0	1	0	0	1	0	0	0	0	0	0	0	0	0
6	0	0	1	0	0	0	0	0	0	0	0	0	0	0	0	0	0	0	1	0
7	0	0	1	0	0	0	0	0	0	0	0	1	0	0	0	0	0	1	0	0
8	0	0	0	0	0	0	0	0	0	0	0	0	1	1	1	0	0	0	0	0
9	0	0	0	0	0	0	0	0	0	0	0	0	0	0	0	1	1	0	0	0
10	0	0	0	0	0	0	0	0	0	0	0	0	0	0	0	0	0	0	0	0
11	0	0	0	0	0	0	0	0	0	0	0	0	0	0	0	0	0	0	0	0
13	0	0	0	0	0	0	0	0	0	0	0	0	0	0	0	0	0	1	1	0
14	0	0	0	0	0	0	0	0	0	0	0	0	0	0	0	0	0	0	1	0
15	0	0	0	0	0	0	0	0	0	0	0	0	0	0	0	0	0	0	0	0
16	0	0	0	0	0	0	0	0	0	0	0	0	0	0	0	0	0	0	0	0
17	0	0	0	0	0	0	0	0	0	0	0	0	0	0	0	0	0	0	0	0
18	0	0	0	0	0	0	0	0	0	0	0	0	0	0	0	0	0	0	0	0
19	0	0	0	0	0	0	0	0	0	0	0	0	0	0	0	0	0	0	1	1
20	0	0	0	0	0	0	0	0	0	0	0	0	0	0	0	0	0	0	0	0
21	0	0	0	0	0	0	0	0	0	0	0	0	0	0	0	0	0	1	1	0
23	0	0	0	0	0	0	0	0	0	0	0	0	0	0	0	0	0	0	0	0
24	0	0	0	0	0	0	0	0	0	0	0	0	0	0	0	0	0	0	0	0
25	0	0	0	0	0	0	0	0	0	0	0	0	0	0	0	0	0	0	0	0
27	0	0	0	0	0	0	0	0	0	0	0	0	0	0	0	0	0	0	0	0
28	0	0	0	0	0	0	0	0	0	0	0	0	0	0	0	0	0	0	0	0
34	0	0	0	0	0	0	0	0	0	0	0	0	0	0	0	0	0	0	0	0
35	0	0	0	0	0	0	0	0	0	0	0	0	0	0	0	0	0	0	0	0
37	0	0	0	0	0	0	0	0	0	0	0	0	0	0	0	0	0	0	0	0
38	0	0	0	0	0	0	0	0	0	0	0	0	0	0	0	0	0	0	0	0
40	0	0	0	0	0	0	0	0	0	0	0	0	0	0	0	0	0	0	0	0
43	0	0	0	0	0	0	0	0	0	0	0	0	0	0	0	0	0	0	0	0

1 grass *Schismus barbatus*
2 cacti (including seeds and fruits)
3 seeds of other plants
4 creosote bush
5 Palo Verde
6 mistletoe (berries)
7 leaves
8 nectar
9 animal carcasses
10 pocket mouse
11 doves, Palmer's thrasher, sage sparrow, Lark bunting, House finch, goldfinch, Gambel sparrow
12 white-tailed deer
13 harvester ants
14 crickets, grasshoppers, caterpillars
15 cottontails
16 ground squirrels
17 wood rats
18 kangaroo rats
19 cactus beetles, cactus weevils
20 Palo Verde weevil
21 leaf-cutting ants
22 hummingbirds
23 honeybees
24 butterflies, moths

(Web 99 cont.)

30	31	32	33	34	35	36	37	38	39	40	41	42	43	44	45	46	47	48
0	0	0	0	0	0	0	0	0	0	0	1	0	0	0	0	0	0	1
0	0	0	0	0	0	0	0	0	0	0	1	0	0	0	0	0	0	0
0	0	0	0	1	0	0	0	0	0	0	1	0	0	0	1	0	0	1
0	0	0	0	0	0	0	0	0	0	0	0	0	0	0	0	0	0	0
0	0	0	0	0	0	0	0	0	0	0	0	0	0	0	0	0	0	0
0	0	0	0	1	1	0	0	0	0	0	1	0	0	0	0	0	0	0
0	0	0	0	0	0	0	0	0	0	0	0	0	0	0	0	0	0	1
0	0	0	0	0	0	0	0	0	0	0	0	0	0	0	0	0	0	0
0	0	0	0	0	0	0	0	0	0	0	0	0	0	0	1	0	0	1
1	0	0	0	0	0	1	0	0	0	1	1	0	0	0	1	0	0	1
1	0	0	0	0	0	1	0	0	0	1	1	0	0	0	1	0	1	0
0	0	0	0	1	0	0	0	1	0	0	0	0	0	0	1	0	0	0
0	1	1	0	1	0	0	0	0	0	0	0	1	0	0	1	0	0	0
0	0	0	0	0	0	1	0	0	0	0	1	0	0	1	0	0	0	1
1	0	0	0	0	0	0	0	0	0	1	0	1	0	1	0	0	1	1
1	0	0	1	0	0	0	0	0	0	1	1	1	0	0	1	0	1	1
1	0	0	1	0	0	1	0	0	0	1	1	1	0	0	1	0	1	1
0	1	1	0	1	1	0	1	1	0	0	1	0	0	0	1	0	0	0
0	0	1	0	0	0	0	0	1	0	0	0	0	0	0	0	0	0	0
0	0	0	0	0	0	0	0	0	0	0	1	0	0	0	0	0	0	0
0	0	0	0	0	0	0	1	0	0	0	0	0	0	0	0	0	0	0
0	0	0	0	1	0	0	1	0	0	0	0	0	0	0	0	0	0	0
0	0	1	0	0	0	0	1	1	0	0	0	0	0	0	1	0	0	0
0	0	0	0	0	0	0	0	0	1	1	0	0	0	1	1	0	0	0
0	0	0	0	0	0	0	0	0	0	0	1	0	0	0	1	0	1	1
0	0	0	0	0	0	0	0	0	0	0	0	0	0	0	0	0	1	1
0	0	0	0	0	0	0	0	0	0	0	0	0	0	0	0	0	1	1
0	0	0	0	0	0	0	0	0	0	0	0	0	0	0	0	0	1	1
0	0	0	0	0	0	0	0	0	0	1	1	1	1	0	1	0	0	0
0	0	0	0	0	0	0	0	0	0	0	0	1	0	0	0	1	0	0
0	0	0	0	0	0	0	0	0	0	0	0	0	0	0	0	0	0	1

25 flies
26 black vulture, turkey vulture, raven
27 uta, horned lizard, gecko
28 sage thrasher, robin
29 cactus woodpecker
30 badger
31 cactus wren
32 flycatcher
33 Horned owl
34 backbirds, mockingbird, oriole, cardinal
35 Blue bird
36 Sparrow hawk
37 swift
38 scorpion, spiders
39 Race runner
40 Bull snake, Red racer, rattlesnakes
41 Gray fox
42 skunk
43 grasshopper mice
44 Red-tailed hawk
45 Road runner
46 Coral snake
47 bobcat
48 coyote

100 Rajasthan Desert, India

100	6	7	8	9	10	11	12	13	14	15	16	17	18	19	20	21	22
1	1	1	1	1	0	0	1	0	0	0	0	1	0	0	0	0	0
2	0	1	1	0	0	0	1	0	0	1	0	0	0	0	0	1	0
3	0	0	1	0	0	0	0	0	0	0	0	0	0	0	0	0	0
4	0	0	0	0	1	0	0	0	0	1	0	0	0	0	0	0	1
5	0	0	0	0	0	0	0	1	0	0	0	0	0	0	0	0	0
6	0	0	0	0	0	1	1	0	1	0	1	0	0	1	0	0	0
7	0	0	0	0	0	0	0	0	0	1	0	0	1	1	1	1	0
8	0	0	0	0	0	0	0	0	0	0	0	0	0	0	0	1	1
9	0	0	0	0	0	0	0	0	1	0	1	0	1	0	1	0	0
11	0	0	0	0	0	0	0	1	1	1	1	1	0	1	0	0	0
12	0	0	0	0	0	0	0	0	1	0	1	0	1	1	1	1	1
13	0	0	0	0	0	0	0	0	1	0	0	0	1	1	0	0	1
14	0	0	0	0	0	0	0	1	0	0	0	0	1	0	0	1	1
15	0	0	0	0	0	0	0	0	0	0	0	0	0	0	0	1	0
16	0	0	0	0	0	0	0	0	0	0	0	0	1	0	0	0	0
17	0	0	0	0	0	0	0	0	0	0	0	0	1	0	0	0	1
18	0	0	0	0	0	0	0	0	0	0	0	0	0	0	0	0	1

1 *Cyperus, Cenchrus, Eleucine*
2 *Crotalaria, Zizyphus*
3 *Prosopis cineraria*
4 animal carcasses
5 animal dung
6 hoppers, ants, termites, beetles
7 gerbils, hares
8 antelopes, gazelle, backbuck, nilgai
9 doves, larks, sandgrouse
10 vultures, crows
11 spiders, wasps, tiger beetles, carpenter ants
12 rodents
13 partridge, peafowl, babbler
14 snakes, varanids
15 jackal *Canis aureaus*
16 hedgehog
17 bultul, shrike, Indian robin
18 desert cat *Felis libyca*
19 fox *Vulpes vulpes*
20 shikra
21 wolf *Canis lupus*
22 stray dog *Canis familiaris*

101 Temporary freshwater rockpool, France

101	3	4	5	6
1	1	1	0	0
2	0	0	1	0
4	0	0	0	1
5	0	0	0	1

1 organic material
2 small algae
3 *Heterocypris, Herpetocypris*
4 *Chironomus, Anopheles, Culex, Theobaldia*
5 *Sigara, Haliplus*
6 *Meladema, Graptodytes, Bidessus, Notonecta*

102 Plankton, oligotrophic tropical Pacific

102	3	4	5	6	7	8	9
1	1	1	1	1	0	0	0
2	0	0	1	1	1	1	0
3	0	1	1	1	1	0	0
4	0	0	1	1	1	1	1
5	0	0	0	1	1	1	1
6	0	0	0	0	1	1	1
7	0	0	0	0	0	1	1
8	0	0	0	0	0	0	1

1 detritus
2 phytoplankton
3 bacteria
4 protozoa
5 nauplii, small sized Euphausiacea,
   Mysidacea, Hyperiidea, Ostracoda
6 large sized Euphausiacea, Mysidacea,
   Hyperiidea, Ostracoda
7 Cyclopoida
8 carnivorous Calanoida
9 Chaetognatha, Polychaeta

103 Tropical plankton community, Pacific

103	5	6	7	8	9	10	11	12	13	14	15	16	17	18	19	20	21	22	23
1	1	1	1	1	1	1	1	1	1	1	0	1	1	1	1	1	1	0	1
2	0	1	1	1	1	1	1	1	1	0	0	0	0	0	0	1	1	0	0
3	0	1	1	0	0	1	1	1	1	0	0	0	0	0	0	1	1	0	0
4	0	1	0	0	0	0	1	0	0	1	0	1	1	1	1	1	0	0	0
5	0	1	1	1	1	1	1	1	1	1	0	1	1	1	1	0	0	0	1
6	0	0	0	0	0	0	0	0	1	0	0	1	0	1	0	0	1	0	0
7	0	0	0	1	0	0	0	1	1	1	1	1	1	1	1	0	1	0	0
8	0	0	0	0	1	0	0	0	1	0	0	0	1	0	1	0	1	0	0
9	0	0	0	0	0	0	0	0	1	0	0	0	1	0	0	1	1	0	0
10	0	0	0	0	0	0	0	0	1	0	1	0	1	0	1	1	0	0	0
11	0	0	0	0	0	0	0	0	1	1	1	1	1	1	0	1	1	1	1
12	0	0	0	0	0	0	0	0	1	1	1	0	1	1	1	1	0	1	1
13	0	0	0	0	0	0	0	0	0	1	1	1	0	0	1	0	1	1	0
14	0	0	0	0	0	0	0	0	0	0	1	0	1	0	0	0	0	1	1
15	0	0	0	0	0	0	0	0	0	0	0	0	0	0	1	1	0	1	1
16	0	0	0	0	0	0	0	0	0	1	0	0	0	0	0	0	0	0	0
17	0	0	0	0	0	0	0	0	0	0	0	0	0	1	1	1	1	1	1
18	0	0	0	0	0	0	0	0	0	0	0	0	0	0	0	0	0	1	1
19	0	0	0	0	0	0	0	0	0	0	0	0	0	0	0	0	0	1	1
20	0	0	0	0	0	0	0	0	0	0	0	0	0	0	0	0	0	1	1
21	0	0	0	0	0	0	0	0	0	0	0	0	0	0	0	0	0	1	1

1 detritus
2 small-size phytoplankton
3 medium-size phytoplankton
4 large-size phytoplankton
5 bacteria
6 large-size Appendicularia
7 nauplii of Copepoda
8 small-size Appendicularia

9 Infusoria
10 Radiolaria
11 copepodites
12 small-size 'Calanus type'
13 small-size 'Acartia type'
14 small-size 'Oithona-Oncaea type'
15 large-size 'Oithona-Oncaea type'
16 large-size 'Acartia type'

17 large-size 'Calanus type'
18 'Euchaeta type'
19 'Centropages type'
20 'Amphipoda type'
21 'Euphausia type'
22 Chaetognatha
23 Medusae, Ctenophora

Note: For technical reasons the sequence of food webs 104-107 has been changed.

106 Rocky shore, Monterey Bay, California

106	6	7	8	9	10	11	12	13	14	15	16	17	18	19	20	21	22	23	24
1	0	0	0	0	0	0	0	0	0	0	1	1	1	1	1	1	1	1	0
2	1	1	1	1	1	1	1	1	1	1	1	0	0	0	0	0	0	0	0
3	0	0	0	0	0	0	0	0	0	0	1	0	0	0	0	0	0	0	0
4	0	0	0	0	0	0	0	0	0	1	0	0	0	0	0	0	0	0	1
5	0	0	0	0	0	0	0	0	0	0	0	0	1	1	1	1	1	1	0
6	0	0	0	0	0	0	0	0	0	0	0	0	0	0	0	0	0	0	0
7	0	0	0	0	0	0	0	0	0	0	0	0	0	0	0	0	0	0	0
8	0	0	0	0	0	0	0	0	0	0	0	0	0	0	0	0	0	0	0
9	0	0	0	0	0	0	0	0	0	0	0	0	0	0	0	0	0	0	0
10	0	0	0	0	0	0	0	0	0	0	0	0	0	0	0	0	0	0	0
11	0	0	0	0	0	0	0	0	0	0	0	0	0	0	0	0	0	0	0
12	0	0	0	0	0	0	0	0	0	0	0	0	0	0	0	0	0	0	0
13	0	0	0	0	0	0	0	0	0	0	0	0	0	0	0	0	0	0	0
14	0	0	0	0	0	0	0	0	0	0	0	0	0	0	0	0	0	0	0
15	0	0	0	0	0	0	0	0	0	0	0	0	0	0	0	0	0	0	0
16	0	0	0	0	0	0	0	0	0	0	0	0	0	0	0	0	0	0	0
18	0	0	0	0	0	0	0	0	0	0	0	0	0	1	1	1	1	1	0
19	0	0	0	0	0	0	0	0	0	0	0	0	0	0	0	0	0	0	0
20	0	0	0	0	0	0	0	0	0	0	0	0	0	0	0	0	0	0	0
21	0	0	0	0	0	0	0	0	0	0	0	0	0	0	0	0	0	0	0
22	0	0	0	0	0	0	0	0	0	0	0	0	0	0	0	0	0	0	0
25	0	0	0	0	0	0	0	0	0	0	0	0	0	0	0	0	0	0	0
28	0	0	0	0	0	0	0	0	0	0	0	0	0	0	0	0	0	0	0
30	0	0	0	0	0	0	0	0	0	0	0	0	0	0	0	0	0	0	0
35	0	0	0	0	0	0	0	0	0	0	0	0	0	0	0	0	0	0	0

1 detritus, plankton
2 diatoms, blue-green algae
3 Gigartina agardhii
4 Endocladia muricata
5 other phytoplankton
6 Suidasia sp.
7 Tegula funebralis
8 Littorina planaxis
9 Littorina scutulata
10 Acmaea digitalis

11 Acmaea pelta
12 Acmaea scabra
13 Cyanoplax dientens
14 Dynamenella glabra
15 Allochertes ptilocerus, Hyale sp.
16 Diaulota densissima
17 Pagurus samuelis
18 zooplankton
19 Chthamalus dalli, C. microtretus
20 Lasaea cistula

107 Bay pilings community, New Jersey

107	3	4	5	6	7	8	9	10
1	1	1	1	0	1	0	0	0
2	0	0	0	1	0	0	0	0
3	0	0	0	0	0	1	0	0
4	0	0	0	0	0	1	1	1
5	0	0	0	0	0	0	1	0
6	0	0	0	0	0	1	0	1
8	0	0	0	0	0	0	0	1
9	0	0	0	0	0	0	0	1

1 plankton, suspended detritus
2 seaweeds
3 Balanus balanoides, B. eburneus
4 Modiolus demissus
5 Molgula manhattensis

6 Littorina littorea
7 Schizoporella, Hydroides, Haliplanella, Bugula
8 Urosalpinx cinerea
9 Neopanope texana sayi
10 Callinectes sapidus

(Web 106 cont.)

25	26	27	28	29	30	31	32	33	34	35	36	37
0	0	0	0	0	0	0	0	0	0	0	0	0
0	0	0	1	0	1	0	0	0	0	0	0	0
1	0	0	0	0	0	0	0	0	0	0	0	1
1	0	0	0	0	0	0	0	0	1	0	0	1
0	0	0	0	0	0	0	0	0	0	0	0	0
0	0	0	0	0	0	0	0	1	0	0	0	0
0	0	0	0	0	0	0	1	0	1	0	0	0
0	0	0	0	0	0	0	0	0	1	0	0	0
0	0	0	0	0	0	1	1	0	1	0	0	1
0	1	1	0	0	0	0	0	0	1	0	0	1
0	0	0	0	0	0	0	0	0	1	0	0	1
0	0	1	0	0	0	0	0	0	1	0	0	1
0	0	0	0	0	0	0	0	0	1	0	0	1
0	0	0	0	0	0	0	0	0	1	0	0	1
0	0	0	0	0	0	0	0	0	0	0	0	1
0	0	0	0	0	0	0	0	0	0	0	0	1
0	0	0	0	0	0	0	0	0	0	0	0	0
0	0	0	0	0	0	0	1	0	0	0	0	0
1	0	0	0	1	0	1	0	0	0	0	0	0
0	0	0	0	0	0	1	0	0	0	0	0	0
0	0	0	0	0	1	1	1	0	0	0	1	1
0	0	0	0	0	0	0	0	0	0	0	1	0
0	0	0	0	0	0	0	0	0	1	1	0	0
0	0	0	0	0	0	0	0	0	1	1	0	0
0	0	0	0	0	0	0	0	0	0	0	0	1

21 *Mytilus californianus*
22 *Balanus glandula*
23 *Filicrisia franciscana, Musculus*
24 *Perinereis monterea*
25 *Nereis grubei*
26 *Notoplana acticola*
27 Oyster catcher
28 *Syllis vittata*
29 *Nemertopsis gracilis*
30 *Syllis spenceri*

31 *Thais emarginata*
32 *Acanthina spirata*
33 *Agauopsis* sp., *Rhombognathus*
34 Tipulidae
35 *Pachygrapsus crassipes* (juv.)
36 *Emplectonema gracilis*
37 Black Turnstone

105 Rocky shore, Gulf of Maine, USA

105	3	4	5	6	7	8	9	10
1	1	1	0	0	0	0	0	0
2	0	0	1	1	0	0	0	0
3	0	0	0	0	1	1	1	1
4	0	0	0	0	1	1	1	0
5	0	0	0	0	1	1	1	1
6	0	0	0	0	1	0	0	1
7	0	0	0	0	0	1	1	0
8	0	0	0	0	0	0	1	1
9	0	0	0	0	0	0	0	1

1 plankton, detritus
2 algae
3 *Mytilus*
4 *Balanus*
5 *Littorina*

6 *Acmaea*
7 *Thais*
8 *Carcinus*
9 *Tautogolabrus*
10 fish, birds, mammals

104 Rocky shore, Bay of Panama

104	3	4	5	6	7	8	9	10	11	12	13	14	15	16	17	18	19	20	21	22	23	24	25	26	27
1	1	1	1	1	1	0	0	0	0	0	0	0	1	1	0	0	0	0	0	0	0	0	0	0	1
2	0	0	0	0	0	1	1	1	1	1	1	1	0	0	0	0	0	0	0	0	0	0	0	0	0
3	0	0	0	0	0	0	0	0	0	0	0	0	0	0	0	0	0	0	1	0	0	0	0	1	0
4	0	0	0	0	0	0	0	0	0	0	0	0	0	0	0	0	1	0	1	0	0	1	0	0	1
5	0	0	0	0	0	0	0	0	0	0	0	0	0	0	0	0	0	0	1	0	0	1	1	1	1
6	0	0	0	0	0	0	0	0	0	0	0	0	0	0	0	0	0	0	0	0	0	0	0	0	0
7	0	0	0	0	0	0	0	0	0	0	0	0	0	0	0	0	0	0	0	0	0	0	0	0	1
8	0	0	0	0	0	0	0	0	0	0	0	0	0	0	1	1	0	0	0	1	0	0	0	0	1
9	0	0	0	0	0	0	0	0	0	0	0	0	0	0	1	1	1	0	0	1	0	1	0	1	1
10	0	0	0	0	0	0	0	0	0	0	0	0	0	0	1	0	1	1	0	0	1	1	0	1	1
11	0	0	0	0	0	0	0	0	0	0	0	0	0	0	1	0	1	0	0	1	0	1	0	0	0
12	0	0	0	0	0	0	0	0	0	0	0	0	0	1	1	0	0	0	0	0	0	1	0	0	1
13	0	0	0	0	0	0	0	0	0	0	0	0	1	0	0	0	0	0	0	1	0	0	0	0	0
14	0	0	0	0	0	0	0	0	0	0	0	0	0	0	0	0	0	0	0	0	0	0	0	0	0
15	0	0	0	0	0	0	0	0	0	0	0	0	0	0	0	0	0	0	0	0	1	0	1	0	1
17	0	0	0	0	0	0	0	0	0	0	0	0	0	0	0	0	0	0	0	0	0	1	0	0	1
19	0	0	0	0	0	0	0	0	0	0	0	0	0	0	0	0	0	0	0	0	0	1	0	0	0
20	0	0	0	0	0	0	0	0	0	0	0	0	0	0	0	0	0	0	0	0	0	1	0	0	0
22	0	0	0	0	0	0	0	0	0	0	0	0	0	0	0	0	0	0	0	0	0	1	1	1	1
24	0	0	0	0	0	0	0	0	0	0	0	0	0	0	0	0	0	0	0	0	0	0	0	1	1
25	0	0	0	0	0	0	0	0	0	0	0	0	0	0	0	0	0	0	0	0	0	0	0	0	1

1 algae
2 plankton, detritus
3 nerites
4 chitons
5 limpets
6 littorines
7 crabs
8 solitary tunicates
9 bivalves

10 barnacles
11 sipunculids
12 vermetids
13 colonial sessile invertebrates
14 serpulid polychaete
15 sea urchins
16 herbivorous fish
17 Thais triangularis
18 Muricanthus

19 Acanthina
20 Thais biserialis
21 Purpura
22 Opeatostoma
23 Leucozonia
24 Thais melones
25 predaceous crabs
26 Heliaster
27 predaceous fish

108 Rocky shore, Cabrillo Point, California

108	5	6	7	8	9	10	11	12	13	14
1	1	0	0	0	0	1	1	0	0	0
2	0	1	1	0	0	0	0	0	0	0
3	0	0	0	0	1	0	0	0	0	0
4	0	0	0	1	0	0	0	0	0	0
5	0	0	0	0	0	1	1	0	0	0
6	0	0	0	0	0	0	0	1	1	1
7	0	0	0	0	0	0	0	0	1	1
8	0	0	0	0	0	0	0	0	0	1
9	0	0	0	0	0	0	0	0	0	1
10	0	0	0	0	0	0	0	0	0	1
11	0	0	0	0	0	0	0	1	1	1

1 phytoplankton, suspended detritus
2 encrusting and mat-forming algae
3 seaweeds
4 detritus
5 zooplankton
6 *Littorina, Tegula*
7 *Acmaea, Amphissa, Columbella*
8 *Pagurus, Pachygrapsus, Halosydna,*
  Cirolana, Pachycheles, Hemigrapsus

9 *Pugettia*
10 *Petrolisthes, Crepidula, Spirorbis*
11 *Balanus*
12 *Acanthina*
13 *Leptasterias*
14 *Cribina*, fishes

109 Rocky shore, central Chile

109	4	5	6	7	8	9	10	11	12	13	14	15	16	17	18	19	20	21
1	1	1	1	0	0	0	0	0	0	0	0	0	0	0	0	0	0	0
2	0	0	0	1	1	1	1	1	0	0	0	0	0	0	0	0	0	0
3	0	0	0	0	0	0	0	0	1	1	0	0	0	0	0	0	0	0
4	0	0	0	0	0	0	0	0	0	0	1	1	1	1	1	0	0	0
5	0	0	0	0	0	0	0	0	0	1	1	1	1	1	0	0	0	0
6	0	0	0	0	0	0	0	0	0	0	1	0	0	0	0	0	0	0
7	0	0	0	0	0	0	0	0	0	0	0	0	0	1	1	1	1	1
8	0	0	0	0	0	0	0	0	0	0	1	0	0	1	1	1	1	1
9	0	0	0	0	0	0	0	0	0	0	1	0	0	1	1	1	0	0
10	0	0	0	0	0	0	0	0	0	0	0	0	0	1	1	1	0	0
11	0	0	0	0	0	0	0	0	0	0	0	0	0	1	1	0	0	0
12	0	0	0	0	0	0	0	0	0	0	1	0	0	1	0	1	0	1
13	0	0	0	0	0	0	0	0	0	0	1	0	0	1	0	0	0	0
14	0	0	0	0	0	0	0	0	0	0	0	0	0	1	1	0	0	0
15	0	0	0	0	0	0	0	0	0	0	0	0	0	1	1	1	1	1
16	0	0	0	0	0	0	0	0	0	0	0	0	0	1	0	1	0	0
17	0	0	0	0	0	0	0	0	0	0	0	0	0	0	0	0	0	1

1 plankton, suspended detritus
2 benthic algae
3 detritus
4 mussels
5 barnacles
6 ascidians
7 keyhole limpets

8 limpets
9 herbivorous snails
10 chitons
11 sea urchins
12 crabs
13 polychaetes, isopods
14 *Crassilabrum crassilabrum*

15 *Concholepas concholepas*
16 *Acanthocyclus* spp.
17 *Sicyases sanguineus*
18 *Heliaster helianthus*
19 *Larus dominicanus*
20 *Haematopus ater*
21 *Lutra felina*

110 Rocky shore, Cape Ann, Massachusetts

110	4	5	6	7	8	9	10	11	12	13
1	1	1	0	0	0	0	0	0	0	0
2	1	0	1	1	0	0	0	0	0	1
3	0	0	0	0	1	1	1	0	1	1
4	0	0	0	0	0	0	0	0	0	1
5	0	0	0	0	0	0	0	0	0	1
6	0	0	0	0	0	0	0	1	1	1
7	0	0	0	0	0	0	0	0	1	1
8	0	0	0	0	0	0	0	0	1	0
10	0	0	0	0	0	0	0	0	1	1
11	0	0	0	0	0	0	0	0	0	1
12	0	0	0	0	0	0	0	0	0	1

1 *Ascophyllum, Fucus, Ulva*, other algae
2 plankton, suspended detritus
3 organic debris
4 *Acmaea, Crepidula*
5 *Littorina littorea, L. obtusata, L. saxatilis*
6 *Mytilus, Balanus*
7 *Sertularia, Metridium, Obelia, Clava*
8 *Spirorbis*
9 *Anurida*
10 *Gammarus, Orchestia*
11 *Thais, Asterias*
12 *Carcinides, Cancer*
13 *Myoxocephalus, Tautogolabrus, Fundulus*

111 Mudflat, Cape Ann, Massachusetts

111	5	6	7	8	9	10	11	12	13	14	15	16	17	18	19
1	1	0	0	0	0	0	0	1	0	0	0	0	0	0	0
2	0	1	1	0	0	0	0	0	0	0	0	0	1	0	0
3	0	0	0	1	1	1	0	0	0	0	0	0	0	0	0
4	0	0	0	0	0	0	1	0	0	0	0	0	0	0	0
5	0	0	0	0	0	0	0	1	1	0	0	1	1	1	0
7	0	0	0	0	0	0	0	0	0	0	1	1	0	0	0
8	0	0	0	0	0	0	0	0	1	0	1	1	1	1	1
9	0	0	0	0	0	0	0	0	0	0	1	1	0	0	1
10	0	0	0	0	0	0	0	0	0	1	0	0	0	0	0
11	0	0	0	0	0	0	0	0	0	1	0	1	0	0	0
12	0	0	0	0	0	0	0	0	0	0	0	1	1	0	0
13	0	0	0	0	0	0	0	0	0	0	1	1	1	0	0
15	0	0	0	0	0	0	0	0	0	0	0	1	1	1	0

1 organic matter
2 organic debris
3 plankton and detritus
4 *Chaetomorpha, Ulva*, other algae
5 *Lumbrinereis, Clymenella*
6 *Anurida*
7 *Gammarus, Talorchestia*
8 *Mya, Ensis, Macoma, Solemya*
9 *Gemma*
10 *Fundulus*, fish fry
11 *Littorina littorea, Onoba*
12 *Crago*
13 *Glycera, Nereis, Cerebratulus*
14 *Megaceryle, Sterna*
15 *Polinices, Nassarius*
16 *Limulus*
17 *Pagurus, Cancer, Carcinides, Myoxocephalus*
18 *Tautogolabrus, Pseudopleuronectes*
19 *Asterias*

112 Low salt marsh, Cape Ann, Massachusetts

112	5	6	7	8	9	10	11	12	13	14
1	1	0	0	0	0	0	0	0	0	0
2	0	1	0	0	0	0	0	0	0	0
3	0	0	1	1	0	0	1	1	0	0
4	0	0	0	0	1	0	0	0	1	0
5	0	0	0	0	0	1	0	0	0	0
6	0	0	0	0	0	0	1	0	0	1
7	0	0	0	0	0	0	1	0	0	0
9	0	0	0	0	0	0	1	1	1	1
13	0	0	0	0	0	0	0	0	0	1

1 *Spartina glabra*
2 *Ascophyllum, Fucus*, other algae
3 plankton, detritus
4 organic debris
5 marsh insects
6 *Littorina littorea, L. obtusata, L. saxatilis*
7 *Mytilus, Balanus*
8 *Brachidontes*
9 *Anurida, Orchestia, Gammarus*
10 spiders, song birds
11 *Fundulus*
12 fish fry
13 *Carcinides, Cancer*
14 *Butorides, Corvus*

113 High salt marsh, Cape Ann, Massachusetts

113	5	6	7	8	9	10	11
1	1	0	0	0	0	0	0
2	0	1	0	0	0	0	0
3	0	0	0	0	1	0	0
4	1	0	1	0	0	0	0
5	0	0	0	1	0	0	1
6	0	0	0	0	1	1	0
7	0	0	0	0	1	1	0
8	0	0	0	0	0	0	1

1 *Spartina patena*
2 algae
3 plankton, detritus
4 organic debris
5 marsh insects
6 *Melampus, Littorina littorea, L. saxatilis*
7 *Orchestia, Philoscia, Anurida, Cylisticus*
8 marsh spiders
9 *Fundulus*, fish fry
10 *Corvus, Pisobia*
11 song birds

References

Auerbach MJ (1984) Stability, probability and the topology of food webs. In: Strong Jr DR, Simberloff D, Abele LG, Thistle AB (eds) Ecological Communities: Conceptual issues and the evidence. Princeton University Press, Princeton, pp 412-436

Aulio K, Jumppanen K, Molsa H, Nevalainen J, Rajasilta M, Vuorinen I (1981) Litoraalin merkitys Pyhajarven kalatuotannolle, Sakylan Pyhajarven Tila Ja Biologinen Tuotanto (Lounais-Suomen Vesiensuojeluyhdistys RY, Turku, Finland) *47*:173-176

Bai ZD, Yin YQ (1988) Convergence to the semicircle law. Annals of Probability *16*:863-875

Baril A (1983) Effect of the water mite Piona constricta on planktonic community structure. M.Sc. Thesis, University of Ottawa, Canada

Beaver RA (1983) The communities living in Nepenthes pitcher plants: fauna and food webs. In: Frank JH, Lounibos LP (eds) Phytotelmata: Terrestrial plants as hosts for aquatic insect communities. Plexus, Medford, NJ, pp 129-159

Beaver RA (1985) Geographical variation in food web structure in Nepenthes pitcher plants. Ecol Entomol *10*:241-248

Bird RD (1930) Biotic communities of the Aspen Parkland of central Canada. Ecology *11*:356-442

Bollobás B (1985) Random graphs. Academic Press, London New York

Bollobás B, Erdös P (1976) Cliques in random graphs. Math Proc Camb Phil Soc *80*:419-427

Booth KS (1975) PQ-tree algorithms. Ph.D. thesis, Computer Science Division, University of California, Berkeley. Lawrence Livermore Laboratory UCRL-51953, November 14, 1975

Bradstreet SW, Cross WE (1982) Trophic relationship at High Arctic ice edges. Arctic *35*:1-12

Briand F (1983) Environmental control of food web structure. Ecology *64*:253-263

Briand F (1983a) Biogeographic patterns in food web organization. In: DeAngelis DL, Post WM, Sugihara G (eds) Current trends in food web theory: Report on a food web workshop. Oak Ridge National Laboratory, Oak Ridge, TN, ORNL-5983, pp 37-39

Briand F, Cohen JE (1984) Community food webs have scale-invariant structure. Nature (Lond) *307*:264-266

Briand F, Cohen JE (1987) Environmental correlates of food chain length. Science *238*:956-960 and (1989) *243*:239-240

Brooks JL, Deevey ES (1963) New England. In: Frey DG (ed) Limnology in North America. University of Wisconsin Press, Madison, Wisconsin, pp 117-162

Brown AC (1964) Food relationships on the intertidal sandy beaches of the Cape Peninsula. S Afr J Sci *60*:35-41

Brown J (1971) The structure and function of the tundra ecosystem. U.S. Tundra biome 1971 Progress Rept., vol 1

Brown J (1975) Ecological investigations of the Tundra Biome in the Prudhoe Bay Region, Alaska, Special Report, no 2. Biol Pap Univ Alaska

Budyko MI (1980) Global Ecology. Progress Publishers, Moscow

Burgis MJ, Dunn IG, Ganf GG, McGowan LM, Viner AB (1972) Lake George, Uganda: Studies on a tropical freshwater ecosystem. In: Kajak Z, Hillbricht-Ilkowska A (eds) Productivity problems of freshwaters. Polish Scientific, Warsaw, pp 301-309

Cable C, Jones KF, Lundgren JR, Seager S (1988) Niche graphs. Manuscript

Calder WA (1984) Size, function, and life history. Harvard University Press, Cambridge, MA

Camerano L (1880) Dell'equilibrio dei viventi mercè la reciproca distruzione. Atti della Reale Accademia delle Scienze di Torino *15*:393-414

Carlson CA (1968) Summer bottom fauna of the Mississippi River above Dam 19, Keokuk, Iowa. Ecology *49*:162-168

Carpenter SR, Kitchell JF, Hodgson JF (1985) Cascading trophic interactions and lake productivity. BioScience *35*(10):634-639

Castilla JC (1981) Perspectivas de investigacion en estructura y dinamica de communidades intermareales rocosas de Chile Central.II. Depredadores de alto nivel trofico. Medio Ambiente *5*:190-215

Clarke TA, Flechsig AO, Grigg RW (1967) Ecological studies during Project Sealab II. Science *157*:1381-1389

Cody ML (1968) On the methods of resource division in grassland bird communities. Am Nat *102*:107-147

Cohen JE (1968) Interval graphs and food webs: a finding and a problem. RAND Corporation Document 17696-PR. Rand Corp, Santa Monica, CA

Cohen JE (1977) Ratio of prey to predators in community food webs. Nature (Lond) *270*:165-167

Cohen JE (1977) Food webs and the dimensionality of trophic niche space. Proc Nat Acad Sci USA *74*:4533-4536

Cohen JE (1978) Food webs and niche space. Princeton University Press, Princeton, NJ, 189 pp

Cohen JE (1983) Recent progress and problems in food web theory. In: DeAngelis DL, Post WM, Sugihara G (eds) Current trends in food web theory, ORNL-5983. Oak Ridge National Laboratory, Oak Ridge, Tennessee, pp 17-24

Cohen JE (1987) Lotka, Alfred James. In: Eatwell J, Milgate M, Newman P (eds) The new Palgrave: A dictionary of economic theory and doctrine, vol 3 Macmillan, London/Stockton Press, New York, pp 245-247

Cohen JE, Briand F (1984) Trophic links of community food webs. Proc Nat Acad Sci USA *81*:4105-4109

Cohen JE, Briand F, Newman CM (1986) A stochastic theory of community food webs. III. Predicted and observed lengths of food chains. Proc R Soc Lond B *228*:317-353

Cohen JE, Komlós J, Mueller T (1979) The probability of an interval graph, and why it matters. In: Ray-Chaudhuri DK (ed) Proceedings of symposia in pure mathematics. Relations between combinatorics and other parts of mathematics, vol 4. American Mathematical Society, Providence, RI, pp 97-115

Cohen JE, Newman CM (1984) The stability of large random matrices and their products. Annals of Probability *12*(2):283-310

Cohen JE, Newman CM (1985) When will a large complex system be stable? J Theoret Biol *113*:153-156

Cohen JE, Newman CM (1985) A stochastic theory of community food webs. I. Models and aggregated data. Proc R Soc Lond B *224*:421-448

Cohen JE, Newman CM (1988) Dynamic basis of food web organization. Ecology *69*:1655-1664

Cohen JE, Newman CM, Briand F (1985) A stochastic theory of community food webs. II. Individual webs. Proc R Soc Lond B *224*:449-461

Connell JH (1983) On the prevalence and relative importance of interspecific competition: evidence from field experiments. American Naturalist *122*:661-696

Cooley JH, Golley FB (eds) (1984) Trends in ecological research for the 1980s. Plenum Press, New York London

Copeland BJ, Tenore KR, Horton DB (1974) Oligohaline regime. In: Odum HT, Copeland BJ, McMahan EA (eds) Coastal ecological systems of the United States, vol 2. The Conservation Foundation, Washington, DC, USA, pp 315-357

Cousins SH (1980) A trophic continuum derived from plant structure, animal size and a detritus cascade. J Theoret Biol *82*:607-618

Critchlow RE, Stearns SC (1982) The structure of food webs. American Naturalist *120*:478-499

Cummins KW, Coffman WP, Roff PA (1966) Trophic relationships in small woodland stream. Verh Int Ver Theor Angew Limnol *16*:627-638

Darwin C (1859) On the origin of species. Facsimile of 1st ed. with Intro. by Ernst Mayr. Harvard University Press, Cambridge, MA, 1966

Day JH (1967) The biology of Knysna estuary, South Africa. In: Lauff GH (ed) Estuaries. American Association for the Advancement of Science, Washington, DC, publication no. 83, pp 397-407

DeAngelis DL, Post WM, Sugihara G (eds) (1983) Current trends in food web theory, ORNL-5983. Oak Ridge Laboratory, Oak Ridge, Tennessee, 137 pp

Dexter RW (1947) The marine communities of a tidal inlet at Cape Ann, Massachusetts: a study in bio-ecology. Ecol Monogr *17*:263-294

Diamond JM, Case TJ (eds) (1986) Community ecology. Harper and Row, Cambridge, 665 pp

Diamond JM, May RM (1981) Island biogeography and the design of natural reserves. In: May RM (ed), pp 228-252

Dunaeva T, Kucheruk V (1941) Material on the ecology of the terrestrial vertebrates of the tundra of south Yamal. Byull Mosk Obshch Ispvt Prir *4*(19):1-80

Dunbar MJ (1954) Arctic and subarctic marine ecology: immediate problems. Arctic *7*:213-228

Edwards DC, Conover DO, Sutter F (1982) Mobile predators and the structure of marine intertidal communities. Ecology *63*:1175-1180

Elton C (1927 [1935]) Animal ecology [New impression with additional notes]. Macmillan, New York, 209 pp

Esary J, Proschan F, Walkup D (1967) Association of random variables with applications. Ann math Statist *38*:1466-1474

Erdős P, Rényi A (1960) On the evolution of random graphs. Magyar Tud Akad Mat Kut Int Kozl *5*:16-61. Reprinted in: Spencer J (ed) The art of counting: selected writings. M.I.T Press, Cambridge (1973), pp 574-617

Forbes SA (1977) Ecological investigations of Stephen Alfred Forbes. Arno Press, New York

Fryer G (1959) The trophic interrelationships and ecology of some littoral communities of Lake Nyasa with especial reference to the fishes, and a discussion of the evolution of a group of rock-frequenting Cichlidae. Proc Zool Soc Lond *132*:153-281

Fulkerson DR, Gross OA (1965) Incidence matrices and interval graphs. Pac J Math *15*:835-855

Gallopín GC (1972) Structural properties of food webs. In: Patten BC (ed) Systems analysis and simulation in ecology, vol 2. Academic Press, New York, pp 241-282

Gardner MR, Ashby WR (1970) Connectance of large dynamical systems: critical values for stability. Nature *288*:784

Geman S (1986) The spectral radius of large random matrices. Ann Probab *14*:1318-1328

Gibbons A (1985) Algorithmic graph theory. Cambridge University Press, Cambridge, 259 pp

Glynn PW (1965) Community composition, structure, and interrelationships in the marine intertidal Endocladia muricata-Balanus glandula association in Monterey Bay, California. Beaufortia *148*:1-198

Grant PR (1986) Interspecific competition in fluctuating environments. In: Diamond J, Case TJ (eds) Community Ecology. Harper and Row, New York, pp 173-191

Haigh J, Smith JM (1972) Can there be more predators than prey? Theor Pop Biol *3*:290-299

Hairston NG, Smith FE, Slobodkin LB (1960) Community structure, population control, and competition. American Naturalist *94*:421-425

Harary F (1961) Who eats whom? Gen Systems *6*:41-44

Hardy AC (1924) The herring in relation to its animate environment. Part 1. The food and feeding habits of the herring with special reference to the East Coast of England. Fisheries Investigations Series II 7(3):1-45

Harris LD, Bowman GB (1980) Vertebrate predator subsystem. In: Breymeyer AI, Van Dyne GM (eds) Grasslands, systems analysis and man (International Biological Programme Series, no 19). Cambridge University Press, Cambridge, England, pp 591-607

Harris TE (1960) A lower bound for the critical probability in a certain percolation process. Proc Camb Phil Soc *59*:13-20

Harrison JL (1962) The distribution of feeding habits among animals in a tropical rain forest. J Anim Ecol *31*:53-63

Hart S (1989) Food chains: The carbon link. Science News *136*:168-170

Hartley PHT (1948) Food and feeding relationships in a community of freshwater fishes. J Anim Ecol *17*:1-14

Hatanaka MA (1977) Sendai Bay. In: Hogetsu K, Horanaka M, Hatanaka T, Kawamura T (eds) Productivity of biocenoses in coastal regions of Japan, vol 14. Japanese Committee for the International Biological Program Synthesis, Tokyo, pp 173-221

Hazen WE (ed) (1964) Readings in population and community ecology. Saunders, Philadelphia

Hewatt WG (1937) Ecological studies on selected marine intertidal communities of Monterey Bay, California. Am Midl Nat *18*:161-206

Hiatt R, Strasburg DW (1960) Ecological relationships of the fish fauna on coral reefs of the Marshall Islands. Ecol Monogr *30*:65-127

Hogetsu K (1979) Biological productivity of some coastal regions of Japan. In: Dunbar J (ed) Marine production mechanisms (International Biological Programme Series, no 20). Cambridge University Press, Cambridge, England, pp 71-87

Holm E, Scholtz CH (1980) Structure and pattern of the Namib Desert dune ecosystem at Gobabed. Madoqua *12*:3-39

Holt RD (1977) Predation, apparent competition and the structure of prey communities. Theoretical Population Biology 11:197-229

Howes PG (1954) The Giant Cactus Forest and Its World. Duell, Sloan and Pearce, New York

Hutchinson GE (1944) Limnological studies in Connecticut. VII. A critical examination of the supposed relationship between phytoplankton periodicity and chemical changes in lake waters. Ecology 25:3-26

Hutchinson GE (1959) Homage to Santa Rosalia or why are there so many kinds of animals? Am Nat 93:145-159

Hutchinson GE (1965) The ecological theater and the evolutionary play. Yale University Press, New Haven London

Jeffries MJ, Lawton JH (1984) Enemy free space and the structure of ecological communities. Biol J Linn Soc 23:269-286

Johnston RF (1956) Predation by short-eared owls on a Salicornia salt-marsh. Wilson Bull 68:91-102

Jones JR (1949) A further ecological study of calcareous streams in the "Black Mountain" district of South Wales. J Anim Ecol 18:142-159

Jones JR (1950) A further ecological study of the River Rheidol: the food of the common insects of the main stream. J anim Ecol 19:159-174

Kawanabe H (1986) Cooperative study on the ecology of Lake Tanganyika between Japanese and African scientists, with special reference to mutual interactions among fishes. Physiol Ecol Jap 23:119-128

Kawanabe H (1987) Niche problems in mutualism. Physiol Ecol Jap 24 (special no):75-80

Kemp WM et al (1977) Energy cost-benefit analysis applied to power plants near Crystal River, Florida. In: Hall CA, Day Jr JW (eds) Ecosystem modeling in theory and practice: An introduction with case histories. Wiley, New York, pp 507-543

Kendall DG (1969) Incidence matrices, interval graphs and seriation in archeology. Pac J Math 28(3):565-570

Kendall MG, Stuart A (1968) The advanced theory of statistics. Vol 3: Design and analysis, and time series, 2nd edn. Hafner, New York/Griffin, London, 557 pp

Kendall MG, Stuart A (1973) The advanced theory of statistics. Vol 2: Inference and relationship, 3rd edn. Hafner, New York/Griffin, London, 723 pp

Kikkawa J, Anderson DJ (1986) Community ecology: Pattern and process. Blackwell Scientific, Melbourne Oxford, 432 pp

Kingsland SE (1985) Modeling nature: Episodes in the history of population ecology. University of Chicago Press, Chicago

Kitazawa Y (1977) Ecosystem metabolism of the subalpine coniferous forest of the Shigayama IBP area, central Japan. In: Kitazawa Y (ed) Ecosystem analysis of the subalpine coniferous forest of Shigayama IBP Area, vol 15. Japanese Committee for the International Biological Program Synthesis, Tokyo, pp 181-196

Kitching JA, Ebling FJ (1967) Ecological studies at Lough Ine. Adv Ecol Res 4:197-291

Kitching RL (1987) Spatial and temporal variation in food webs in water-filled treeholes. Oikos (Copenhagen) 48:280-288

Kitching RL, Pimm SL (1985) The length of food chains: phytotelmata in Australia and elsewhere. Proc Ecol Soc Aust 14:123-139

Klee V (1969) What are the intersection graphs of arcs in a circle? Am Math Mon 76:810-813

Knox GA (1970) Antarctic marine ecosystems. In: Holdgate MW (ed) Antarctic ecology. Academic Press, New York, NY, pp 69-96

Kohn AJ (1959) The ecology of Conus in Hawaii. Ecol Monogr 29:47-90

Kremer JN, Nixon SW (1978) A coastal marine ecosystem: Simulation and analysis. Ecological Studies, vol 24. Springer, Berlin Heidelberg

Kuusela K (1979) Early summer ecology and community structure of the macrozoobenthos on stones in the Javajankoski rapids on the river Lestijoki. Acta Universitatis Ouluensis (Ser. A, no 87, Oulu, Finland)

Lawlor LR (1978) A comment on randomly constructed model ecosystems. Am Nat 112:445-447

Lawton JH (1989) Food webs. In: Cherrett JM (ed) Ecological concepts. Blackwell Scientific, Oxford, pp 43-78

Lekkerkerker CG, Boland JC (1962) Representation of a finite graph by a set of intervals on the real line. Fund Math Polska Akad Nauk 51:45-64

Levin SA (1978) Population models and community structure in heterogeneous environments. In: Levin SA (ed) Studies in mathematical biology 2:439-476. Mathematical Association of America, Washington, DC

Levin SA, Paine RT (1974) Disturbance, patch formation and community structure. Proc Nat Acad Sci USA 71:2744-2747

Lindemann RL (1942) The trophic-dynamic aspect of ecology. Ecology 23:399-418

Lotka AJ (1925) Elements of physical biology. Williams and Wilkins, Baltimore. Reprinted 1956: Elements of Mathematical Biology. Dover, New York

Lundgren JR, Maybee JS (1985) Food webs with interval competition graphs. In: Harary F, Maybee JS (eds) Graphs and applications, pp 245-256

MacArthur RH, Wilson EO (1967) The theory of island biogeography. Princeton University Press, Princeton

MacDonald N (1979) Simple aspects of foodweb complexity. J Theoret Biol 80:577-588

MacDonald N (1983) Trees and networks in biological models. Wiley, Chichester New York

MacGinitie GE (1935) Ecological aspects of a Californian marine estuary. Am Midl Nat 16:629-765

Mackintosh NA (1964) A survey of antarctic biology up to 1945. In: Carrick R, Holdgate M, Prevost J (eds) Biologie antarctique. Hermann, Paris, France, pp 3-38

Mann KH, Britton RH, Kowalczewski A, Lack TJ, Mathews CP, McDonald I (1972) Productivity and energy flow at all trophic levels in the River Thames, England. In: Kajak Z, Hillbricht-Ilkowska A (eds) Productivity problems of freshwaters. Polish Scientific, Warsaw, pp 579-596

Mantel N (1982) Simpson's paradox in reverse. Am Stat 36:395

Margalef R (1984) Simple facts about life and the environment not to forget in preparing school-books for our grandchildren. In: Cooley and Golley, pp 299-320

May RM (1972) Will a large complex system be stable? Nature 238:413-414

May RM (1973) Stability and complexity in model ecosystems. Princeton University Press, Princeton, NJ

May RM (1981) Patterns in multi-species communities. In: May RH (ed) Theoretical ecology. Principles and applications. Sinauer, Sunderland, Massachusetts, USA, pp 197-227

May RM (1983) The structure of food webs. News and views. Nature (Lond) 301:566-568

May RM (1986) The search for patterns in the balance of nature: advances and retreats. Ecology 67:1115-1126

May RM, Rubenstein DI (1985) Reproductive strategies. In: Austin CR, Short RV (eds) Reproduction in mammals. Book 4: Reproductive fitness, 2nd ed. Cambridge University Press, Cambridge

McNeil DR (1977) Interactive data analysis. A practical primer (Probability and mathematical statistics series). Wiley, New York

Mehta ML (1986) Random matrices in nuclear physics and number theory. In: Cohen JE, Kesten H, Newman CM (eds) Random matrices and their applications. Contemporary mathematics, vol 50. American Mathematical Society, Providence, RI, pp 295-303

Menge BA, Lubchenco J, Gaines SD, Ashkenas LR (1986) A test of the Menge-Sutherland model of community organization in a tropical rocky intertidal food web. Oecologica (Berlin) 71:75-89

Menge BA, Sutherland JP (1976) Species diversity gradients: synthesis of the roles of predation, competition and temporal heterogeneity. American Naturalist 110:351-369

Miller RS (1967) Pattern and process in competition. In: Cragg JB (ed) Advances in ecological research, vol 4. Academic Press, London, pp 1-74

Milne H, Dunnet GM (1972) Standing crop, productivity and trophic relations of the fauna of the Ythan estuary. In: Barnes RSK, Green J (eds) The estuarine environment. Elsevier, New York, pp 86-106

Minshall GW (1967) Role of allochthonous detritus in the trophic structure of a woodland springbrook community. Ecology 48:139-149

Mithen SJ, Lawton JH (1986) Food-web models that generate constant predator-prey ratios. Oecologia (Berlin) 69:542-550

Mizuno T, Furtado JI (1982) in: Furtado JI, Mori S (eds) Tasek Bera. Junk, The Hague, Netherlands, pp 357-359

Morgan NC, McLusky DS (1972) A summary of the Loch Leven IBP results in relation to lake management and future research. Proc R Soc Edinburgh Ser. B 74:407-416

National Research Council, Commission on Life Sciences, Committee on the Applications of Ecological Theory to Environmental Problems (1986) Ecological knowledge and environmental problem-solving. National Academy Press, Washington, DC

Newman CM (1980) Normal fluctuations and the FKG inequalities. Comm Math Phys *74*:119-128

Newman CM (1984) Asymptotic independence and limit theorems for positively and negatively dependent random variables. In: Tong YL (ed) Inequalities in statistics and probability. Institute of Mathematical Statistics Lecture Notes - Monograph Series, vol 5, pp 127-140

Newman CM, Cohen JE (1986) A stochastic theory of community food webs. IV. Theory of food chain lengths in large webs. Proc R Soc Lond B *228*:355-377

Newman CM, Rinott Y, Tversky A (1984) Nearest neighbors and Voronoi regions in certain point processes. Adv Appl Probab *15*:726-751

Newman CM, Schulman LS (1985) Infinite chains and clusters in one dimensional directed and undirected percolation. In: Durrett R (ed) Particle systems, random media and large deviations. Contemporary Mathematics vol 41. American Mathematical Society, Providence, RI, pp 297-313

Niering WA (1963) Terrestrial ecology of Kapingamarangi Atoll, Caroline Islands. Ecol Monogr *33*:131-160

Nixon SW, Oviatt CA (1973) Ecology of a New England salt marsh. Ecol Monogr *43*:463-498

Nybakken JW (1982) Marine biology: An ecological approach. Harper and Row, New York

Ohm P, Remmert H (1955) Etudes sur les rockpools des Pyrenees-Orientales. Vie Milieu *6*:194-209

Osmolovskaya VI (1948) Geographical distribution of raptors in Kazakhstan plains and their importance for pest control. Tr Acad Sci USSR Inst Geogr *41*:5-77

Paine RT (1980) Food webs: linkage, interaction strength and community infrastructure. J Anim Ecol *49*:667-685

Paine RT (1988) On food webs: roadmaps of interactions or grist for theoretical development? Ecology *69*:1648-1654

Parin NV (1970) Ichthyofauna of the Epipelagic Zone. Israel Program for Scientific Translations, Jerusalem

Patten BC et al (1975) Total ecosystem model for a cove in Lake Texoma. In: Patten BC (ed) System analysis and simulation in ecology, vol 3. Academic Press, New York, pp 205-421

Patten BC, Finn JT (1979) Systems approach to continental shelf ecosystems. In: Halfon E (ed) Theoretical systems ecology: Advances and case studies. Academic Press, New York, pp 184-212

Pattie DL, Verbeek NAM (1966) Alpine birds of the Beartooth Mountains. Condor *68*:167-176; (1967) Alpine mammals of the Beartooth Mountains, Northwest Sci *41*:110-117

Paviour-Smith K (1956) The biotic community of a salt meadow in New Zealand. Trans R Soc N.Z. *83*:525-554

Pechlaner R, Bretschko G, Gollmann P, Pfeifer H, Tilzer M, Weissenbach HP (1972) Ein Hochgebirgssee (Vorderer Finstertaler See, Kubtai, Tirol) als Modell des Energietransportes durch ein limnisches Ökosystem. Verh Dtsch Zool Ges *65*:47-56

Peters RH (1983) The ecological implications of body size. Cambridge University Press, Cambridge

Peters RH, Raelson JV (1984) Relations between individual size and mammalian population density. American Naturalist *124*:498-517

Peterson CH (1979) The importance of predation and competition in organizing the intertidal epifaunal communities of Barnegat Inlet, New Jersey. Oecologia (Berlin) *39*:1-24

Peterson RO, Page RE, Dodge KM (1984) Wolves, mouse, and the allometry of population cycles. Science *224*:1350-1352

Petipa TS (1979) Trophic relationships in communities and the functioning of marine ecosystems. In: Dunbar M (ed) Marine production mechanisms. Cambridge University Press, Cambridge, pp 233-250

Petipa TS, Pavlova EV, Mironov GN (1970) The food web structure, utilization transport of energy by trophic levels in the planktonic communities. In: Steele JH (ed) Marine food chains. Oliver and Boyd, Edinburgh, Scotland, pp 142-167

Pianka ER (1976) Competition and niche theory. In: May RM (ed) Theoretical ecology: principles and applications. Blackwell Scientific Publications, Oxford, pp 114-141

Pimm SL (1980) Bounds on food web connectance. Nature (Lond) *284*:591

Pimm SL (1982) Food webs. Chapman and Hall, London

Pimm SL (1984) The complexity and stability of ecosystems. Nature (Lond) *307*:321-326

Pimm SL (1988) The geometry of niches. In: Hastings A (ed) Community ecology. Lecture
    Notes in Biomathematics 77:92-111
Pimm SL, Kitching RL (1987) The determinants of food chain lengths. Oikos, Special Issue
    on Trophic Exploitation, 50:302-307
Pimm SL, Lawton JH (1977) The number of trophic levels in ecological communities. Nature
    268:329-331
Qazim SZ (1970) Some problems related to the food chain in a tropical estuary. In: Steele JH
    (ed) Marine food chains. Oliver and Boyd, Edinburgh, Scotland, pp 46-51
Rasmussen DI (1941) Biotic communities of Kaibab Plateau, Arizona. Ecol Monogr 11:228-275
Rejmánek M, Starý P (1979) Connectance in real biotic communities and critical values for
    stability of model ecosystems. Nature 280:311-313
Rektorys K (1969) Survey of applicable mathematics. M.I.T. Press, Cambridge
Remmert H (1984) And now? Ecosystem research! In: Cooley and Golley, pp 179-191
Richards OW (1926) Studies on the ecology of English heaths. III. Animal communities of the
    felling and burn successions at Oxshott Heath, Surrey. J Ecol 14:244-281
Riedel D (1962) Der Margheritensee (Sudabessinien) – Zugleich ein Beitrag zur Kenntnis der
    Abessinischen Graben-Seen. Arch Hydrobiol 58:435-466
Roberts FS (1969) On the boxicity and cubicity of a graph. In: Tutte WT (ed) Recent progress
    in combinatorics. Academic Press, New York, pp 301-310
Roberts FS (1969) Indifference graphs. In: Harary F (ed) Proof techniques in graph theory.
    Academic Press, New York
Roberts FS (1988) Applications of combinatorics and graph theory to the biological and social
    sciences: seven fundamental ideas. In: Roberts FS (ed) Proceedings of the Workshop on
    Applications of Combinatorics and Graph Theory to the Biological and Social Sciences.
    Institute of Mathematics and Its Applications, University of Minnesota, Minneapolis
Robinson DF, Foulds LR (1980) Digraphs: theory and techniques. Gordon and Breach, New
    York
Rose DJ, Tarjan RE, Lueker GS (1976) Algorithmic aspects of vertex elimination on graphs.
    SIAM J Comput 5(2):266-283
Rosenthal RJ, Clarke WD, Dayton PK (1974) Ecology and natural history of a stand of giant
    kelp, Macrocystis pyrifera, off Del Mar, California. Fish Bull (Dublin) 72:670-684
Ruciński A (1988) Enumeration of "up-and-down" labelings of a cycle. Submitted
Saldanha L (1980) Estudio Ambiental do Estuario do Tejo. CNA/Tejo, Lisbon
Sarvala J (1974) Paarjaven energiatalous. Luonnon Tutkija 78:181-190
Schiemer F (1979) The benthic community of the open lake. In: Loeffler H (ed) Neusiedlersee:
    The limnology of a shallow lake in Central Europe. Junk, The Hague, Netherlands, pp
    337-384
Schiemer F, Bobek M, Gludovatz P, Ioschenkohl A, Zweimuller I, Martinetz M (1982) Trophi-
    sche Interaktionen im Pelagial des Hafnersees. Sitzungsber Akad Wiss Wien Math Natur-
    wiss Kl Abt 1:191-209
Schoener TW (1967) The ecological significance of sexual dimorphism in size in the lizard
    Anolis conspersus. Science 155:474-477
Schoener TW (1974) Resource partitioning in ecological communities. Science 185:27-39
Schoener TW (1976) The species-area relation within archipelagos: models and evidence from
    island land birds. Proceedings of the 16th International Ornithological Congress, Canberra,
    pp 629-642
Schoener TW (1982) The controversy over interspecific competition. Am Scient 70:586-594
Schoener TW (1983) Field experiments on interspecific competition. American Naturalist
    122:240-285
Schoener TW (1985) Some comments on Connell's and my my reviews of field experiments on
    interspecific competition. American Naturalist 125:730-740
Schoener TW (1986) Patterns in terrestrial vertebrate versus arthropod communities: do sys-
    tematic differences in regularity exist? In: Diamond and Case (eds), pp 556-586
Sharma IK (1980) A study of ecosystems of the Indian desert. Trans Indian Soc Desert Technol
    Univ Center Desert Stud 5:51-55; A study of agro-ecosystems in the Indian desert. Ibid.
    5:77-82
Shelford VE (1913) Animal communities in temperate America as illustrated in the Chicago
    region: A study in animal ecology. Geographic Society of Chicago Bulletin 5, University
    of Chicago Press, Chicago
Shure DJ (1973) Radionuclide tracer analysis of trophic relationships in an old-field ecosystem.
    Ecol Monogr 43:1-19

Shushkina EA, Vinogradov ME (1979) Trophic relationships in communities and the functioning of marine ecosystems. II. In: Dunbar MJ (ed) Marine production mechanisms. International Biological Programme Series, no 20. Cambridge University Press, Cambridge, pp
   251-268

Silvert W (1984) Particle size spectra in ecology. In: Levin SA, Hallam TG (eds) Mathematical
   ecology. Lecture Notes in Biomathematics 54:154-162

Simenstad CA, Estes JA, Kenyon KW (1978) Aleuts, sea otters, and alternate stable-state
   communities. Science 200:403-411

Smirnov NN (1961) Food cycles in sphagnous bogs. Hydrobiologia 17:175-182

Snedecor GW, Cochran WG (1967) Statistical methods, 6th edn. Iowa State University Press,
   Ames, Iowa, 593 pp

Sorokin YI (1972) Biological productivity of the Rybinsk reservoir. In: Kajak Z, Hillbricht-
   Ilkowska A (eds) Productivity problems of freshwaters. Polish Scientific, Warsaw, pp 493-
   503

Stein GJ (1988) Biological science and the roots of Nazism. American Scientist 76:50-58

Strong DR, Simberloff D, Abele LG, Thistle AB (eds) (1984) Ecological communities: Conceptual issues and the evidence. Princeton University Press, Princeton

Sugihara G (1982) Niche hierarchy: structure, organization, and assembly in natural communities. Ph.D. dissertation, Department of Biology, Princeton University

Sugihara G (1983) Holes in niche space: a derived assembly rule and its relation to intervality.
   In: DeAngelis et al (eds), pp 25-35

Sugihara G (1984) Graph theory, homology and food webs. Proc Symp App Math 30:83-101.
   American Mathematical Society, Providence, RI

Sullivan AL, Shaffer ML (1975) Biogeography of the megazoo. Science 189:13-17

Summerhayes VS, Elton CS (1923) Contributions to the ecology of Spitzbergen and Bear
   Island. J Ecol 11:214-286

Summerhayes VS, Elton CS (1928) Further contributions to the ecology of Spitzbergen. J Ecol
   16:193-268

Swan LW (1961) The ecology of the high Himalayas. Sci Am 205:68-78

Taylor SR (1987) The origin of the moon. American Scientist 75:469-477

Teal JM (1962) Energy flow in the saltmarsh ecosystem of Georgia. Ecology 43:614-624

Tilly LJ (1968) The structure and dynamics of Cone Spring. Ecol Monogr 38:169-197

Tilman D (1986) A consumer-resource approach to community structure. American Zoologist
   26:5-22

Tsuda M (1972) Interim results of the Yoshino River productivity survey, especially on benthic
   animals. In: Kajak Z, Hillbricht-Ilkowska A (eds) Productivity problems of freshwaters.
   Polish Scientific, Warsaw, pp 829-841

Tukey JW (1977) Exploratory data analysis. Addison-Wesley, Reading, MA; reprinted by
   University Microfilms International, Ann Arbor, MI

Twomey AC (1945) The bird population of an elm-maple forest with special reference to
   aspection, territorialism, and coactions. Ecol Monogr 15:175-205

Vandermeer JH (1972) Niche theory. Annu Rev Ecol Syst 3:107-132

Van Dover CL, Fry B, Grassle JF, Humphris S, Rona PA (1988) Feeding biology of the shrimp,
   Rimicaris exoculata, at hydrothermal vents on the mid-Atlantic ridge. Marine Biology
   98:209-216

Varley GC (1970) The concept of energy applied to a woodland community. In: Watson A
   (ed) Animal populations in relation to their food resources. Blackwell Scientific, Oxford,
   England, pp 389-401

Vézina AF (1985) Empirical relationships between predator and prey size among terrestrial
   vertebrate predators. Oecologia (Berlin) 67:555-565

Vinogradov ME, Shushkina EA (1978) Some development patterns of plankton communities
   in the upwelling areas of the Pacific Ocean. Mar Biol 48:357-366

Walsh GE (1967) An ecological study of a Hawaiian mangrove swamp. In: Lauff GH (ed) Estuaries, publication no 83. American Association for the Advancement of Science, Washington, DC, pp 420-431

Warren PH, Lawton JH (1987) Invertebrate predator-prey body size relationships: an explanation for upper triangular food webs and patterns in food web structure? Oecologia (Berlin)
   74:231-235

Wiegert RG (ed) (1976) Ecological energetics. Benchmark Papers in Ecology 4. Dowden,
   Hutchinson and Ross, Stroudsberg, Pennsylvania

Wigner EP (1958) On the distribution of the roots of certain symmetric matrices. Ann Math 67:325-327

Wilbur HM (1972) Competition, predation, and the structure of the Ambystoma-Rana sylvatica community. Ecology 53:3-21

Wood TE (1982) Sequences of associated random variables. Ph.D. dissertation, University of Virginia

Woodwell GM (1967) Toxic substances and ecological cycles. Sci Am 216:24-31

Woodwell G (1970) The energy cycle of the biosphere. Scientific American, September. Reprinted in: A Scientific American Book, The Biosphere. Freeman, San Francisco, 1970, pp 26-36; and in: Wilson EO (ed) Ecology, Evolution and Population Biology; Readings from Scientific American. Freeman, San Francisco, 1974, pp 204-214

Yanez-Arancibia A (1978) Taxonomia, ecologia y estructura de las comunidades de peces en lagunas costeras con bocas efimeras del Pacifico de Mexico. Cent Cienc del Mar Univ Nal Auton Mex Publ Espec 2:1-306

Yodzis P (1980) The connectance of real ecosystems. Nature (Lond) 284:544-545

Yodzis P (1981) The structure of assembled communities. J Theor Biol 92:103-117

Yodzis P (1982) The compartmentation of real and assembled ecosystems. American Naturalist 120:551-570

Yodzis P (1983) Community assembly, energy flow and food web structure. In: DeAngelis DL, Post WM, Sugihara G (eds) Current trends in food web theory. Oak Ridge National Laboratory Report 5983, Oak Ridge, TN, pp 41-44

Yodzis P (1984) The structure of assembled communities II. J Theor Biol 107:115-126

Acknowledgements

The research reported in this book has been supported by grants from the U.S. National Science Foundation to J.E.C. (GB-31049, BMS 74-13276, DEB 80-11026, BSR 84-07461, BSR 87-05047), from the Natural Science and Engineering Research Council of Canada to F.B. (A9868), and from the U.S. National Science Foundation to C.M.N. (MCS 80-19384). We acknowledge with thanks fellowships awarded to J.E.C. and C.M.N. by the John Simon Guggenheim Memorial Foundation and to J.E.C. by the John D. and Catherine T. MacArthur Foundation. We are grateful for the hospitality to F.B. of J.-C. Lefeuvre and to J.E.C. of Mr. and Mrs. William T. Golden.

Numerous colleagues assisted us during the work on which this book is based, including skeptical critics who would not welcome the guilt by association of being identified here. We especially thank Donald L. DeAngelis, Stuart Pimm, George Sugihara, and Peter Yodzis for their consistently constructive criticisms. Zoe Sara Cohen helped proofread the book.

We thank the following publishers for permission to use here the essential parts of articles they have published.

Food Webs and Community Structure
Perspectives on Ecological Theory
ed. Jonathan Roughgarden, Robert M. May, Simon Levin, pp. 181-202, 1989
© Princeton University Press 1989

Ratio of Prey to Predators in Community Food Webs
Nature, Vol. 270, No. 5633, pp. 165-167, November 10, 1977
© Macmillan Journals Ltd. 1977

Community Food Webs Have Scale-Invariant Structure
Nature, Vol. 307, pp. 264-267, January 19, 1984
© Macmillan Journals Ltd. 1984

Trophic Links of Community Food Webs
Proc. Natl. Acad. Sci. USA, Vol. 81, pp. 4105-4109, July 1984
with permission of the National Academy of Sciences, USA

Food Webs and the Dimensionality of Trophic Niche Space
Proc. Natl. Acad. Sci. USA, Vol. 74, pp. 4533-4536, October 1977
with permission of the National Academy of Sciences, USA

Environmental Control of Food Web Structure
Ecology, 64 (2), 1983, pp. 253-263
© Ecological Society of America 1983

Environmental Correlates of Food Chain Length
Science, Vol. 238, pp. 956-960, 13 November 1987
© American Association for the Advancement of Science 1987

A Stochastic Theory of Community Food Webs: I. Models and Aggregated Data
Proc. R. Soc. Lond. B. 224, pp. 421-448, 1985
© The Royal Society and J.E. Cohen and C.M. Newman 1985

A Stochastic Theory of Community Food Webs: II. Individual Webs
Proc. R. Soc. Lond. B 224, pp. 449-461, 1985
© The Royal Society and J.E. Cohen, C.M. Newman and F. Briand 1985

A Stochastic Theory of Community Food Webs: III. Predicted and Observed
Length of Food Chains
Proc. R. Soc. Lond. B 228, pp. 317-353, 1986
© The Royal Society and J.E. Cohen, F. Briand and C.M. Newman 1986

A Stochastic Theory of Community Food Webs: IV. Theory of Food Chain
Lengths in Large Webs
Proc. R. Soc. Lond. B 228, pp. 355-377, 1986
© The Royal Society and C.M. Newman and J.E. Cohen 1986

A Stochastic Theory of Community Food Webs: V. Intervality and Triangulation
in the Trophic Niche Overlap Graph
American Naturalist March 1990
© Joel E. Cohen and Zbigniew J. Palka 1990

Authors Index

Subject Index

Volume 19

J. D. Murray, Oxford University

Mathematical Biology

1989. XIV, 767 pp. 262 figs. ISBN 3-540-19460-6

This textbook gives an in-depth account of the practical use of mathematical modelling in several important and diverse areas in the biomedical sciences.
The emphasis is on what is required to solve the real biological problem.
The subject matter is drawn, for example, from population biology, reaction kinetics, biological oscillators and switches, Belousov-Zhabotinskii reaction, neural models, spread of epidemics.
The aim of the book is to provide a thorough training in practical mathematical biology and to show how exciting and novel mathematical challenges arise from a genuine interdisciplinary involvement with the biosciences. It also aims to show how mathematics can contribute to biology and how physical scientists must get involved.
The book also presents a broad view of the field of theoretical and mathematical biology and is a good starting place from which to start genuine interdisciplinary research.

Volume 18

S. A. Levin, Cornell University, Ithaca, NY; **T. G. Hallam**, **L. J. Gross**, University of Tennessee, Knoxville, TN (Eds.)

Applied Mathematical Ecology

1989. XIV, 489 pp. 114 figs. ISBN 3-540-19465-7

Contents: Introduction. – Resource Management. – Epidemiology: Fundamental Aspects of Epidemiology Case Studies. – Ecotoxicology. – Demography and Population Biology. – Author Index. – Subject Index.

This book builds on the basic framework developed in the earlier volume – "Mathematical Ecology", edited by T. G. Hallam and S. A. Levin, Springer 1986, which lays out the essentials of the subject. In the present book, the applications of mathematical ecology in ecotoxicology, in resource management, and epidemiology are illustrated in detail. The most important features are the case studies, and the interrelatedness of theory and application. There is no comparable text in the literature so far. The reader of the two-volume set will gain an appreciation of the broad scope of mathematical ecology.

Springer-Verlag Berlin
Heidelberg New York London
Paris Tokyo Hong Kong

Springer-Verlag
Berlin Heidelberg New York
London Paris Tokyo Hong Kong